The Modern Measuring Circuit Encyclopedia

Other Books in the Series

3889 The Modern Power Supply and Battery Charger Circuit Encyclopedia
3891 The Modern Detector Circuit Encyclopedia
3892 The Modern Converter Circuit Encyclopedia
3893 The Modern Oscillator Circuit Encyclopedia
3894 The Modern Amplifier Circuit Encyclopedia

The Modern Measuring Circuit Encyclopedia

Rudolf F. Graf

To sweet and lovely Allison,
with love, from Popsi

FIRST EDITION
FIRST PRINTING

Published by TAB Books.
TAB Books is a division of McGraw-Hill, Inc.

Library of Congress Cataloging-in-Publication Data

Graf, Rudolf F.
The modern measuring circuit encyclopedia / by Rudolf F. Graf.
p.cm.
Includes index.
ISBN 0-8306-4156-4
1. Electronic circuits. 2. Electronic measurements. I. Title.
TK7867.G67 1992
621.381′5—dc20 92-9564
CIP

TAB Books offers software for sale. For information and a catalog, please contact TAB Software Department, Blue Ridge Summit, PA 17294-0850.

Acquisitions Editor: Roland S. Phelps
Technical Editor: Andrew Yoder
Director of Production: Katherine G. Brown

Contents

Introduction

Like the other volumes in this series, this book contains a wealth of ready-to-use circuits that serve the needs of the engineer, technician, student and, of course, the browser. These unique books contain more practical, ready-to-use circuits focused on a specific field of interest, than can be found anywhere in a single volume.

1

Battery Monitoring Circuits

The sources of the following circuits are contained in the Sources section, which begins on page 217. The figure number in the box of each circuit correlates to the source entry in the Sources section.

BATTERY STATUS INDICATOR

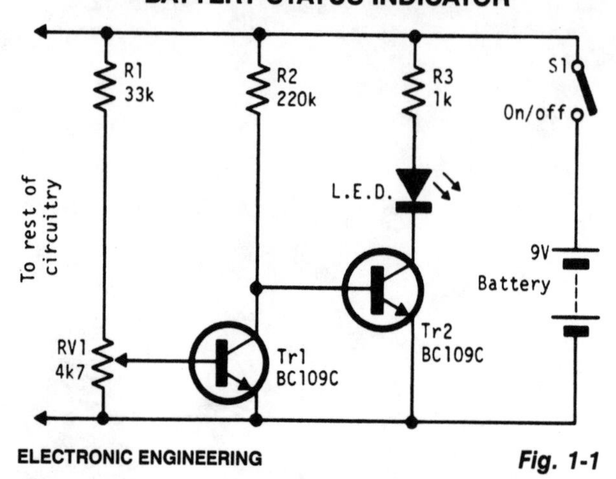

ELECTRONIC ENGINEERING

Fig. 1-1

This indicator continually monitors the battery voltage during use and consumes only about 250 μA (until the end point is reached). Near the end point, Tr1 turns off, allowing Tr2 to illuminate the LED to increase current drain, which further leads to a distinct turn-off point.

LOW-BATTERY INDICATOR I

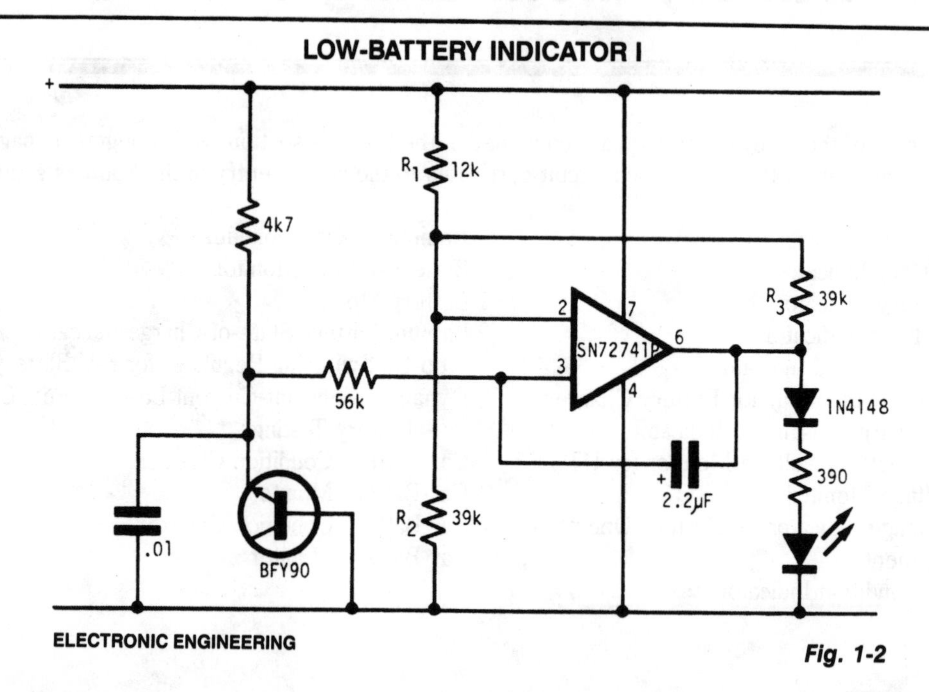

ELECTRONIC ENGINEERING

Fig. 1-2

Under good battery conditions, the LED is off. As the battery voltage falls, the LED begins to flash until, in the low-battery condition, the LED lights continuously. Designed for a 9-V battery, with the values shown the LED flashes from 7.5 to 6.5 V.

LOW-BATTERY INDICATOR II

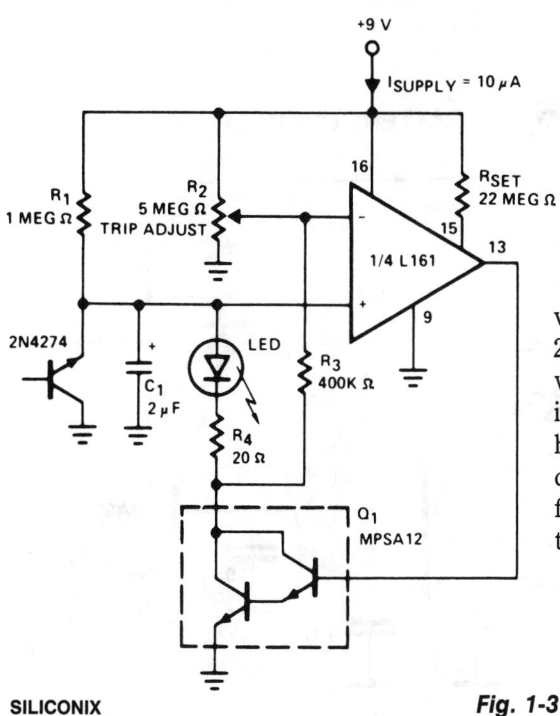

The indicator flashes an LED when the battery voltage drops below a certain threshold. The 2N4274 emitter-base junction serves as a zener, which establishes about 6 V on the L161's positive input. As the battery drops, the L161 output goes high. This turns on the Darlington, which discharges C1 through the LED. The interval between flashes is roughly two seconds and gives a low-battery warning with only 10-μA average power drain.

SILICONIX

Fig. 1-3

BATTERY-LEVEL INDICATOR

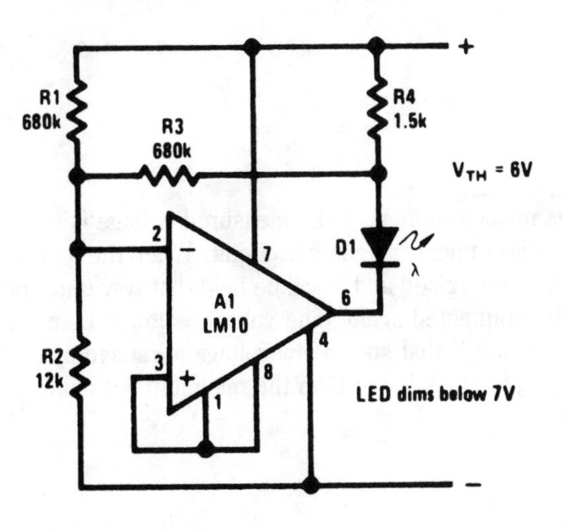

NATIONAL SEMICONDUCTOR

Fig. 1-4

BATTERY-THRESHOLD INDICATOR

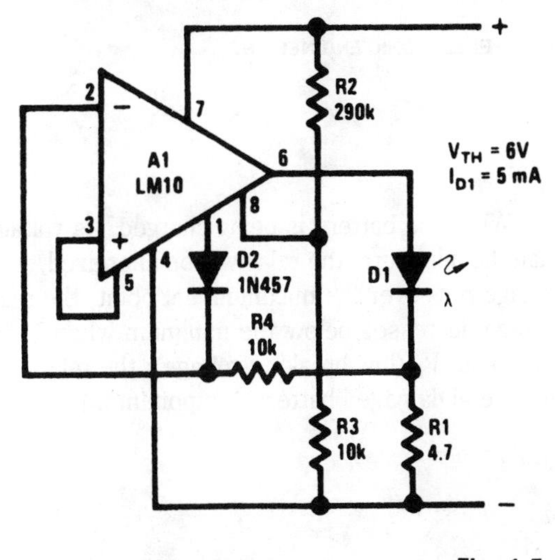

NATIONAL SEMICONDUCTOR

Fig. 1-5

VOLTAGE-DETECTOR RELAY FOR BATTERY CHARGER

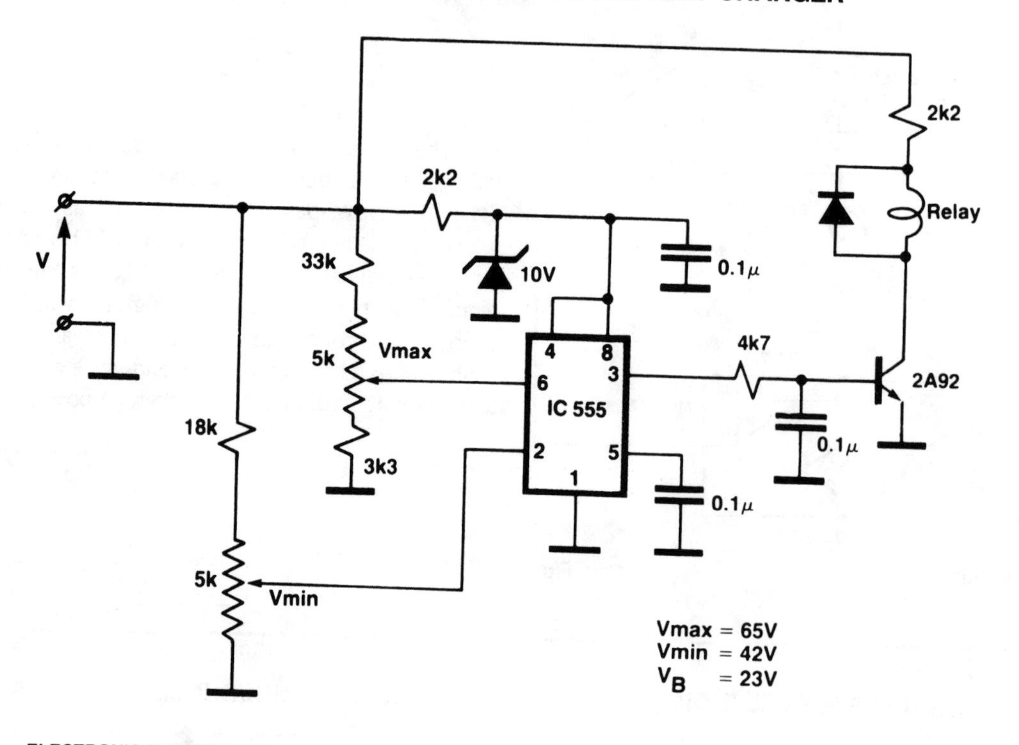

ELECTRONIC ENGINEERING

Fig. 1-6

While the battery is being charged, its voltage is measured at V. If the measured voltage is lower than the minimum, the relay will be energized, which will connect the charger circuit. When the battery voltage runs over the maximum set point, the relay is deenergized and it will be held that way until the voltage decreases below the minimum when it will be connected again. The voltage is lower than the threshold, V_B (low breaking voltage); the relay will "assume" that such a low voltage is caused by one or several damaged battery components. Of course, V_B is much lower than the minimum set point.

BATTERY-CHARGE/DISCHARGE INDICATOR

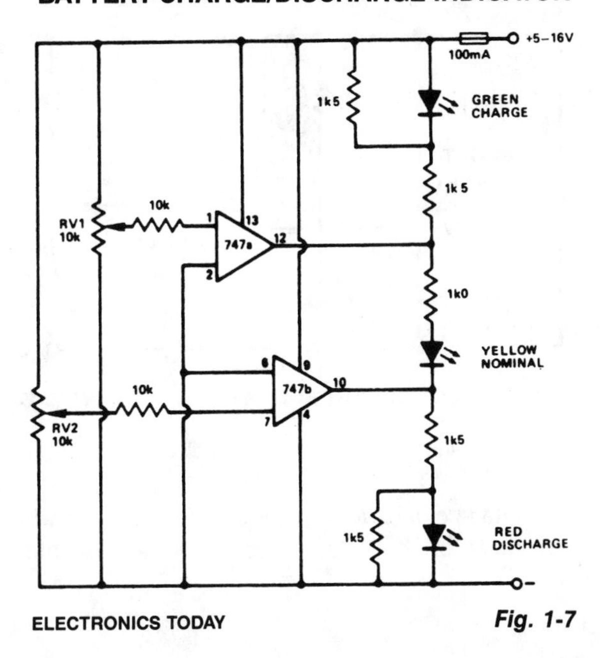

ELECTRONICS TODAY *Fig. 1-7*

This circuit monitors car battery voltage. It provides an indication of nominal supply voltage as well as low or high voltage. RV1 and RV2 adjust the point at which the red/yellow and yellow/green LEDs are on or off. For example, the red LED comes on at 11 V, and the green LED at 12 V. The yellow LED is on between these values.

PRECISION BATTERY-VOLTAGE MONITOR FOR HTS

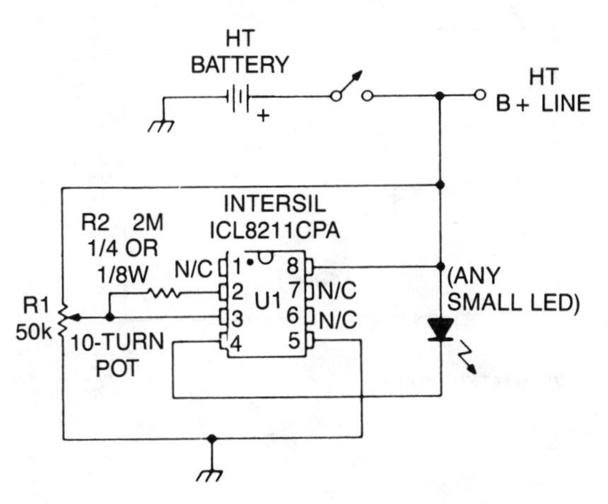

HAM RADIO *Fig. 1-8*

The precision voltage-monitor chip contains a temperature-compensated voltage reference. R1 divides down the battery voltage to match the built-in reference voltage of IC1 (1.15 V). When the voltage at pin 3 falls below 1.15 V, pin 4 supplies a constant current of 7 mA to drive a small LED. About 0.2 V of hysteresis is added with R2. Without hysteresis, the LED could flicker on and off when the monitored voltage varies around the set point, as might be the case on voice peaks during receive.

LOW-VOLTAGE MONITOR

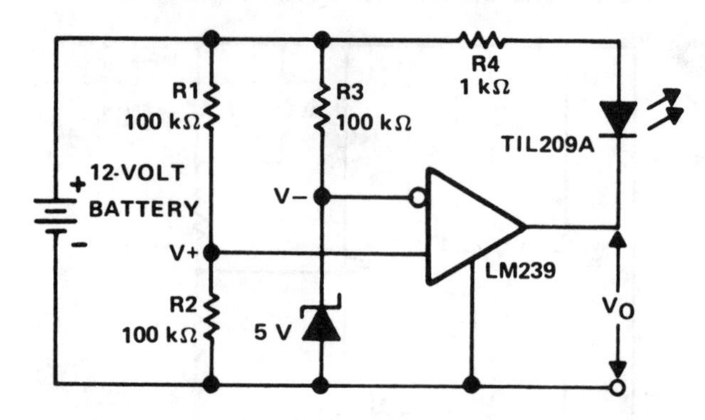

a. SCHEMATIC OF CIRCUIT FOR LOW-VOLTAGE INDICATOR

TEXAS INSTRUMENTS *Fig. 1-9*

This circuit monitors the voltage of a battery and warns the operator when the battery voltage is below a preset level by turning on an LED. The values are set for a 12-V automobile battery. The preset value is 10 V.

UNDERVOLTAGE INDICATOR FOR BATTERY-OPERATED EQUIPMENT

As a result of the low duty cycle of flashing LED, the average current drain is 1 mA or less. The NE555 will trigger the LED on when the monitored voltage falls to 12 V. The ratio of R1 to R2 only needs to be changed if you want to change the voltage point at which the LED is triggered.

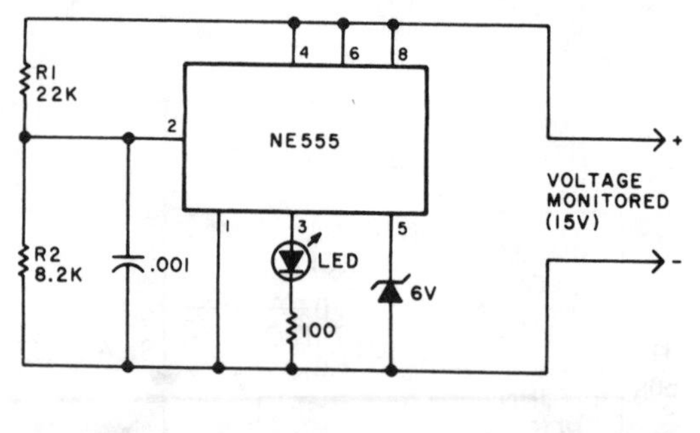

73 AMATEUR RADIO *Fig. 1-10*

BATTERY-CONDITION INDICATOR

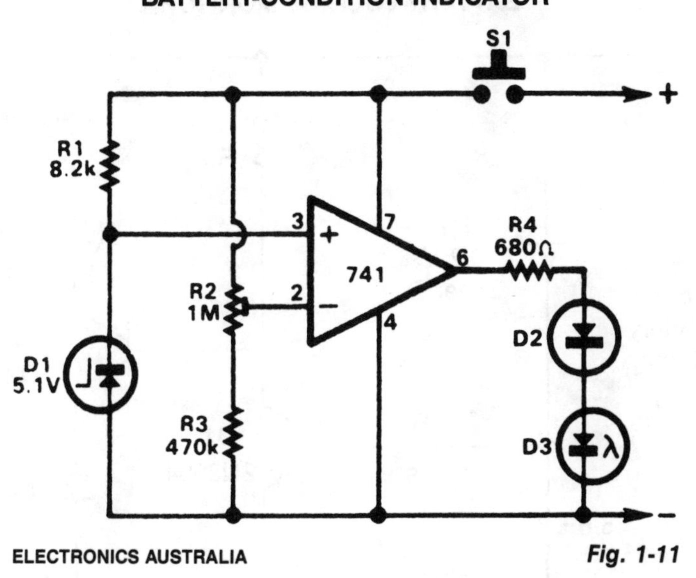

ELECTRONICS AUSTRALIA *Fig. 1-11*

A 741 op amp is used as a voltage comparator. The noninverting input is connected to zener reference source. The reference voltage is 5.1 V. R2 is adjusted so that the voltage at the inverting input is half the supply voltage. When supply is higher than 10.2 V, the LED will not light. When the supply falls just fractionally below the 10.2-V level, the IC inverting input will be slightly negative of the noninverting input, and the output will swing fully positive. The LED will light and indicate that the supply voltage has fallen to the preset threshold level. The LED can be made to light at other voltages by adjusting R2.

EQUIPMENT-ON REMINDER

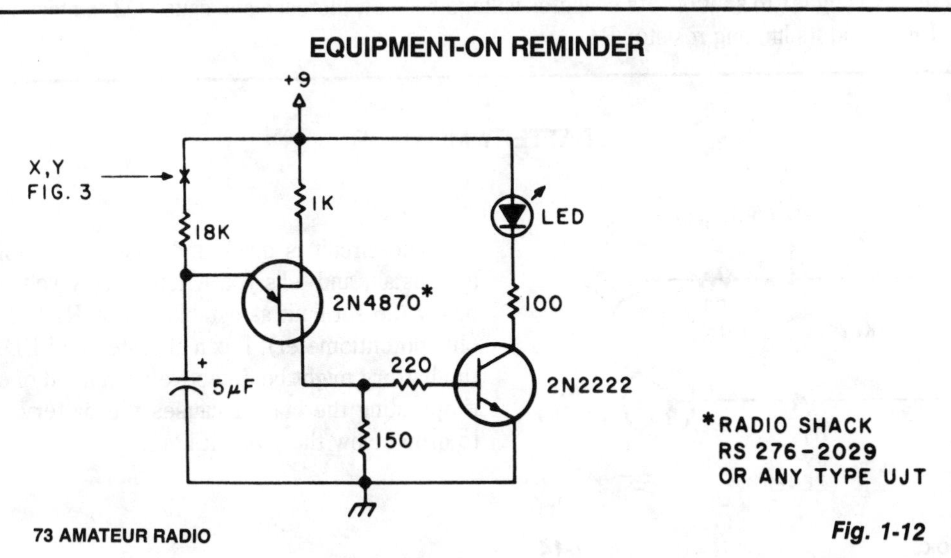

73 AMATEUR RADIO *Fig. 1-12*

As a result of the low duty cycle of the flashing LED, the average current drain is 1 mA or less.

BATTERY-VOLTAGE MONITOR

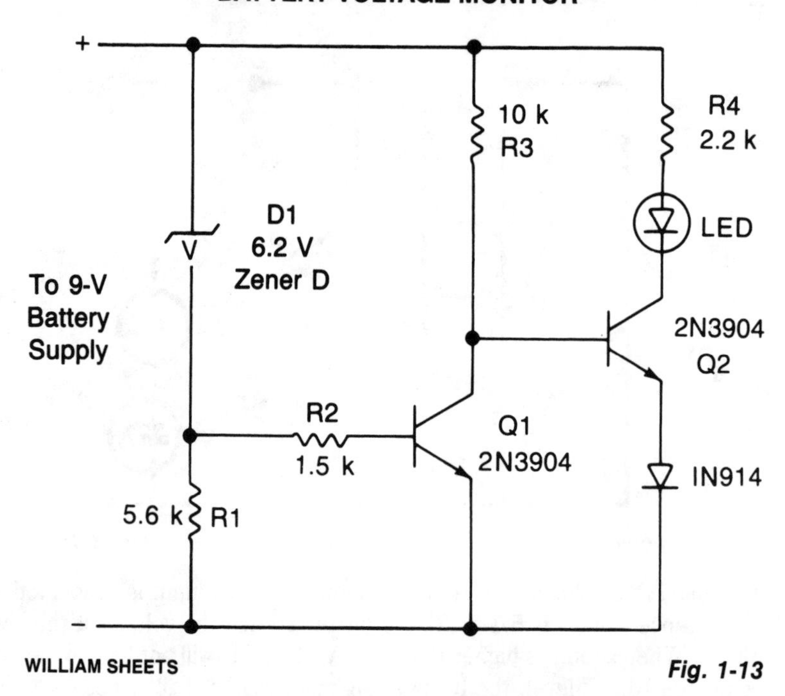

WILLIAM SHEETS

Fig. 1-13

This circuit gives an early warning of the discharge of batteries. Zener diode D1 is chosen for the voltage below which an indication is required (9 V). Should the supply drop to below 7 V, D1 will cease conducting causing Q1 to shut off. Its collector voltage will now increase and cause Q2 to start conducting via LED1 and its limiting resistor R4.

BATTERY MONITOR

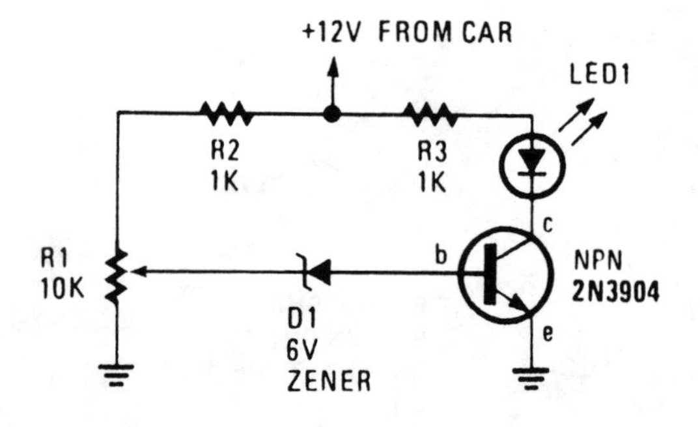

TAB BOOKS

Fig. 1-14

The circuit is quick and easy to put together and install, and tells you when battery voltage falls below the set limit as established by R1 (a 10,000-ohm potentiometer). It can indicate, via LED1, that the battery might be defective or in need of change if operating the starter causes the battery voltage to drop below the present limit.

LITHIUM BATTERY STATE-OF-CHARGE INDICATOR

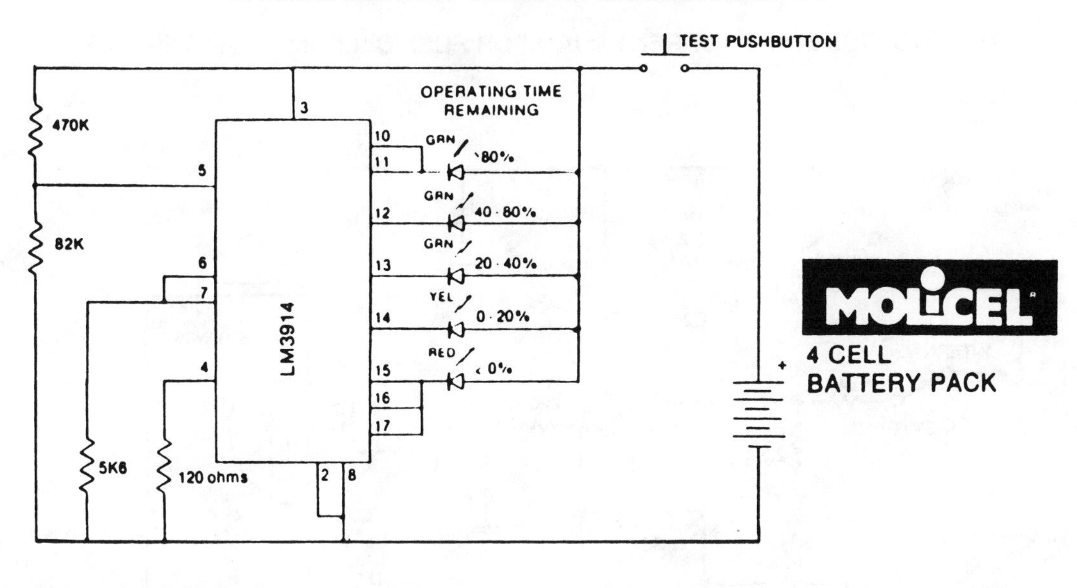

MOLI ENERGY LIMITED ***Fig. 1-15***

State-of-charge indication for a sloping-voltage discharge can be used as a state-of-charge indicator. A typical voltage comparator circuit that gives a visual indication of state-of-charge is shown. Components identified are for a 4-cell input voltage of 9.6 to 5.2 V.

STEP-UP SWITCHING REGULATOR FOR 6-V BATTERIES

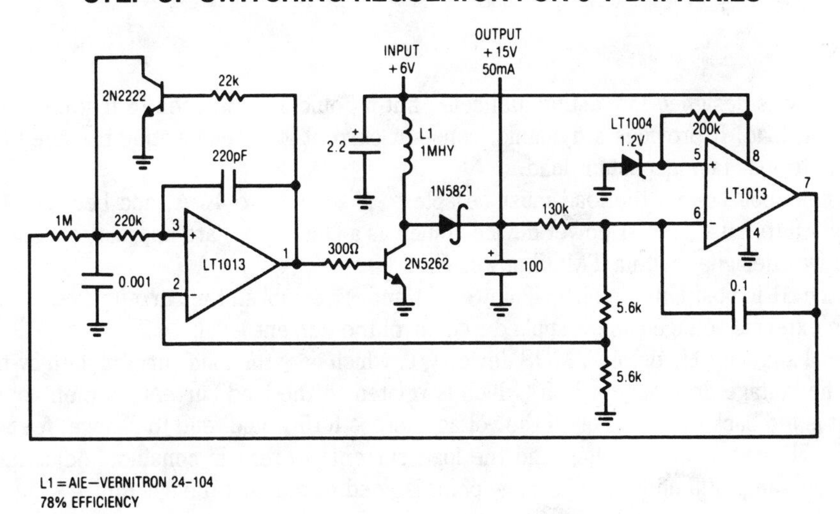

LINEAR TECHNOLOGY CORP. ***Fig. 1-16***

DYNAMIC, CONSTANT-CURRENT LOAD FOR FUEL CELL/BATTERY TESTING

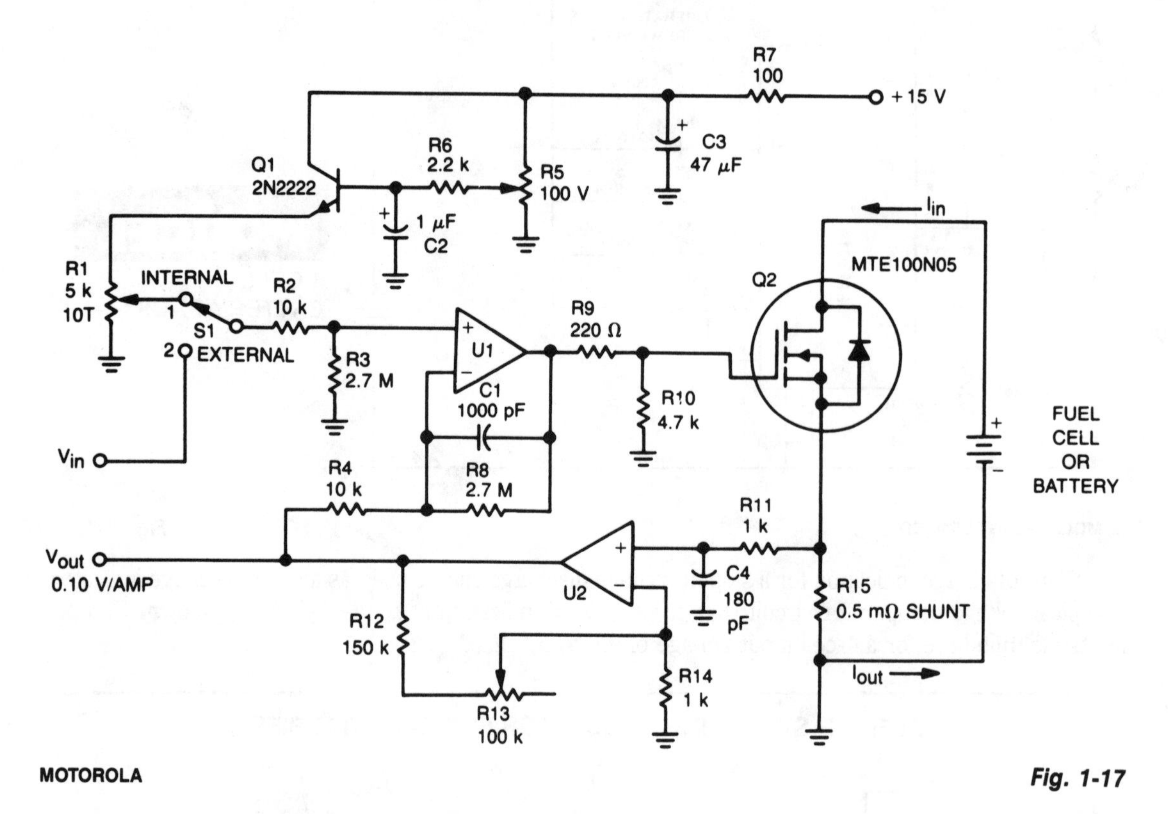

MOTOROLA *Fig. 1-17*

This circuit was designed for testing fuel cells, but it could also be used to test batteries under a constant-current load. It provides a dynamic, constant-current load, eliminating the need to manually adjust the load to maintain a constant load.

For fuel cell applications, the load must be able to absorb 20 to 40 A, and because a single cell develops only 0.5 to 1.0 V, bipolar power devices (such as a Darlington) are impractical. Therefore, this dynamic load was designed with a TMOS power FET (Q2).

With switch S1 in position 1, emitter-follower Q1 and R1 establish the current level for the load. In position 2, an external voltage can be applied to control the current level.

Operational amplifier U1 drives TMOS device Q1, which sets the load current seen by the fuel cell or battery. The voltage drop across R15, which is related to the load current, is then applied to U2, whose output is fed back to U1. Thus, if the voltage across R15 would tend to change, feedback to the minus input of U1 causes the voltage (and the load current) to remain constant. Adjustment of R13 controls the volts/amp of feedback. The V_{OUT} point is used to monitor the system.

CAR BATTERY-CONDITION CHECKER

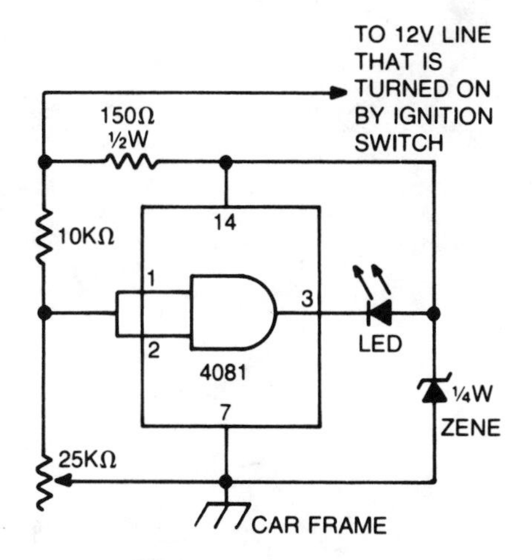

MODERN ELECTRONICS *Fig. 1-18*

This circuit uses an LED and a 4081 CMOS integrated circuit. The variable resistor sets the voltage at which the LED turns on. Set the control so that the LED lights when the voltage from the car's ignition switch drops below 13.8 V. The LED normally will light every now and then for a short period of time. But, if it stays on for very long, your electrical system is in trouble.

CAR-BATTERY MONITOR

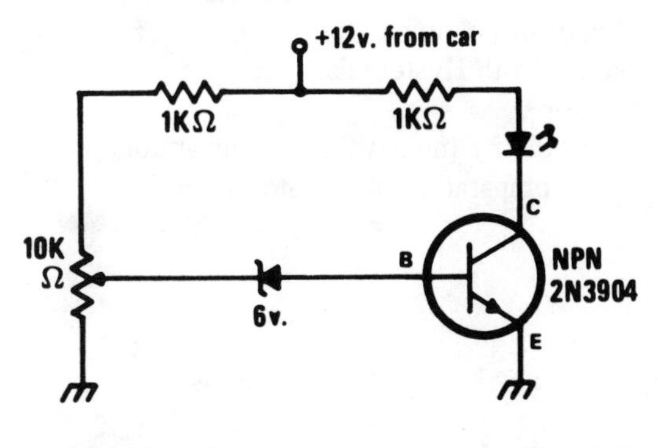

MODERN ELECTRONICS *Fig. 1-19*

The warning light (LED) indicates when the battery voltage falls below level set by 10-kΩ pot. This monitor can indicate that battery is defective or needs charging if cranking drops battery voltage below a preset ''safe'' limit.

2

Comparator Circuits

The sources of the following circuits are contained in the Sources section, which begins on page 217. The figure number in the box of each circuit correlates to the source entry in the Sources section.

DIODE FEEDBACK COMPARATOR

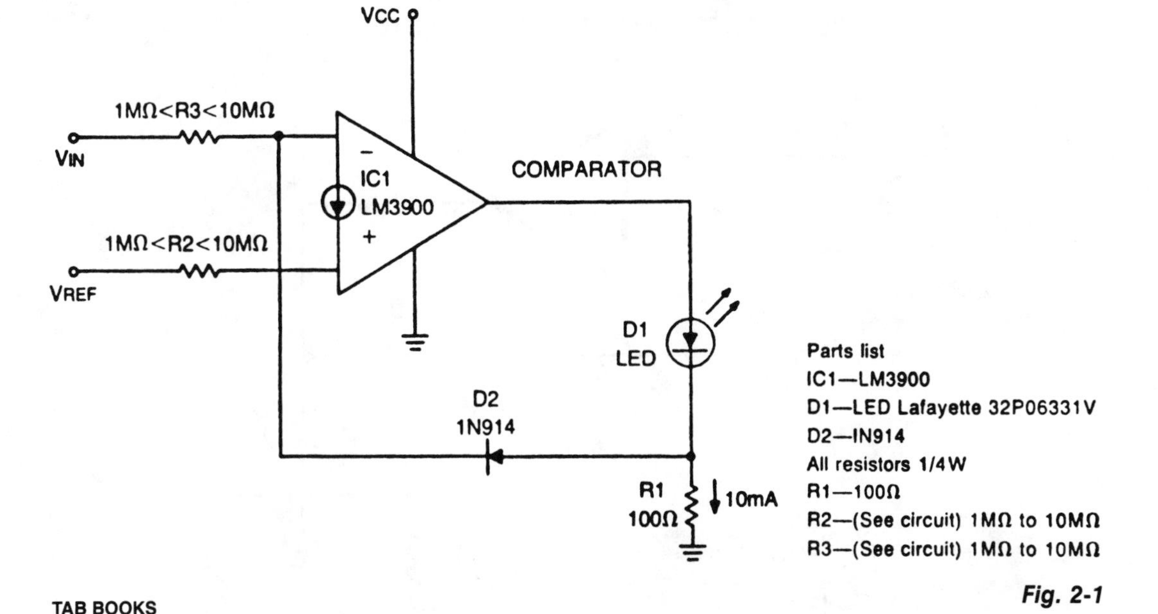

TAB BOOKS

Fig. 2-1

This circuit can drive an LED display with constant current independently of wide power-supply voltage changes. It can operate with a power supply range of at least 4 to 30 V. With 10 MΩ resistances for R2, R3, and the inverting input of the comparator grounded, the circuit becomes an LED driver with very high input impedance. The circuit can also be used in many other applications where a controllable constant-current source is needed.

UNDERVOLTAGE/OVERVOLTAGE INDICATOR

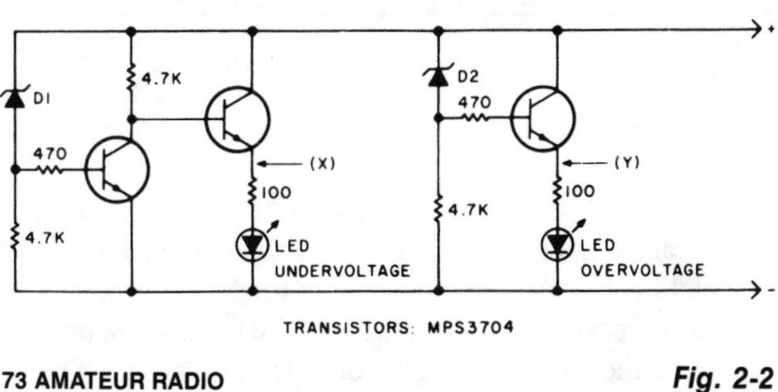

73 AMATEUR RADIO

Fig. 2-2

This circuit will make the appropriate LED glow if the monitored voltage goes below or above the value determined by zener diodes D1 and D2.

COMPARATOR WITH VARIABLE HYSTERESIS

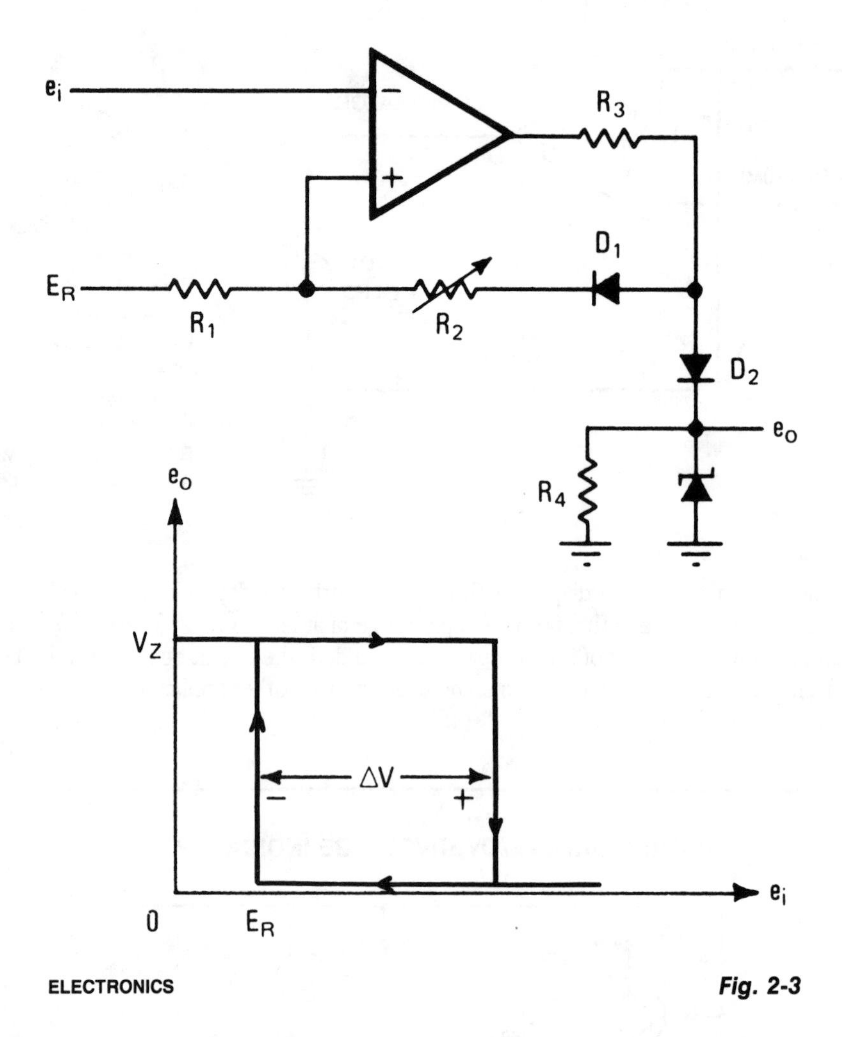

ELECTRONICS

Fig. 2-3

An operational amplifier can be used as a convenient device for analog comparator applications that require two different trip points. The addition of a positive-feedback network introduces a precise variable hysteresis into the usual comparator switching action. Such feedback develops two comparator trip points that are centered about the initial trip point or reference point. The voltage difference, ΔV, between the trip points can be adjusted by varying resistor R2. When the output voltage is taken from the zener diode, as shown, it switches between zero and V_Z, the zener voltage.

LIMIT COMPARATOR I

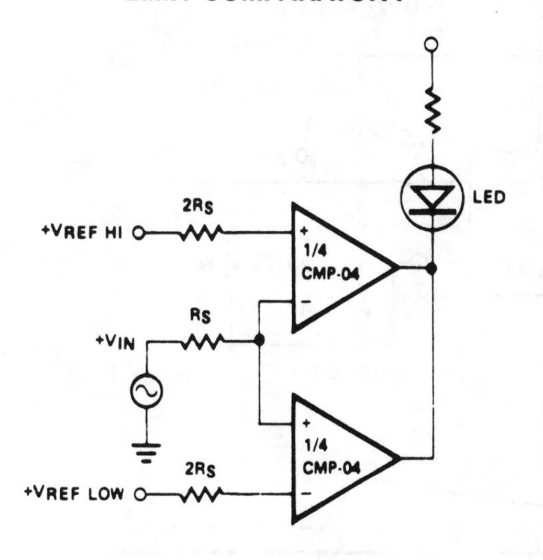

PRECISION MONOLITHICS *Fig. 2-4*

DOUBLE-ENDED LIMIT COMPARATOR

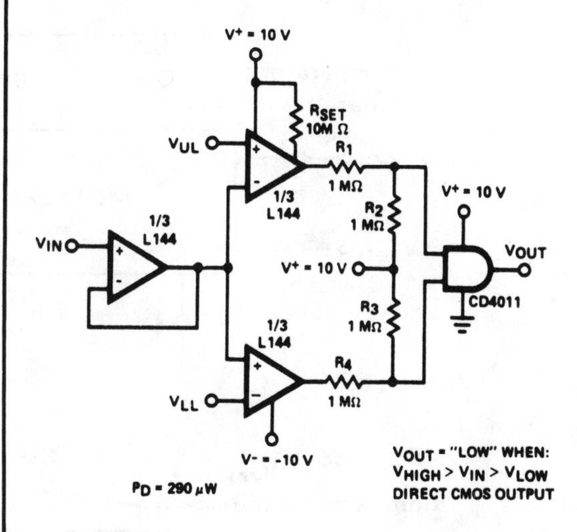

SILICONIX *Fig. 2-6*

COMPARATOR CLOCK CIRCUIT

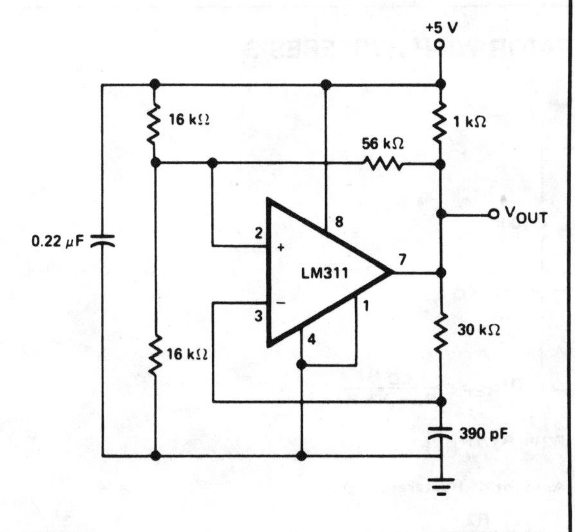

TELEDYNE *Fig. 2-5*

LIMIT COMPARATOR II

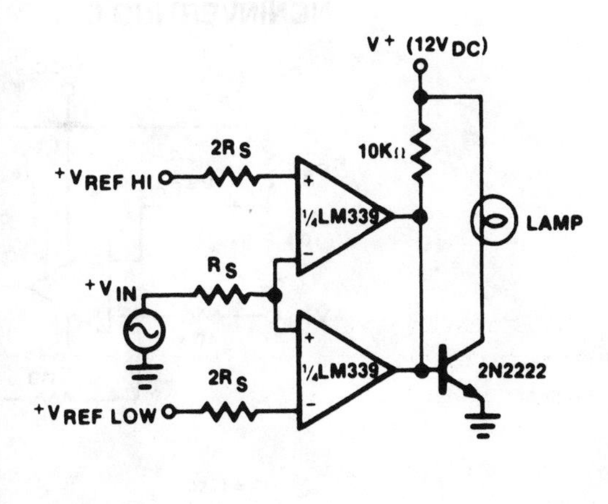

SIGNETICS *Fig. 2-7*

COMPARATOR WITH TIME OUT

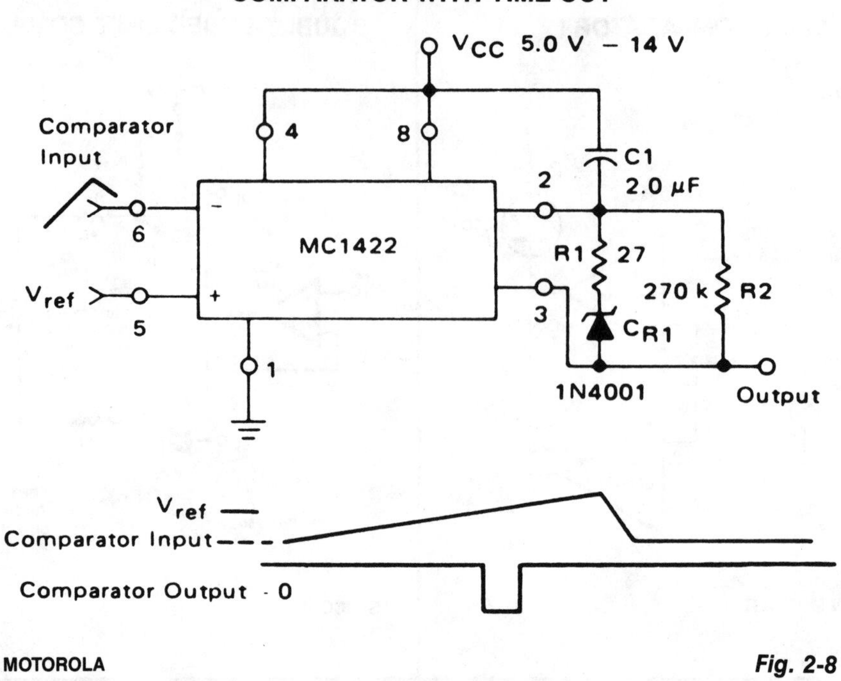

MOTOROLA *Fig. 2-8*

The MC1422 is used as a comparator with the capability of timing output pulse when the inverting input (Pin 6) is ≥ the noninverting input (Pin 5). The frequency of the pulses for the values of R2 and C1 as shown is approximately 2.0 Hz, and the pulse width 0.3 ms.

NONINVERTING COMPARATOR WITH HYSTERESIS

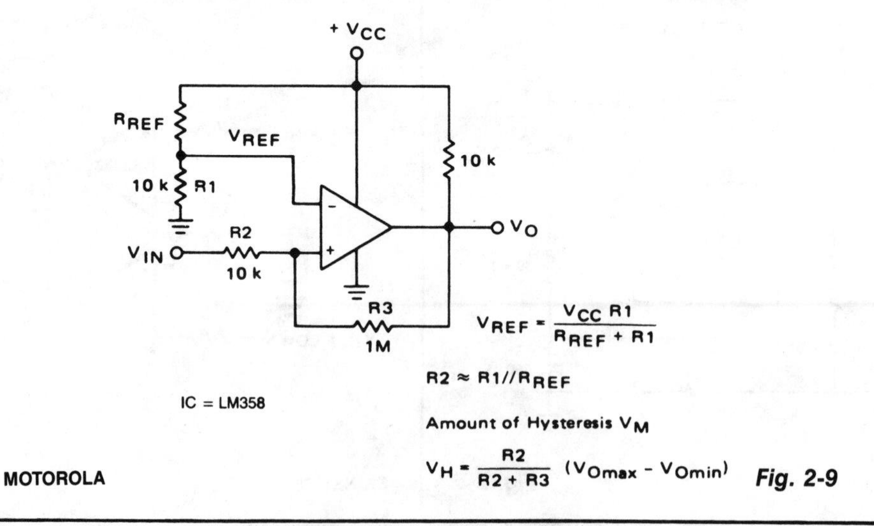

MOTOROLA *Fig. 2-9*

DUAL-LIMIT COMPARATOR

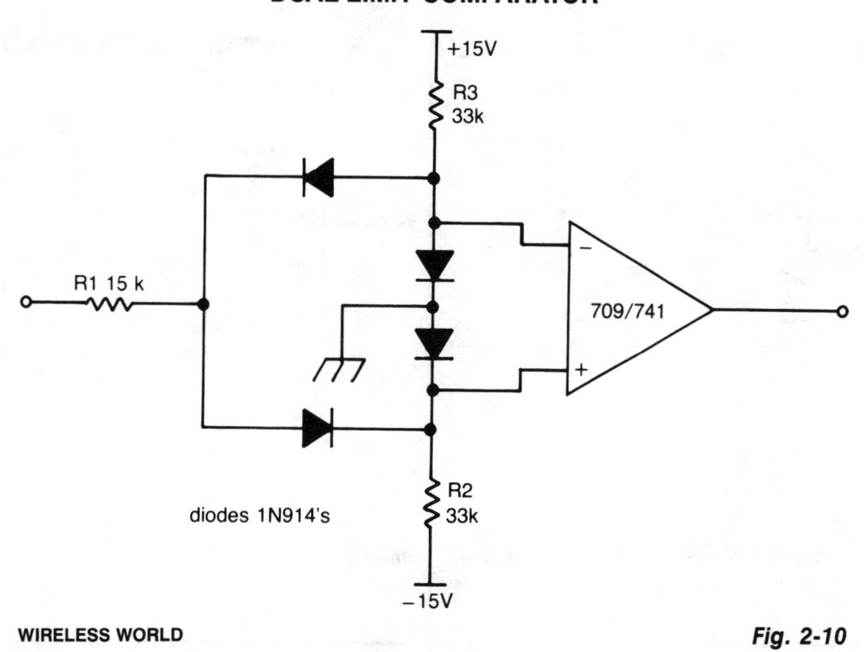

WIRELESS WORLD

Fig. 2-10

This circuit gives a positive output when the input voltage exceeds 8.5 V. Between these limits the output is negative. The positive limit point is determined by the ratio of R1, R2, and the negative point by R1 and R3. The forward voltage drop across the diodes must be allowed for. The output can be inverted by reversing the inputs to the op amp. The 709 is used without frequency compensation.

HIGH-/LOW-LIMIT ALARM

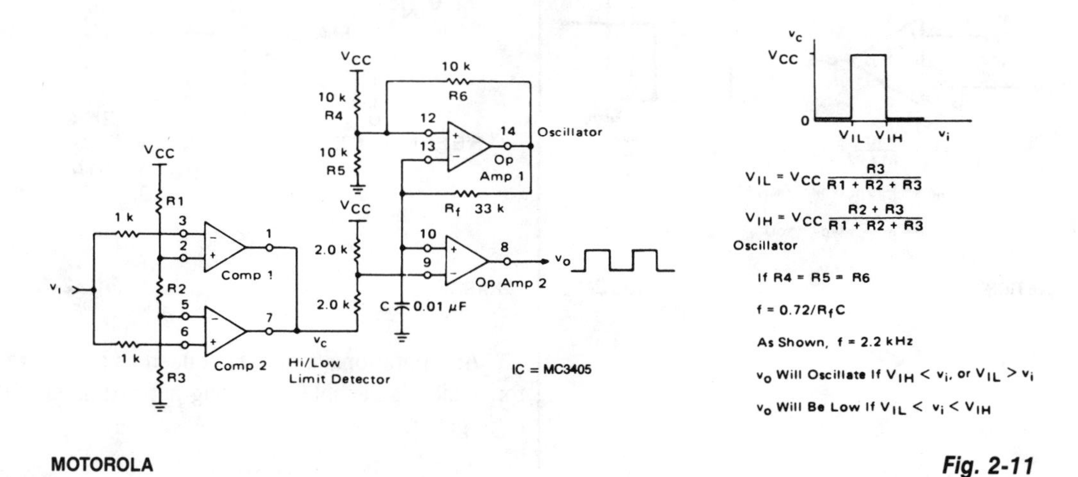

MOTOROLA

Fig. 2-11

PRECISION DUAL-LIMIT GO/NO-GO TESTER

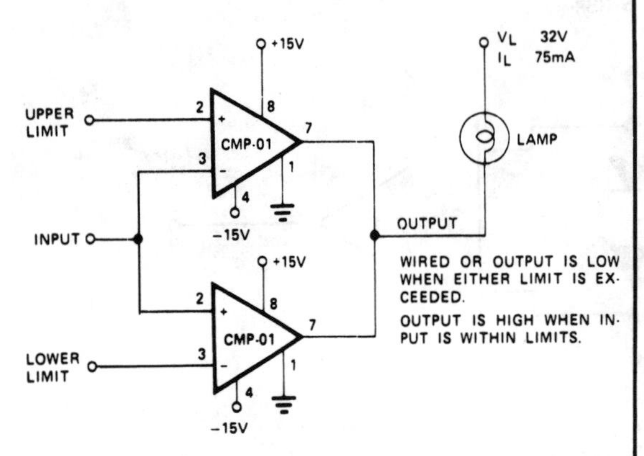

PRECISION MONOLITHICS *Fig. 2-12*

HIGH-IMPEDANCE COMPARATOR

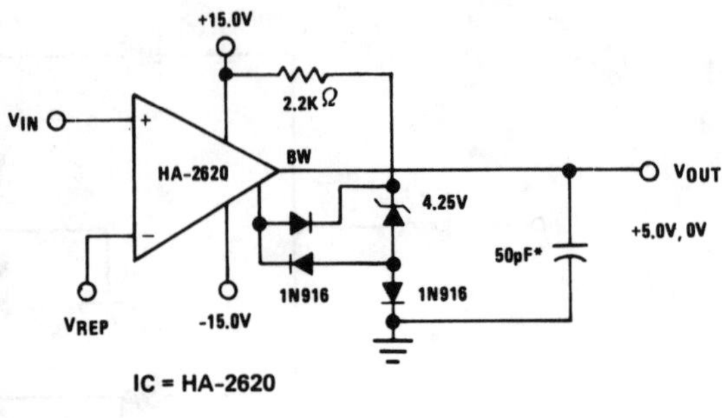

HARRIS *Fig. 2-14*

COMPARATOR WITH HYSTERESIS

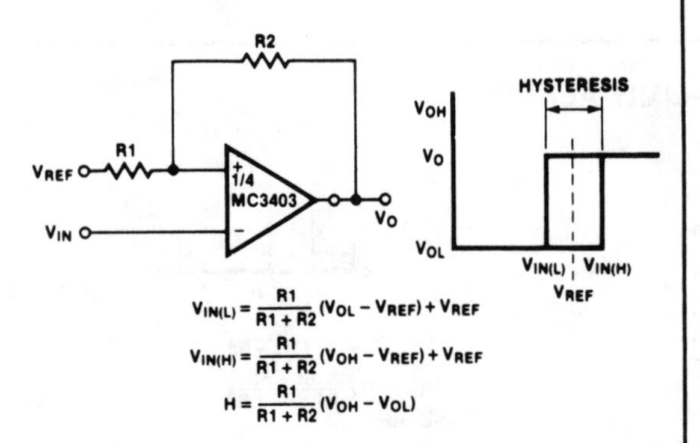

$$V_{IN(L)} = \frac{R1}{R1 + R2}(V_{OL} - V_{REF}) + V_{REF}$$

$$V_{IN(H)} = \frac{R1}{R1 + R2}(V_{OH} - V_{REF}) + V_{REF}$$

$$H = \frac{R1}{R1 + R2}(V_{OH} - V_{OL})$$

SIGNETICS *Fig. 2-13*

COMPARATOR

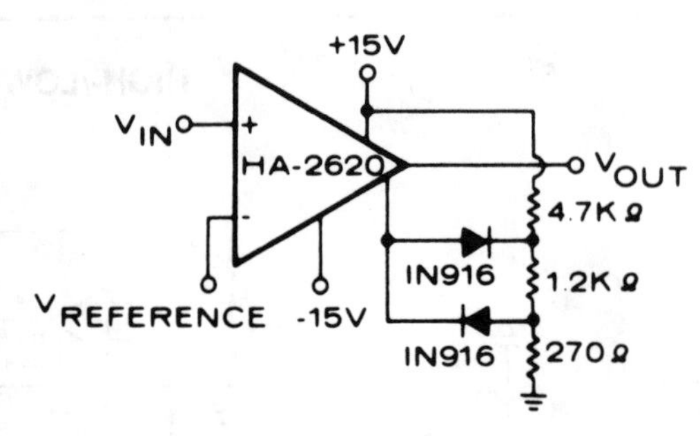

HARRIS *Fig. 2-15*

An operational amplifier is used as a conparator, which is capable of driving approximately 10 logic gates.

OPPOSITE-POLARITY INPUT-VOLTAGE COMPARATOR

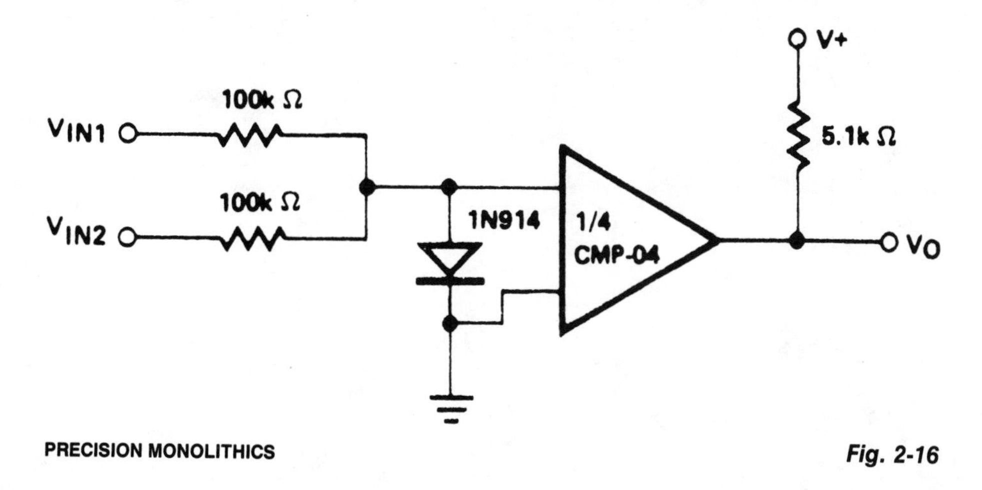

PRECISION MONOLITHICS

Fig. 2-16

INVERTING COMPARATOR WITH HYSTERESIS

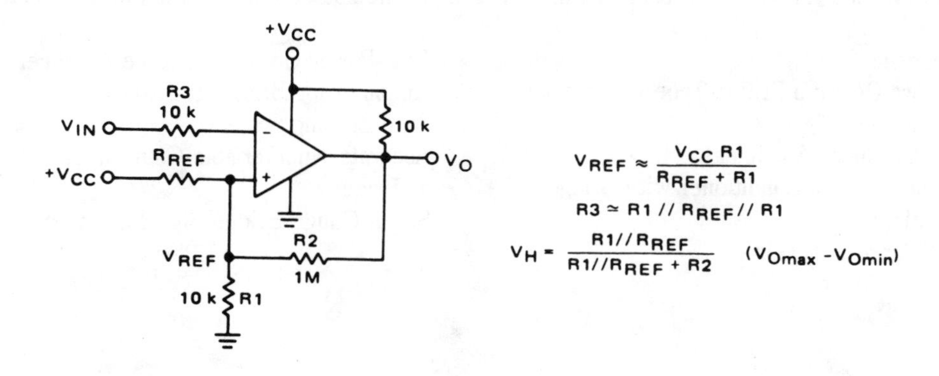

$$V_{REF} \approx \frac{V_{CC}\,R1}{R_{REF} + R1}$$

$$R3 \simeq R1 \,//\, R_{REF} \,//\, R1$$

$$V_H = \frac{R1//R_{REF}}{R1//R_{REF} + R2} \quad (V_{Omax} - V_{Omin})$$

MOTOROLA

Fig. 2-17

3

Bridge Circuits

The sources of the following circuits are contained in the Sources section, which begins on page 217. The figure number in the box of each circuit correlates to the source entry in the Sources section.

BRIDGE CIRCUIT

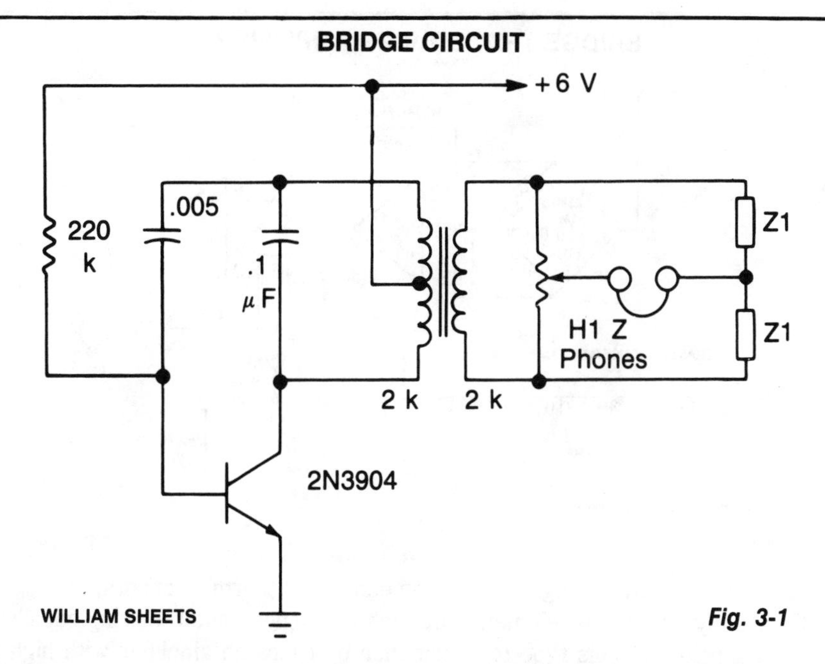

Fig. 3-1

The transistor is connected as an audio oscillator, using an audio transformer in the collector. The secondary goes to a linear pot. The ratio between the two parts of the pot from the slider is proportional to the values of Z_1 and Z_2 when no signal is heard in the phones.

TYPICAL TWO OP-AMP BRIDGE-TYPE DIFFERENTIAL AMPLIFIER

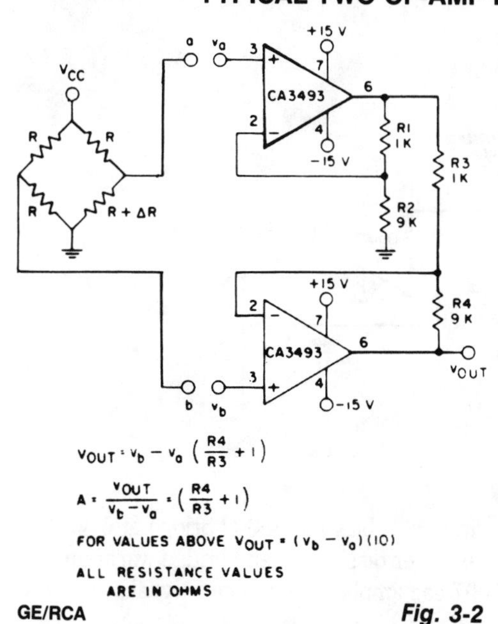

GE/RCA

Fig. 3-2

Using a CA3493 BiMOS op amp to provide high input impedance and good common-mode rejection ratio (depends primarily on matching of resistor networks).

BRIDGE TRANSDUCER AMPLIFIER

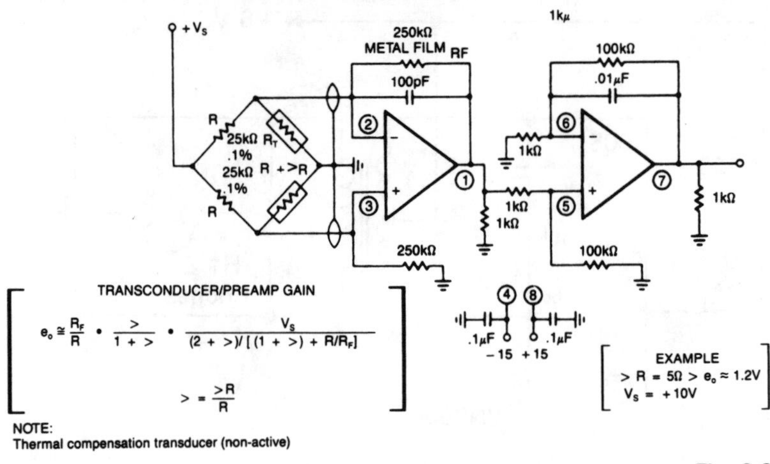

SIGNETICS

Fig. 3-3

In applications involving strain gauges, accelerometers, and thermal sensors, a bridge transducer is often used. Frequently, the sensor elements are high resistance units requiring equally high bridge resistance for good sensitivity. This type of circuit then demands an amplifier with high input impedance, low bias current and low drift. The circuit shown represents a possible solution to these general requirements.

STRAIN GAUGE SIGNAL CONDITIONER WITH BRIDGE EXCITATION

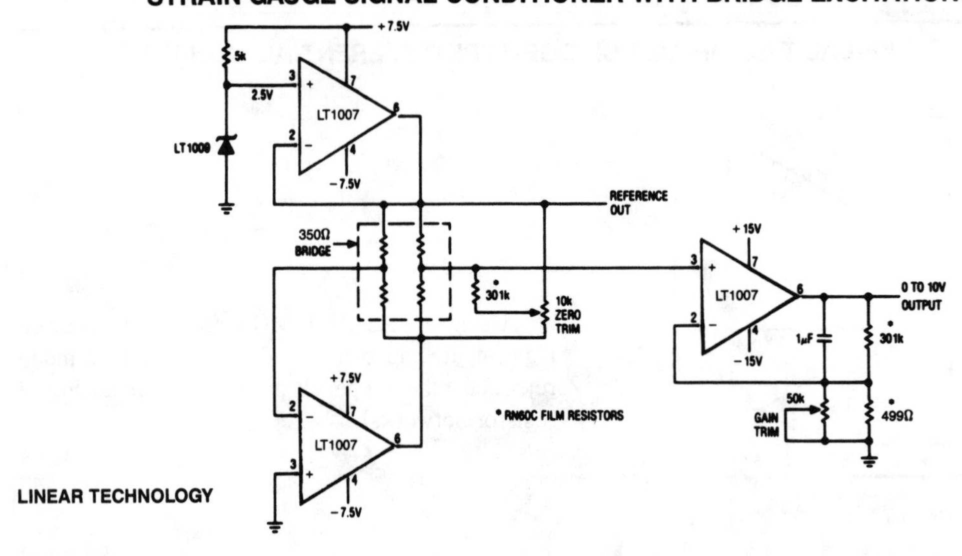

LINEAR TECHNOLOGY

Fig. 3-4

The LT1007 is capable of providing excitation current directly to bias the 350-Ω bridge at 5 V. With only 5 V across the bridge, as opposed to the usual 10 V, total power dissipation and bridge warm-up drift is reduced. The bridge output signal is halved, but the LT1007 can amplify the reduced signal accurately.

BRIDGE BALANCE INDICATOR

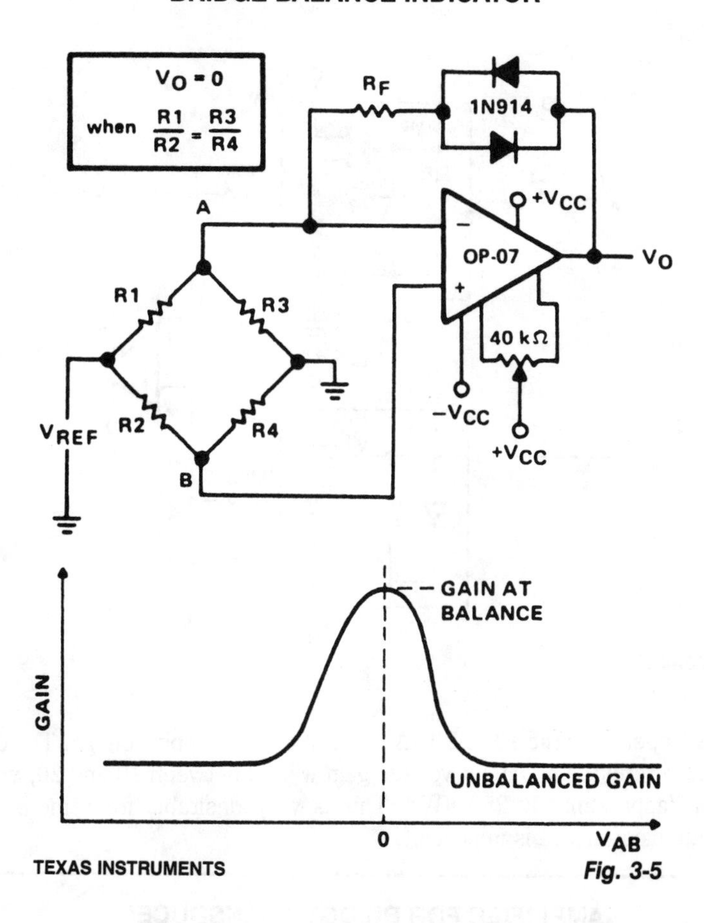

TEXAS INSTRUMENTS

Fig. 3-5

The indicator provides an accurate comparison of two voltages by indicating their degree of balance (or imbalance). Detecting small variations near the null point is difficult with the basic Wheatstone bridge alone. Amplifying voltage differences near the null point will improve circuit accuracy and ease of use.

The 1N914 diodes in the feedback loop result in high sensitivity near the point of balance ($R_1/R_2 = R_3/R_4$). When the bridge is unbalanced the amplifier's closed-loop gain is approximately R_F/r, where r is the parallel equivalent of R_1 and R_3. The resulting gain equation is $G = R_F(1/R_1 + 1/R_3)$. During an unbalanced condition the voltage at point A is different from that at point B. This difference voltage (V_{AB}), amplified by the gain factor G, appears as an output voltage, as the bridge approaches a balanced condition ($R_1/R_2 = R_3/R_4$), V_{AB} approaches zero. As V_{AB} approaches zero the 1N914 diodes in the feedback loop lose their forward bias and their resistance increases, causing the total feedback resistance to increase. This increases circuit gain and accuracy in detecting a balanced condition. The figure shows the effect of approaching balance on circuit gain. The visual indicator used at the output of the OP-07 could be a sensitive voltmeter or oscilloscope.

LOW-POWER COMMON-SOURCE AMPLIFIER

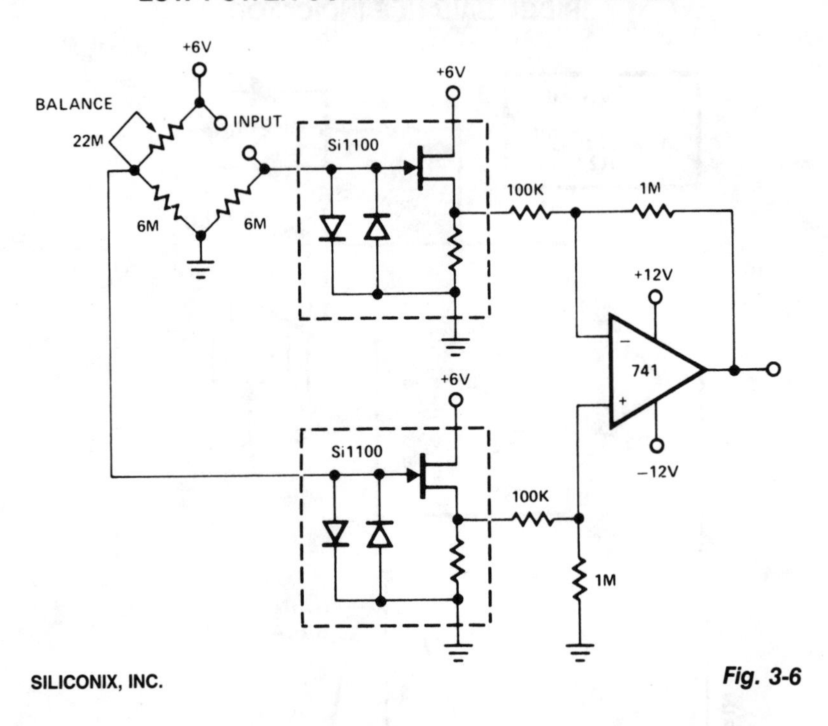

SILICONIX, INC. *Fig. 3-6*

A circuit that will operate in the 10-to-20 μA range at a 12-V supply voltage. The diode protection is available in this configuration. The circuit voltage gain will be between 10 and 20, with extremely low power consumption (approximately 250 μW). This is very desirable for remote or battery operation, where minimum maintenance is important.

AMPLIFIER FOR BRIDGE TRANSDUCERS

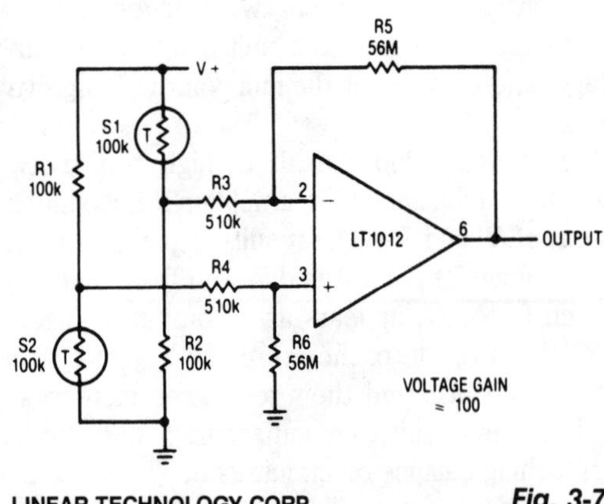

LINEAR TECHNOLOGY CORP. *Fig. 3-7*

AUTO-ZEROING SCALE

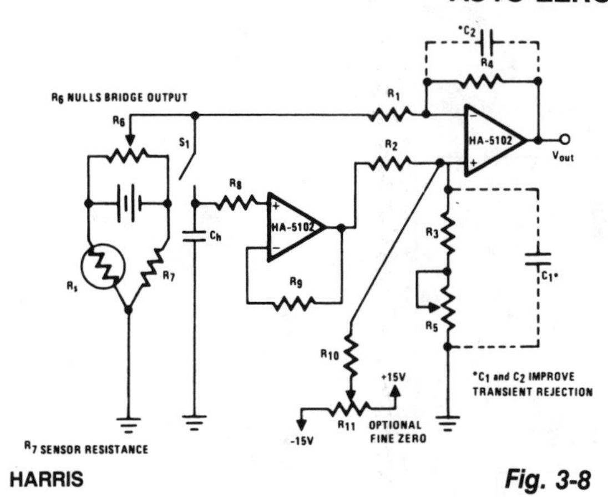

HARRIS

Fig. 3-8

Electronic scales have come into wide use and the HA-510X, as a very low noise device, can improve such designs. This circuit uses a strain-gauge sensing element as part of a resistive Wien-bridge. An auto-zero circuit is also used in this design by including a sample-and-hold-network.

The bridge signal drives the inverting input of a differentially configured HA-5102. The noninverting input is driven by the other half of the HA-5102 used as a buffer for the holding capacitor, CH. This second amplifier and its capacitor CH form the sampling circuit used for automatic output zeroing. The 20-kΩ resistor between the holding capacitor CH and the input terminal, reduces the drain from the bias currents. A second resistor RG is used in the feedback loop to balance the effect of R8. If R7 is approximately equal to the resistance of the strain gauge, the input signal from the bridge can be roughly nulled with R6. With very close matching of the ratio R4/R1 to R3/R2, the output offset can be nulled by closing S1. This will charge CH and provide a 0-V difference to the inputs of the second amplifier, which results in a 0-V output. In this manner, the output of the strain gauge can be indirectly zeroed. R10 and potentiometer R11 provide an additional mechanism for fine tuning V_{OUT}, but they can also increase offset voltage away from the zero point.

ACCURATE-NULL/VARIABLE-GAIN CIRCUIT

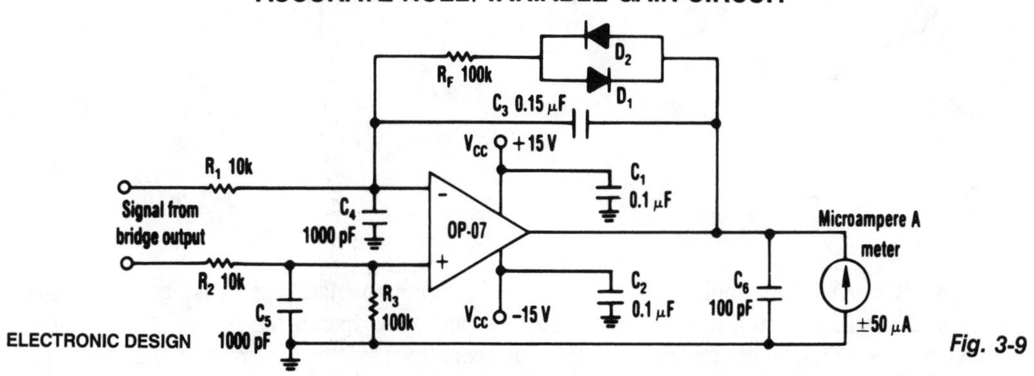

ELECTRONIC DESIGN

Fig. 3-9

The circuit can use any general-purpose, low offset, low-drift op amp, such as the OP-07. The differential signal from the bridge feeds an amplifier that drives an ordinary, rugged ±50-μA meter. Near the null point, however, the drastically reduced signal level from the bridge requires very high gain to acheive a high null resolution. To provide the variable-gain feature, the op amp's feedback path needs a dynamic resistance that increases as the input signal drops. Two common signal diodes, D1 and D2, in an antiparallel configuration in the feedback path supply function for all positive and negative inputs. To stabilize the op amp circuit at high gain, capacitors C3, C5, and C6 reduce response to high frequencies; capacitors C1 and C2 bypass the amplifier's power supplies.

ac BRIDGE

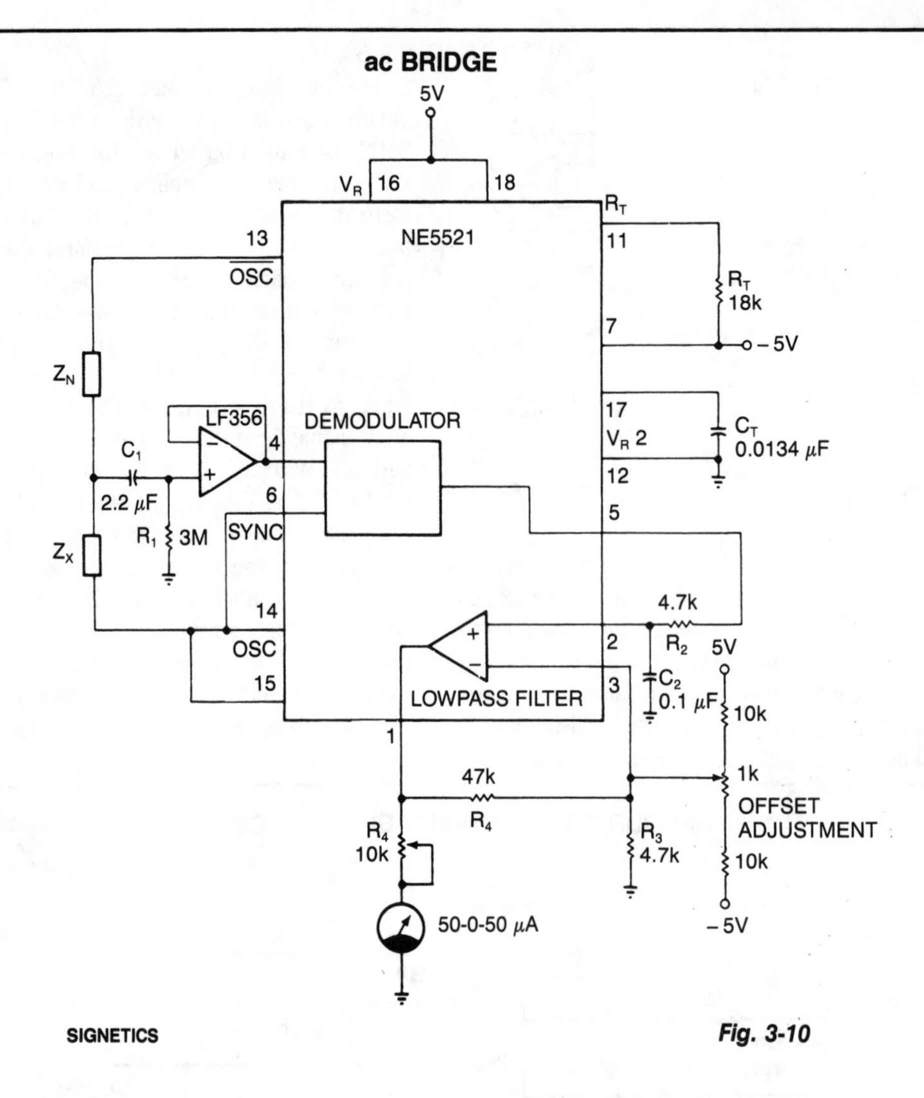

SIGNETICS

Fig. 3-10

The circuit provides a simple and cost-effective solution to matching resistors and capacitors. Impedances Z_R and Z_X form a half-bridge, while OSC and $\overline{\text{OSC}}$ excite the bridge differentially. The external op amp is a FET input amplifier (LF356) with very low input bias current on the order of 30 pA (typical). C1 allows ac coupling by blocking the dc common-mode voltage from the bridge, while R1 biases the output of LF356 to 0 V at dc. Use of FET input op amp ensures that dc offset as a result of bias current through R1 is negligible. Ac output of the demodulator is filtered via the uncommitted amp to provide dc voltage for the meter. The 10-kΩ potentiometer, R5, limits the current into the meter to a safe level. Calibration begins by placing equal impedances at Z_R and Z_X, and the system offset is nulled by the offset adjust circuit so that Pin 1 is at 0 V. Next, known values are placed at Z_X and the meter deviations are calibrated. The bridge is now ready to measure an unknown impedance at Z_X with ±0.05% accuracy or better.

STRAIN-GAUGE BRIDGE-SIGNAL CONDITIONER

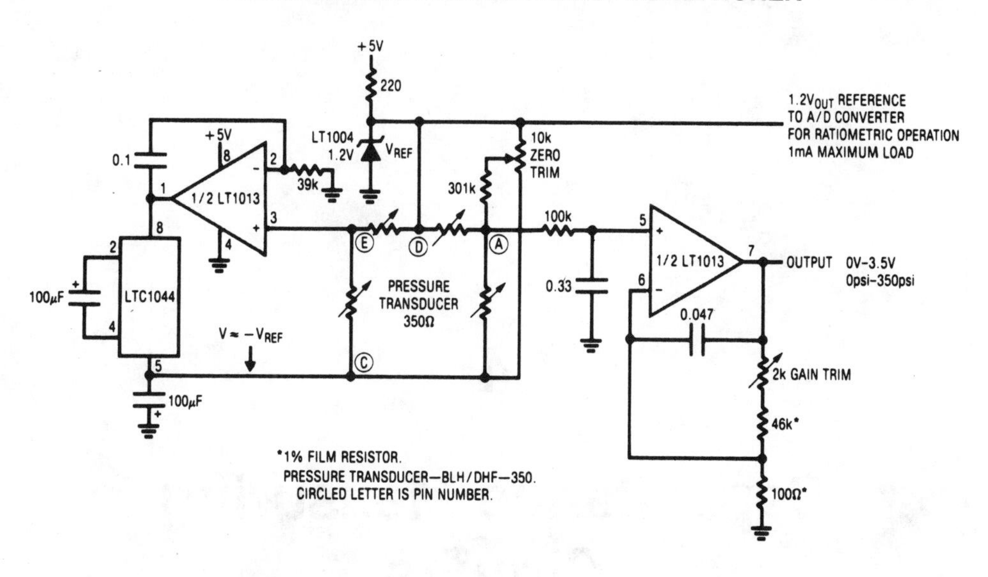

LINEAR TECHNOLOGY CORP. *Fig. 3-11*

4

Capacitance Measuring Circuits

The sources of the following circuits are contained in the Sources section, which begins on page 217. The figure number in the box of each circuit correlates to the source entry in the Sources section.

ACCURATE DIGITAL CAPACITANCE METER

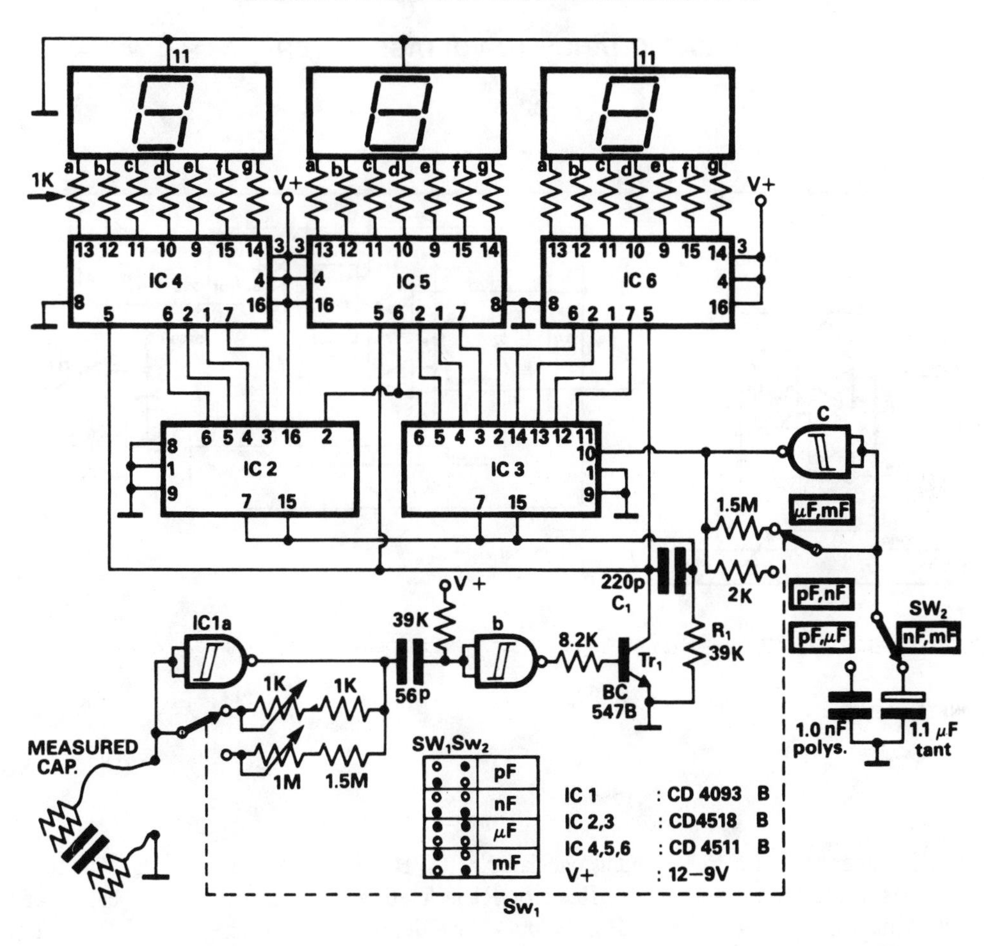

ELECTRONIC ENGINEERING

Fig. 4-1

The principle of operation is counting the pulse number derived from a constant-frequency oscillator during a fixed time interval produced by another lower frequency oscillator. This oscillator uses the capacitor being measured as the timing. The capacitance measurement is proportional during pulse counting during a fixed time interval. The astable oscillator formed by IC1C produces a pulse train of constant frequency. Gate IC1A also forms an oscillator whose oscillation period is given approximately by the equation: $T = 0.7$ RC.

Period T is linearly dependent on the capacitance, C. This period is used as the time interval for one measurement. The differentiator network following the oscillator creates the negative spikes shaped in narrow pulses by IC1B NAND Schmitt Trigger. The differentiator formed by R1 and C1 produces a negative spike which resets the counters. The display shows the number of high-frequency oscillator pulses that enter the counter during the measurement period.

CAPACITANCE-TO-VOLTAGE METER

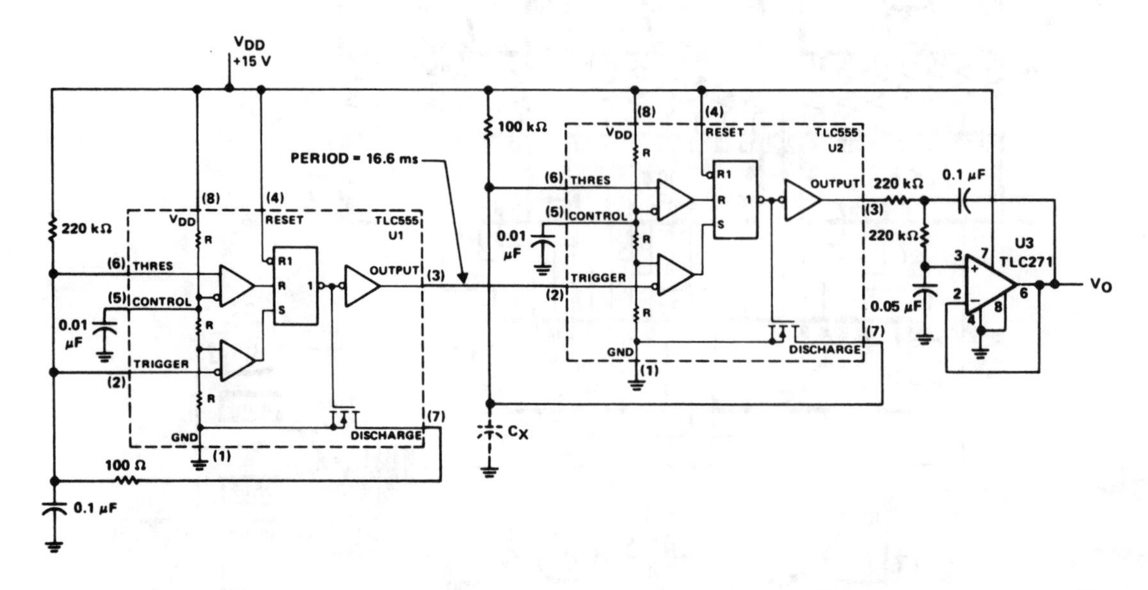

TEXAS INSTRUMENTS

Fig. 4-2

Timer U1 operates as a free-running oscillator at 60 Hz, providing trigger pulses to timer U2, which operates in the monostable mode. Resistor R1 is fixed and capacitor Cx is the capacitor being measured. While the output of U2 is 60 Hz, the duty cycle depends on the value of C_x. U3 is a combination low-pass filter and unity-gain follower whose dc voltage output is the time-averaged amplitude of the output pulses of U2, as shown in the timing diagram.

The diagram shows when the value of C_x is small the duty cycle is relatively low. The output pulses are narrow and produce a lower average dc voltage level at the output of U3. As the capacitance value of C_x increases, the duty cycle increases, which makes the output pulses at U2 wider and the average dc level output at U3 increases. The graph illustrates capacitance values of 0.01 to 0.1μF, plotted against the output voltage of U3. Notice the excellent linearity and direct one-to-one scale calibration of the meter. If this does not occur the 100 kΩ resistor, R1, can be replaced with a potentiometer, which can be adjusted to the proper value for the meter being used.

CAPACITANCE-TO-VOLTAGE METER (*Cont.*)

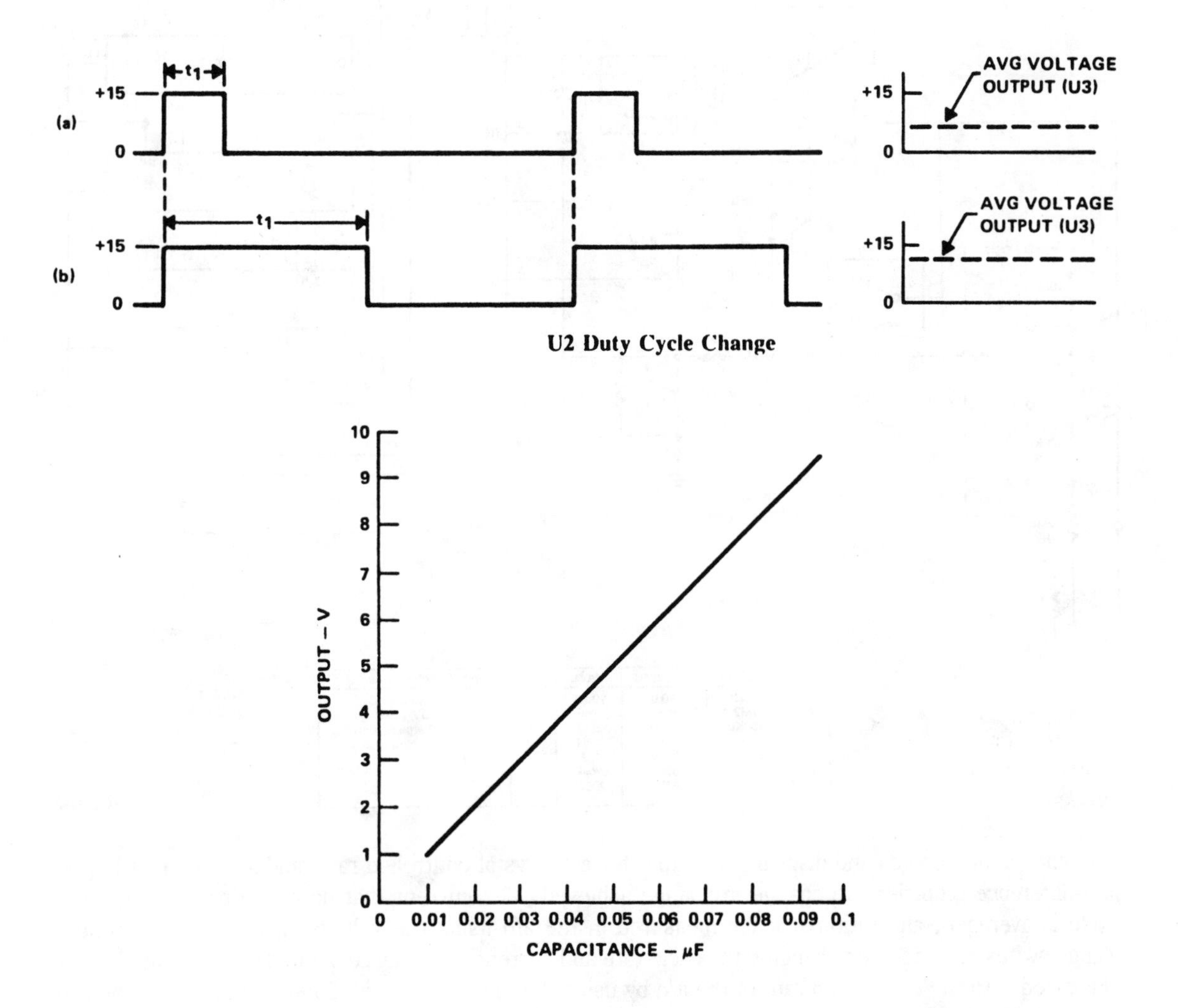

U2 Duty Cycle Change

$3^1/_2$-DIGIT A/D CAPACITANCE METER

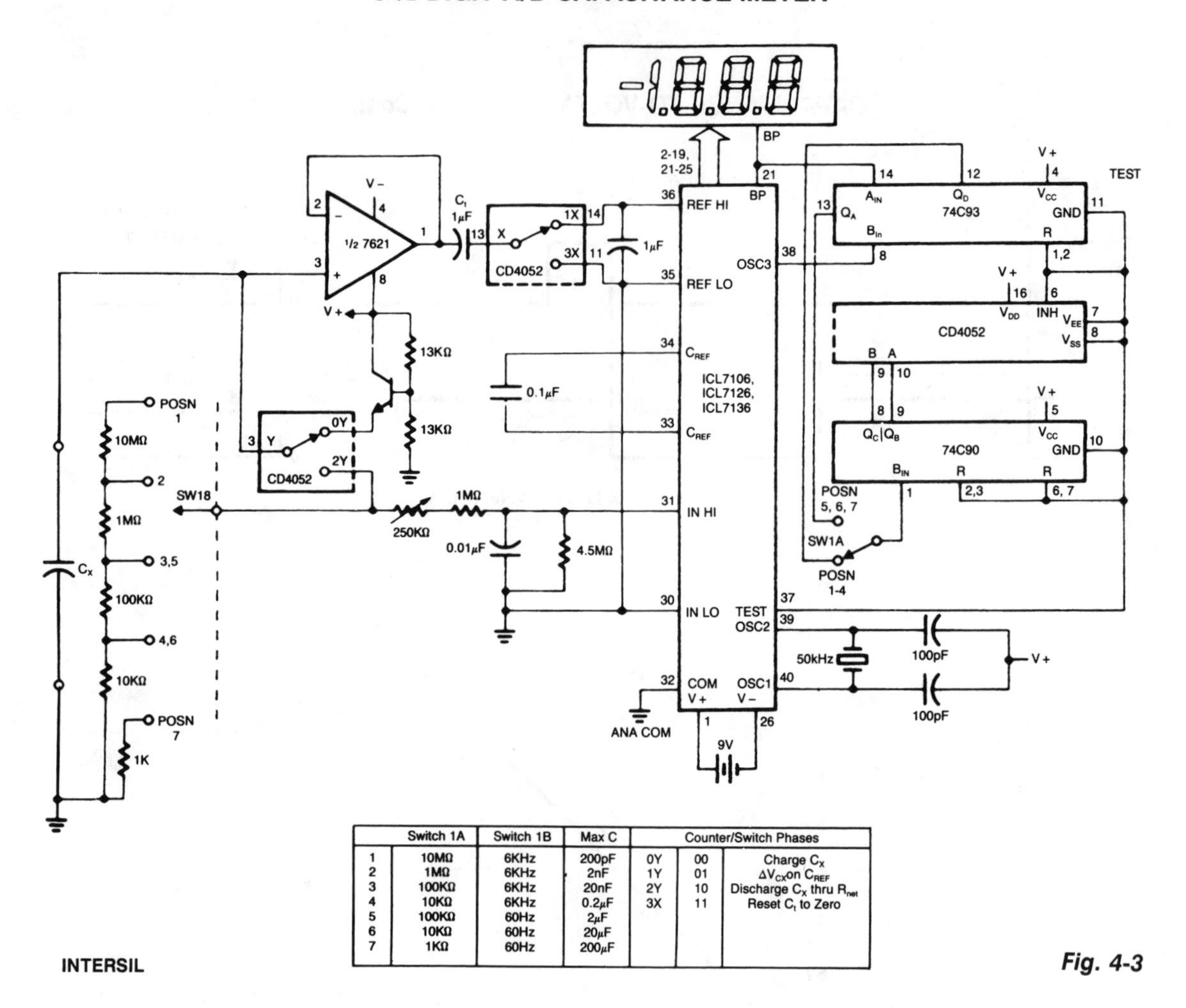

	Switch 1A	Switch 1B	Max C	Counter/Switch Phases		
1	10MΩ	6KHz	200pF	0Y	00	Charge C_X
2	1MΩ	6KHz	2nF	1Y	01	ΔV_{CX} on C_{REF}
3	100KΩ	6KHz	20nF	2Y	10	Discharge C_X thru R_{net}
4	10KΩ	6KHz	0.2μF	3X	11	Reset C_t to Zero
5	100KΩ	60Hz	2μF			
6	10KΩ	60Hz	20μF			
7	1KΩ	60Hz	200μF			

INTERSIL

Fig. 4-3

The circuit charges and discharges a capacitor at a crystal-controlled rate, and stores on a sample-and-difference amplifier the change in voltage achieved. The current that flows during the discharge cycle is averaged, and ratiometrically measured in the a/d using the voltage change as a reference. Range switching is done by changing the cycle rate and current metering resistor. The cycle rate is synchronized with the conversion rate of the a/d by using the externally divided internal oscillator and the internally divided back plane signals. For convenience in timing, the switching cycle takes 5 counter states, although only four switch configurations are used. Capacitances up to 200 μF can be measured, and the resolution on the lowest range is down to 0.1 pF.

The zero integrator time can be set initially at $^1/_3$ to $^1/_2$, the minimum auto-zero time, but if an optimum adjustment is required, look at the comparator output with a scope under worst-case overload conditions. The output of the delay timer should stay low until after the comparator has come off the rail, and is in the linear region (usually fairly noisy).

CAPACITANCE METER I

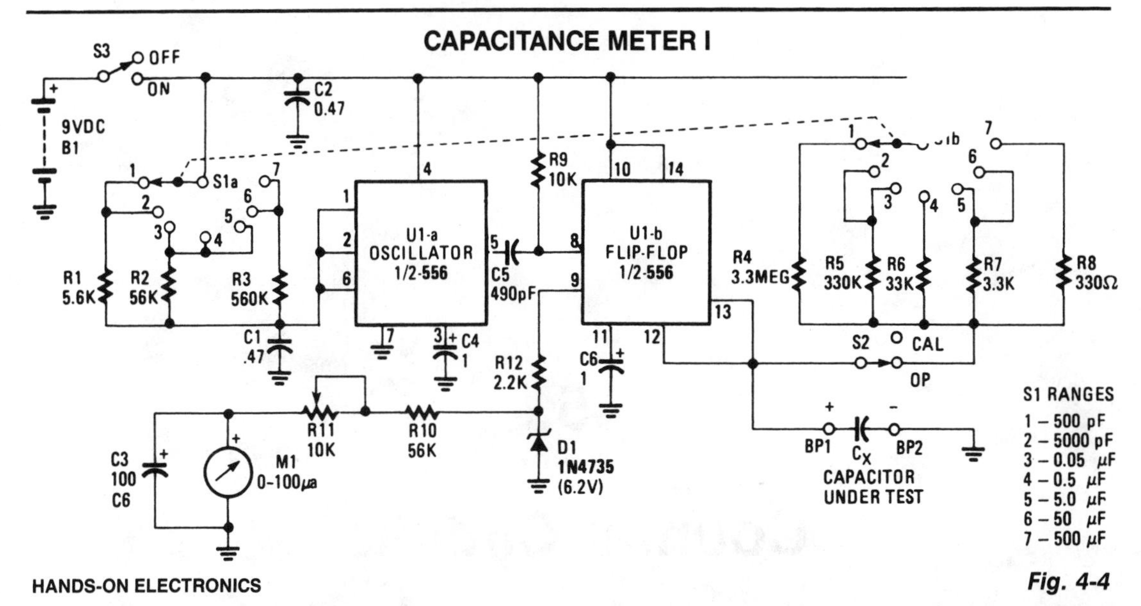

HANDS-ON ELECTRONICS *Fig. 4-4*

U1A is an oscillator and U1B is the measurement part of the circuit. It converts unknown capacity into a pulse-width modulated signal the same way an automotive dwell meter works. The meter is linear, so the fraction or percentage of time that the output is high is directly proportional to the unknown capacitance (*CX* in the schematic). Meter M1 reads the average voltage of those pulses because its mechanical frequency response is low compared to the oscillator frequency of U1A.

CAPACITANCE METER II

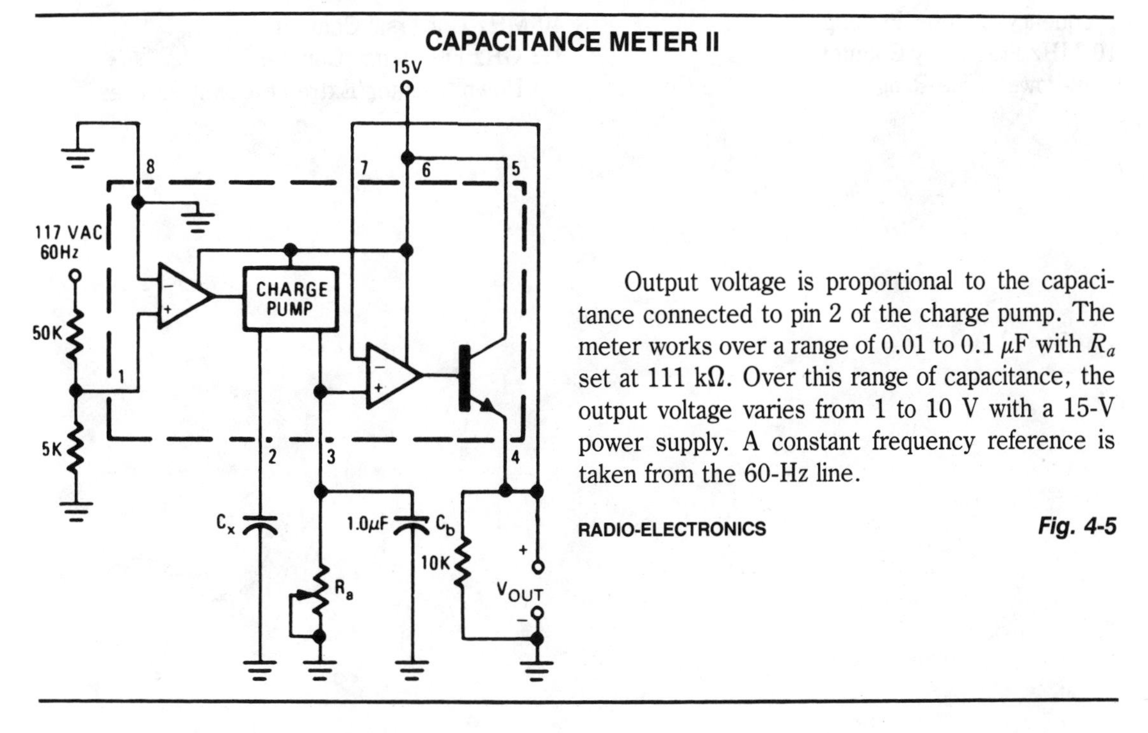

Output voltage is proportional to the capacitance connected to pin 2 of the charge pump. The meter works over a range of 0.01 to 0.1 μF with R_a set at 111 kΩ. Over this range of capacitance, the output voltage varies from 1 to 10 V with a 15-V power supply. A constant frequency reference is taken from the 60-Hz line.

RADIO-ELECTRONICS *Fig. 4-5*

5

Counter Circuits

The sources of the following circuits are contained in the Sources section, which begins on page 217. The figure number in the box of each circuit correlates to the source entry in the Sources section.

FREQUENCY-COUNTER PREAMP

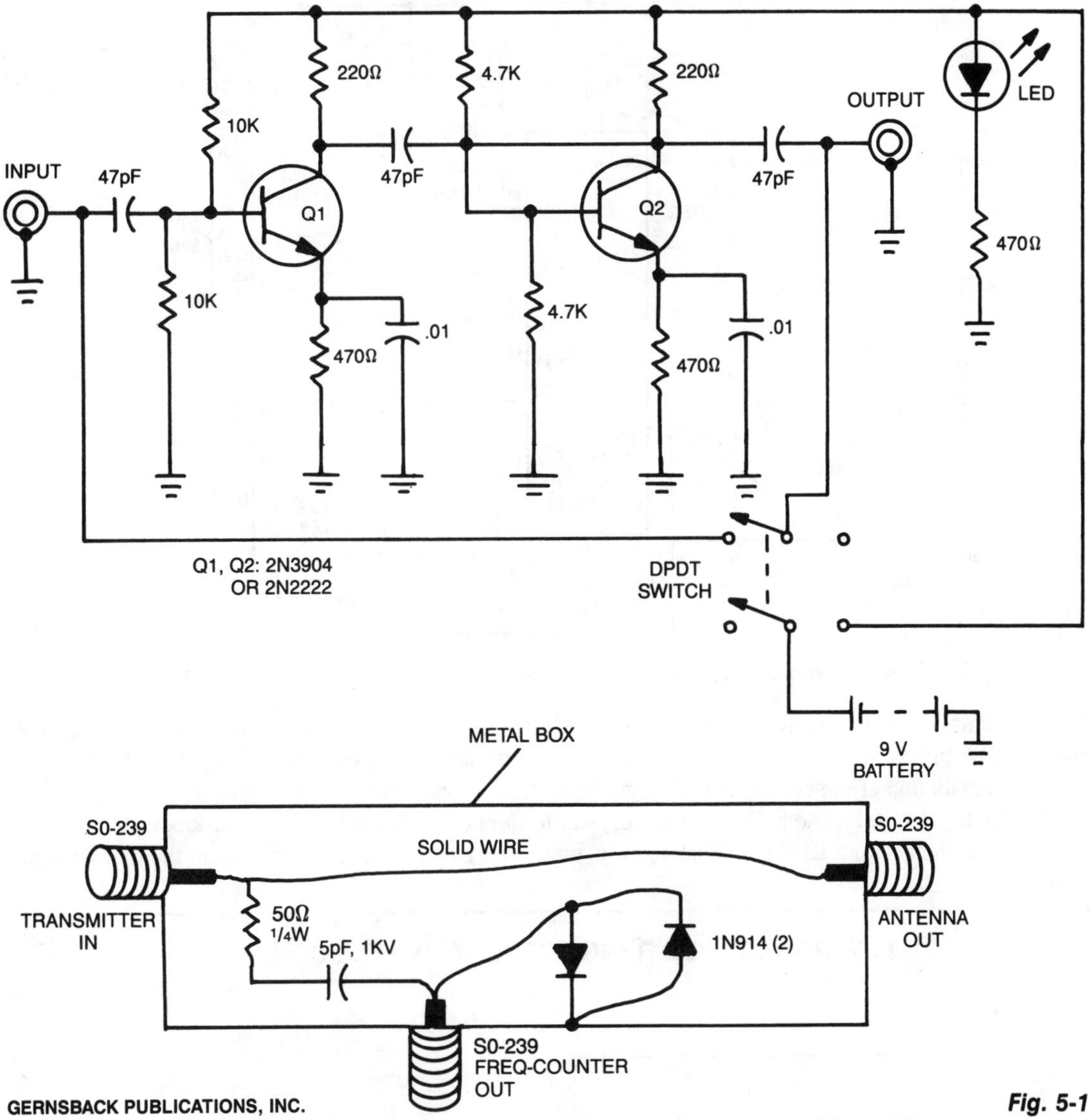

GERNSBACK PUBLICATIONS, INC.

Fig. 5-1

By using the preamplifier with a short length of shielded cable and clip leads, signals that generally could not generate a readout, generate precise and stable readouts on the counter. The DPDT switch is used to bypass the circuit when amplification is not needed. The preamplifier can also be used for other purposes. For example, the unit was also tested as a receiver preamplifier and increased received signal strength about 6 S-units at 30 MHz. A line tap can be used to measure the frequency directly at the output of a transmitter. The entire circuit for that consists of two diodes, one resistor, and one capacitor. The line tap simply picks a low-amplitude signal for measurement by the frequency counter. The tap can be connected to transmitters with an output power of between 1 and 250 W.

10-MHz FREQUENCY COUNTER

HANDS-ON ELECTRONICS/POPULAR ELECTRONICS *Fig. 5-2*

The circuit consists of ICM7208 seven-decade counter U1, ICM7207A oscillator controller U2, and CA3130 biFET op amp U3. IC U1 counts input signals, decodes them to 7-segment format, and outputs signals that are used to drive a 7-digit display. IC U2 provides the timing for U1, while U3 conditions the input to U1. The 5.24288-MHz crystal frequency is divided by U2 to produce a 1280-MHz multiplexing signal at pin 12 of U2. That signal is input to U1 at pin 16 and used to scan the display digits in sequence.

LOW-POWER WIDE-RANGE PROGRAMMABLE COUNTER

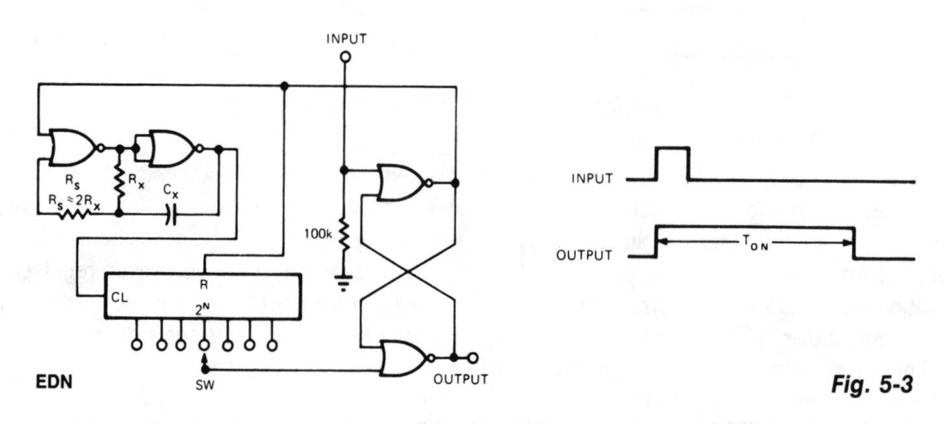

EDN *Fig. 5-3*

LOW-POWER WIDE-RANGE PROGRAMMABLE COUNTER (*Cont.*)

This CMOS circuit can be used as a 1-shot time delay switch and general-purpose timer. The circuit consists of a gated oscillator and a latch made from one CD4001 quad 2-input NOR gate as shown and a CD4020 14-stage counter. T_{ON} is a function of the oscillator frequency from the $R_X C_X$ and the proper 2^N output from the counter. A pulse applied to the latch will "enable" the oscillator and counter. The latch output will remain high until the 2^N count resets the latch and disables the oscillator and counter. The circuit provides μs to the hour interval timing. The extraordinarily long period available from the CMOS oscillator, combined with the 14-stage counter, make this range possible. Further decoding is required for variations finer than a power of two.

40-MHz UNIVERSAL COUNTER

This circuit can be used to measure frequencies up to 40 MHz. To obtain the correct measured value, it is necessary to divide the oscillator frequency and the input frequency by four. In doing this, the time between measurements is also lengthened to 800 ms and the display multiplex rate is decreased to 125 Hz.

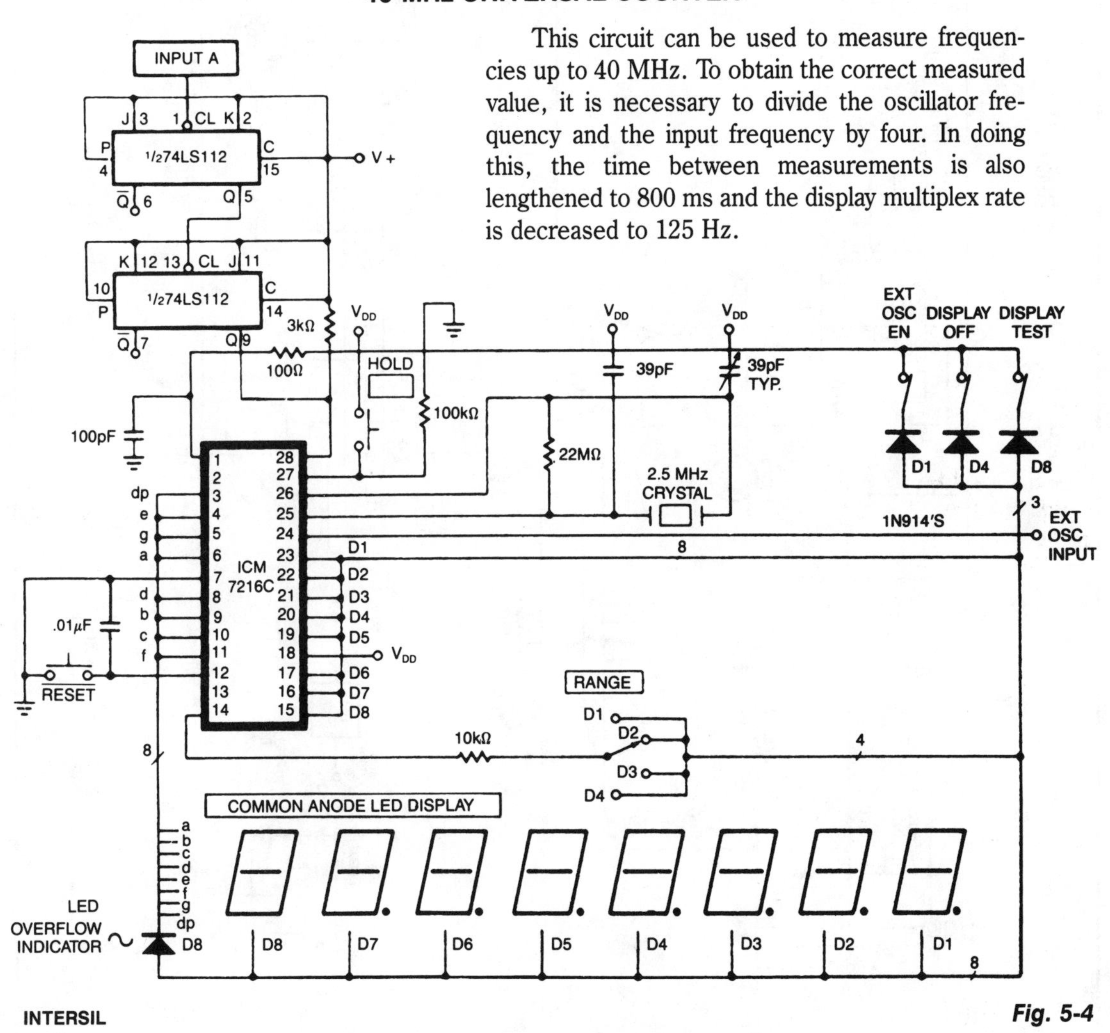

Fig. 5-4

1.2 GHz FREQUENCY COUNTER

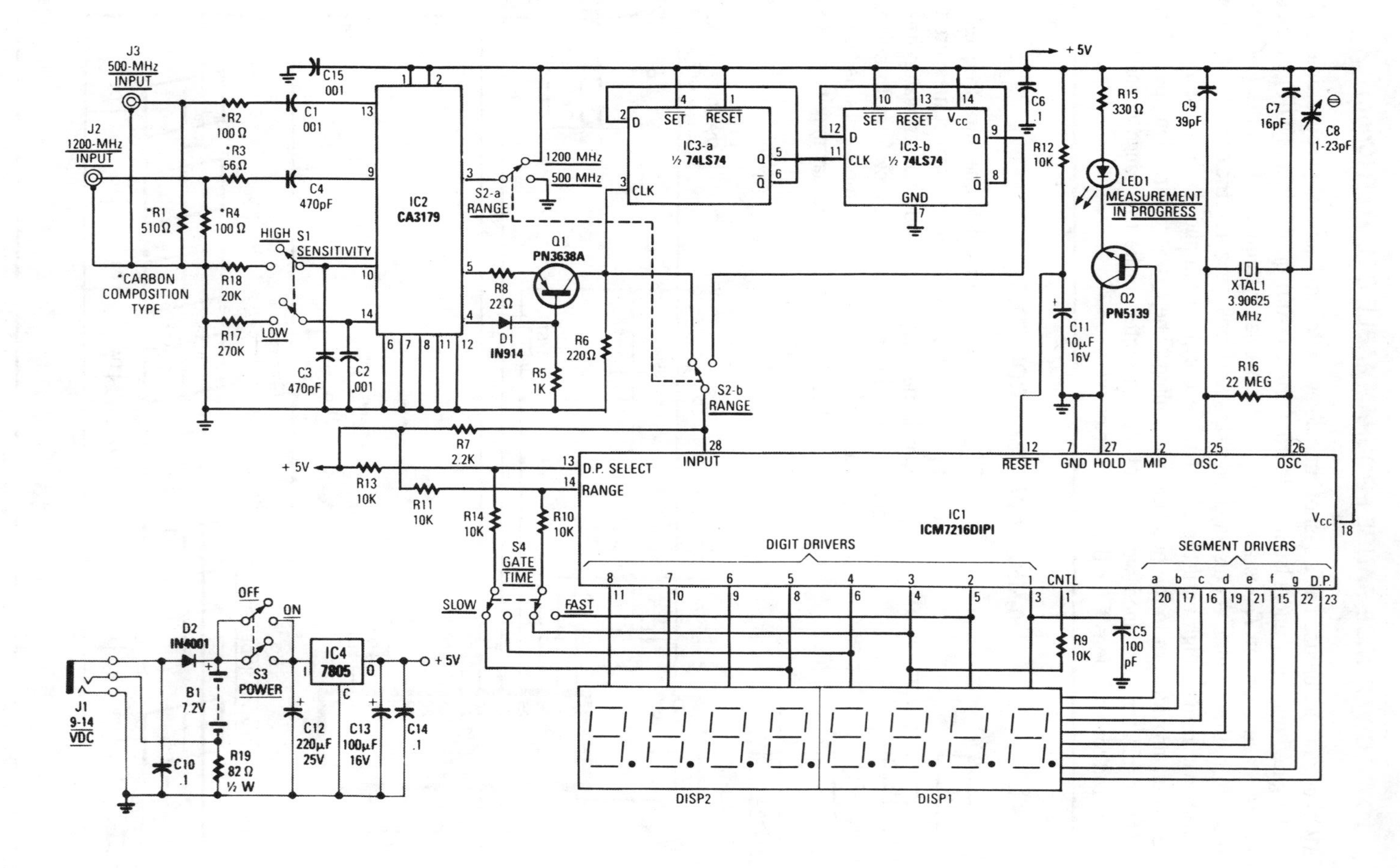

Fig. 5-5

UP/DOWN COUNTER/EXTREME-COUNT FREEZER

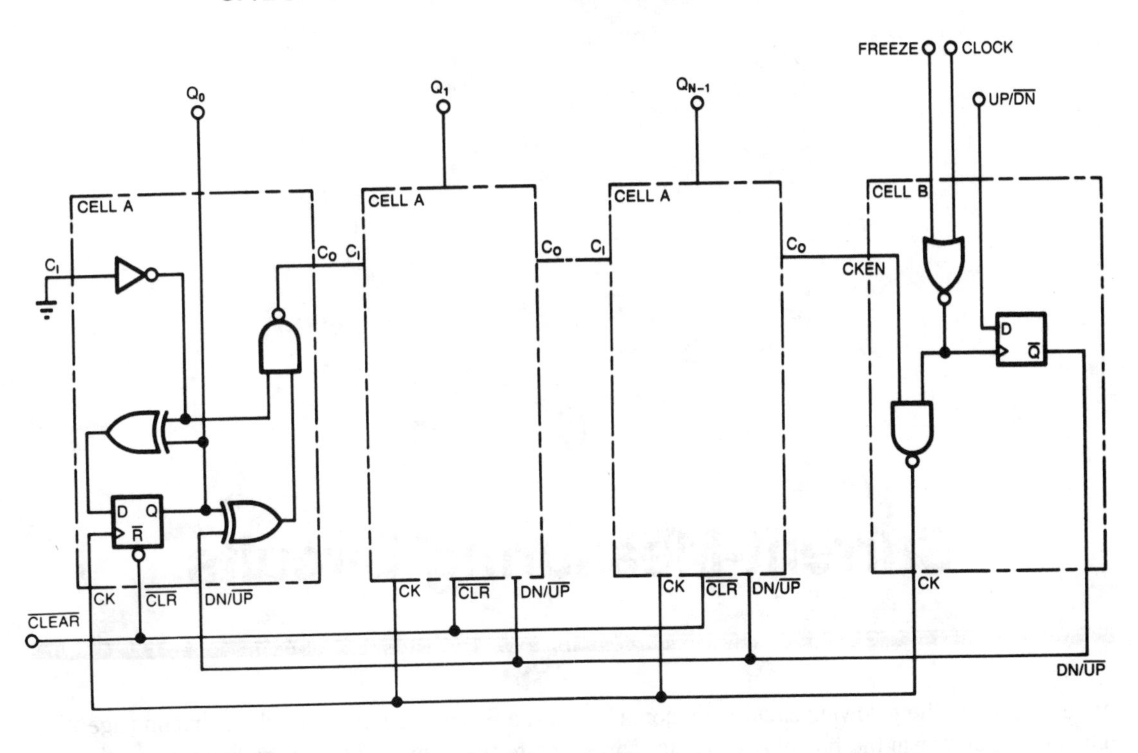

STATE TABLE

FREEZE	UP/$\overline{DN}$	CLOCK	CURRENT STATE $Q_{N-1} \cdots Q_0$	NEXT STATE $Q_{N-1} \cdots Q_0$
L	H	↑	11...10	11...11
L	H	↑	11...11	11...11
L	L	↑	00...01	00...00
L	L	↑	00...00	00...00
H	X	X	Q_X	Q_X

EDN

Fig. 5-6

The discrete-gate up/down-counter design has the unusual property of freezing or saturating when it reaches its lowest count in the down-count mode or its highest count in the count-up mode, instead of rolling over and resetting as do most counters. This property is especially useful in position-control systems. For example, you wouldn't want a robot's arm to slowly move to full extension as the counter counts up, then have it suddenly slam back to its rest position when the counter resets to zero.

You can cascade as many of the A cells as you need because the counter's outputs are synchronous. The B cell accepts the carry bit from the most significant bit's A cell and provides the clock control that stops the counter. Make sure that the freeze input to the B cell doesn't get asserted when the clock input is low; otherwise, the counter might make an extra count.

6

Current-Measuring Circuits

The sources of the following circuits are contained in the Sources section, which begins on page 217. The figure number in the box of each circuit correlates to the source entry in the Sources section.

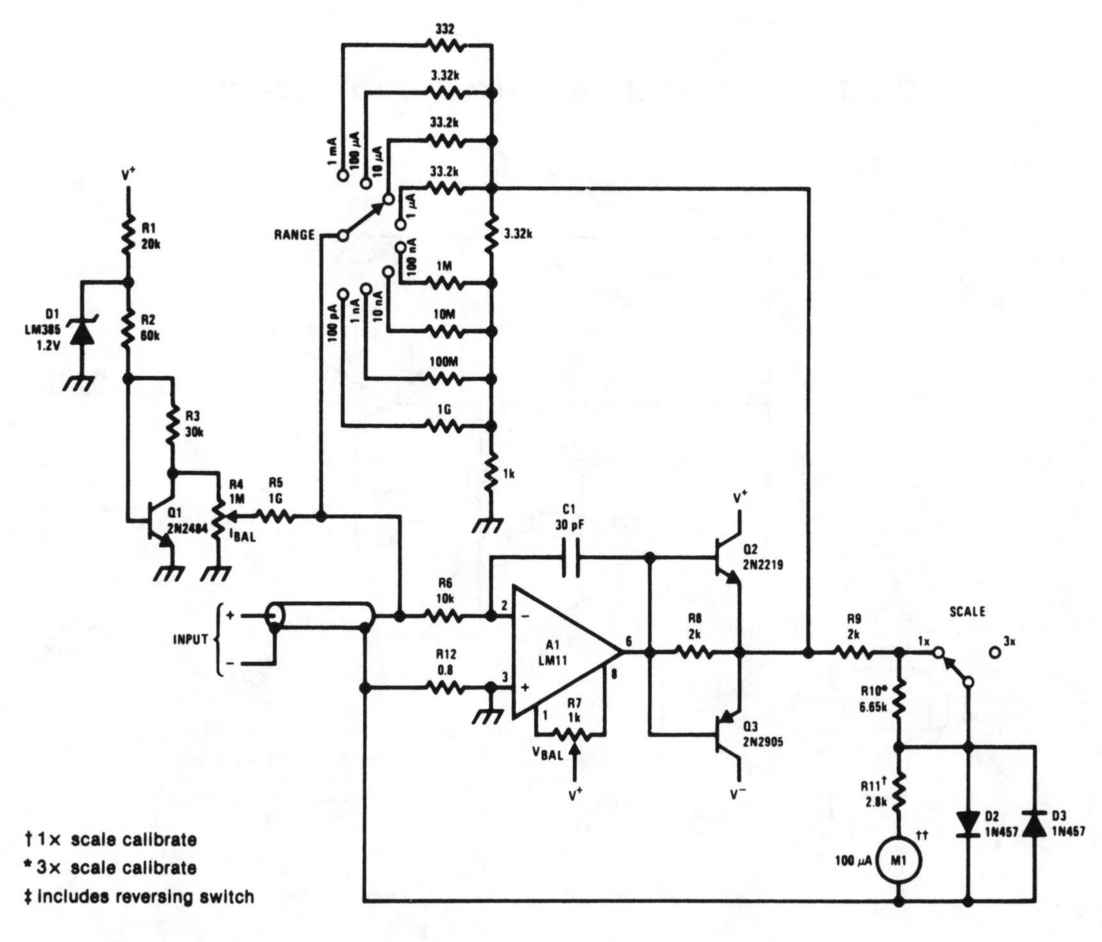

NATIONAL SEMICONDUCTOR

Fig. 6-1

This current meter ranges from 100 pA to 3 mA full scale. Voltage across input is 100 μV at lower ranges rising to 3 mV at 3 mA. The buffers on the op amp are to remove ambiguity with high-current overload. The output can also drive a DVM or a DPM.

ELECTROMETER AMPLIFIER WITH OVERLOAD PROTECTION

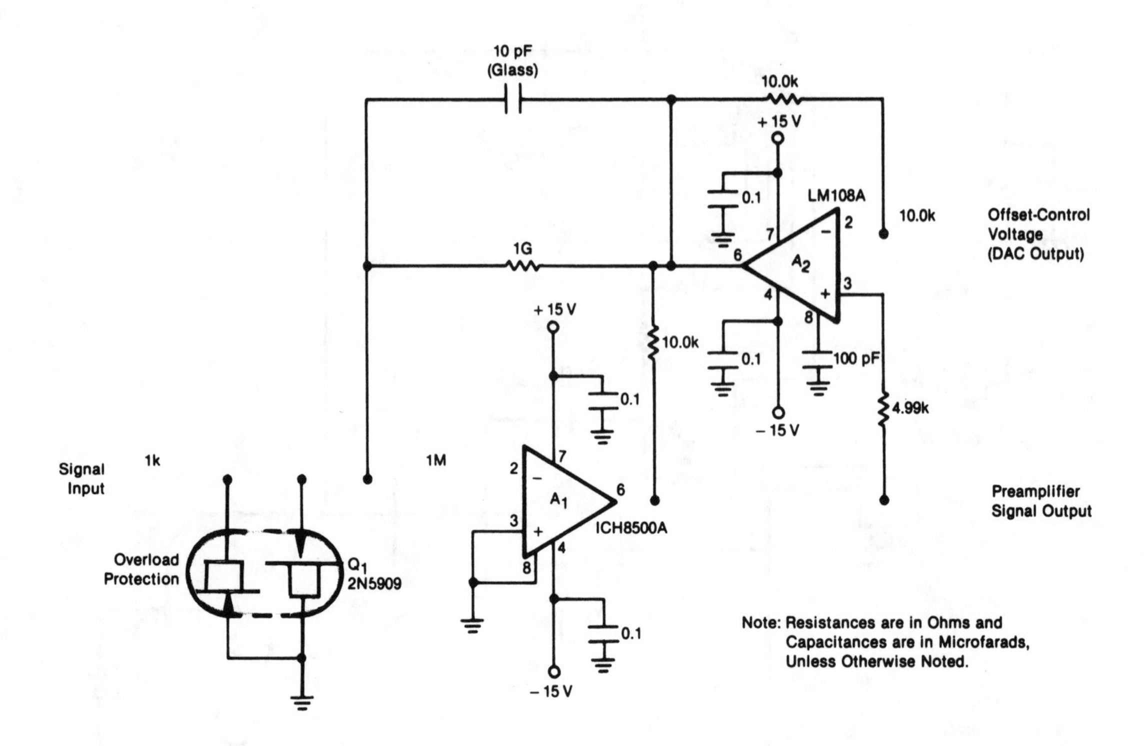

NASA TECH BRIEFS *Fig. 6-2*

The preamplifier is protected from excessive input signals of either polarity by the 2N5909 junction field-effect transistor. A nulling circuit makes it possible to set the preamplifier output voltage to zero at a fixed low level (up to $\pm 10^{-8}$A) of the input current. This level is called the standing current and corresponds to the zero-signal level of the instrumentation. The opposing (offset) current is generated in the 10^9 feedback resistor to buck the standing current. Different current ranges are reached by feeding the preamplifier output to low-gain and high-gain amplifier chains. To reduce noise, each chain includes a 1.5-Hz corner active filter.

PICOAMMETER CIRCUIT

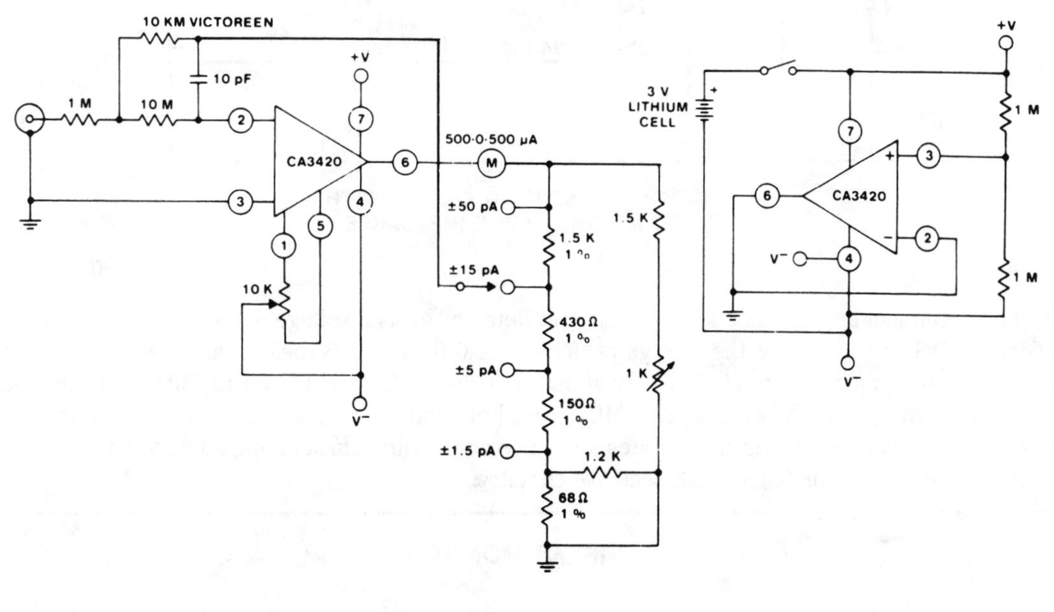

GE/RCA

Fig. 6-3

The circuit uses the exceptionally low input current (0.1 pA) of the CA3420 BiMOS op amp. With only a single 10-MΩ resistor, the circuit covers the range from ±50 pA maximum to a full-scale sensitivity of ±1.5 pA. Using an additional CA3420, a low-resistance center tap is obtained from a single 3-V lithium battery.

NANOAMPERE SENSING CIRCUIT WITH 100-MΩ INPUT IMPEDANCE

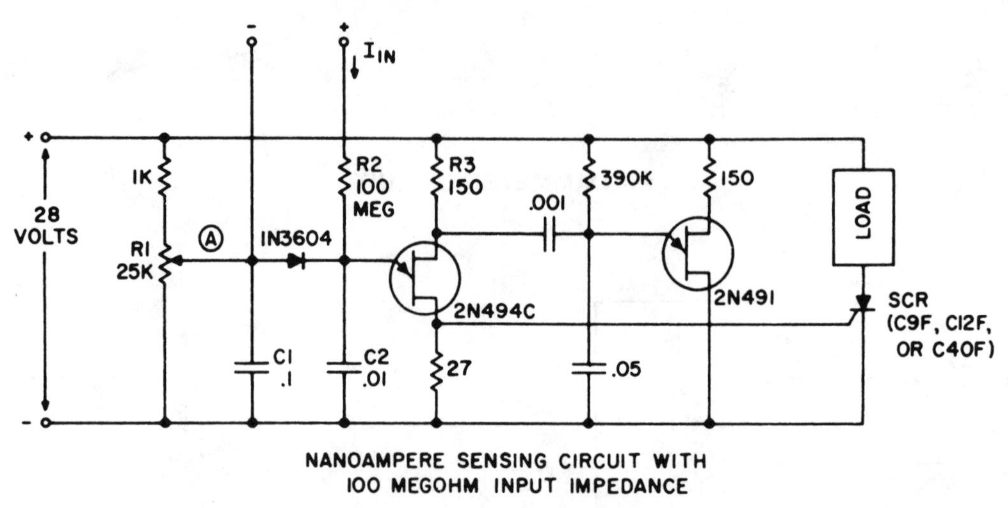

NANOAMPERE SENSING CIRCUIT WITH 100 MEGOHM INPUT IMPEDANCE

GE *Fig. 6-4*

The circuit can be used as a sensitive current detector or as a voltage detector having high input impedance. R1 is set so that the voltage at point A is 0.5 to 0.75 V below the level that fires the 2N494C. A small input current (I_{IN}) of only 40 nA will charge C2 and raise the voltage at the emitter to the firing level. When the 2N494C fires, both capacitors, C1 and C2, are discharged through the 27-Ω resistor, which generates a positive pulse with sufficient amplitude to trigger a controlled rectifier (SCR), or other pulse sensitive circuitry.

CURRENT MONITOR

R1 senses current flow of a power supply. The JFET is used as a buffer because $I_D = I_S$; therefore the output monitor voltage accurately reflects the power supply current flow.

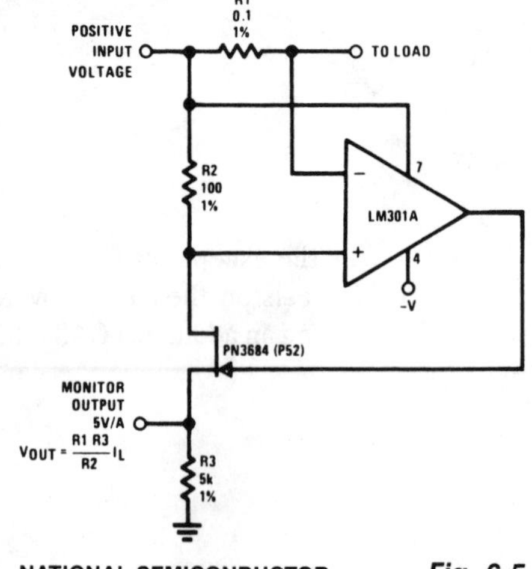

NATIONAL SEMICONDUCTOR *Fig. 6-5*

GUARDED-INPUT PICOAMMETER CIRCUIT

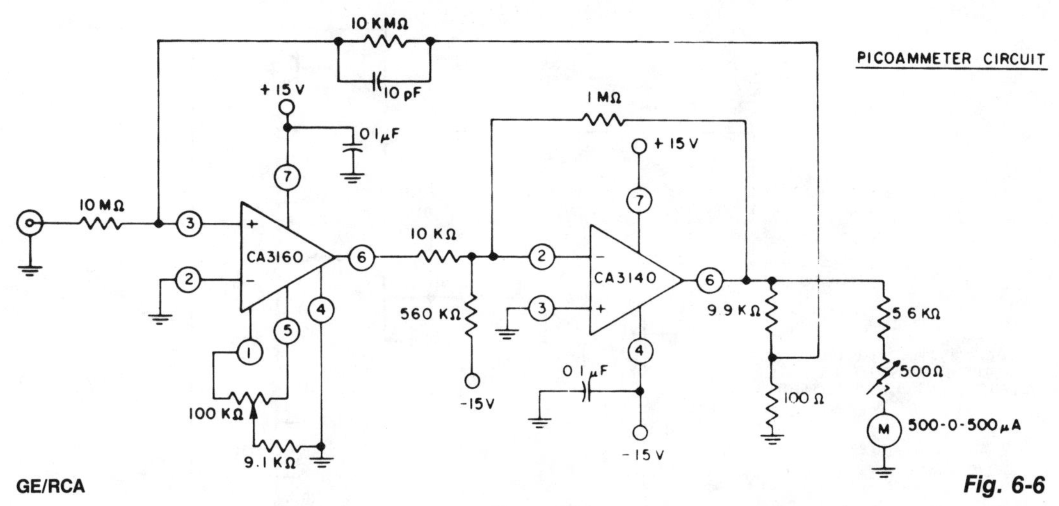

GE/RCA

Fig. 6-6

The circuit utilizes CA3160 and CA3140 BiMOS op amps to provide a full-scale meter deflection of ±3 pA. The CA3140 serves as an X100 gain stage to provide the required plus and minus output swing for the meter and feedback network. Terminals 2 and 4 of the CA3160 are at ground potential, thus its input is operated in the "guarded mode."

AMMETER WITH SIX-DECADE RANGE I

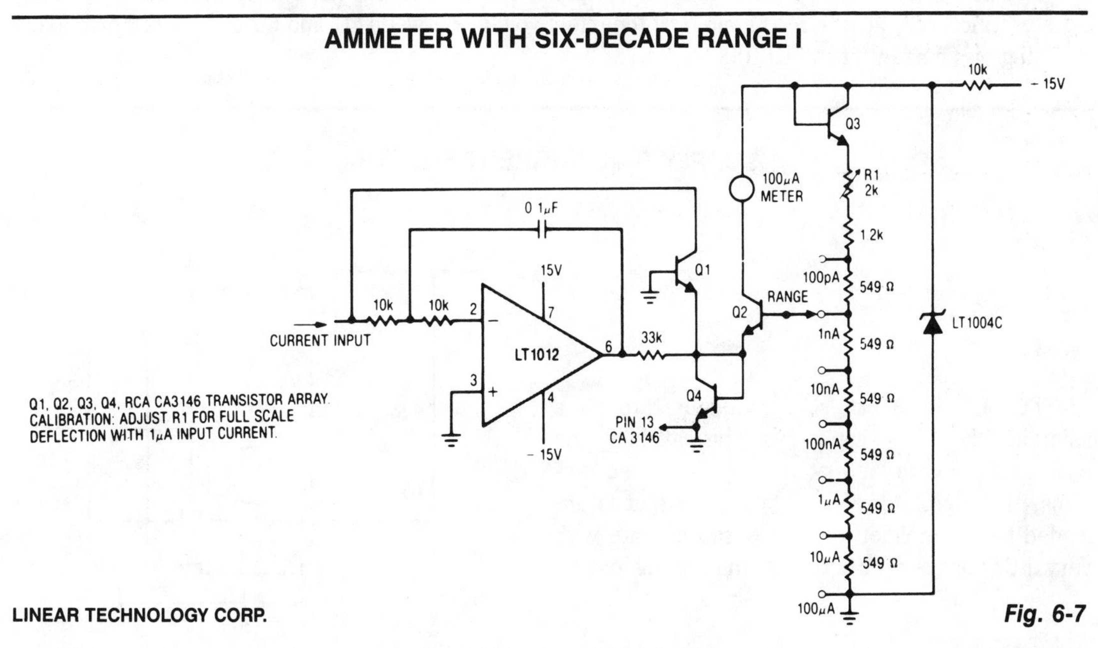

LINEAR TECHNOLOGY CORP.

Fig. 6-7

The ammeter measures currents from 100 pA to 100 μA without the use of expensive high-value resistors. Accuracy at 100 μA is limited by the offset voltage between Q1 and Q2 and, at 100 pA, by the inverting bias current of the LT1008.

AMMETER WITH SIX-DECADE RANGE II

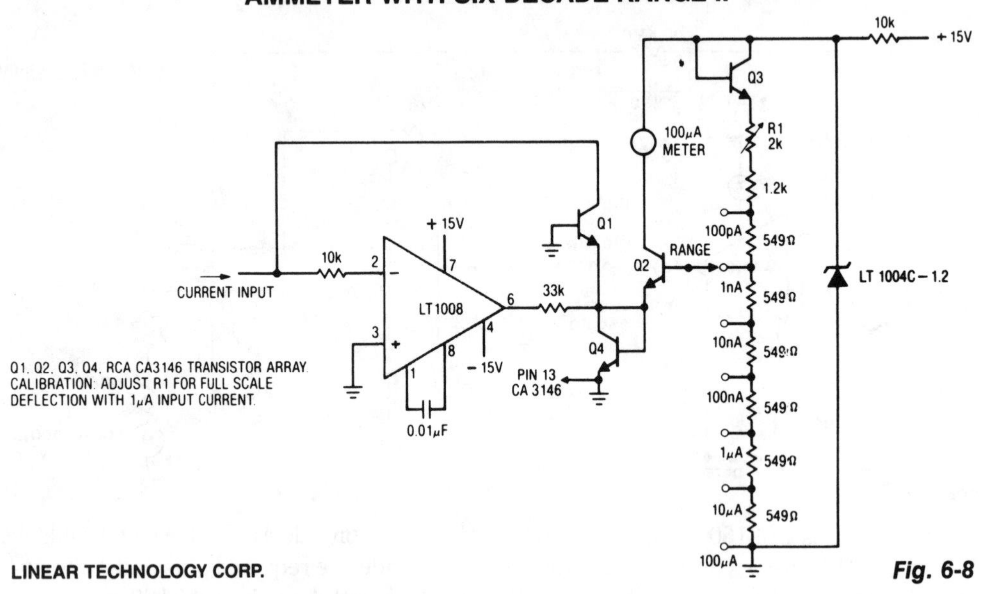

Q1, Q2, Q3, Q4, RCA CA3146 TRANSISTOR ARRAY.
CALIBRATION: ADJUST R1 FOR FULL SCALE DEFLECTION WITH 1μA INPUT CURRENT.

LINEAR TECHNOLOGY CORP.

Fig. 6-8

The ammeter measures currents from 100 pA to 100 μA without the use of expensive high-value resistors. Accuracy at 100 μA is limited by the offset voltage between Q1 and Q2 and, at 100 pA, by the inverting bias current of the LT1008.

SUPPLY RAIL CURRENT SENSOR

The LTC1043 can sense current through a shunt in either of its supply rails. This capability has wide application in battery and solar-powered systems. If the ground-referred voltage output is unloaded by an amplifier, the shunt can operate with very little voltage drop across it, minimizing losses.

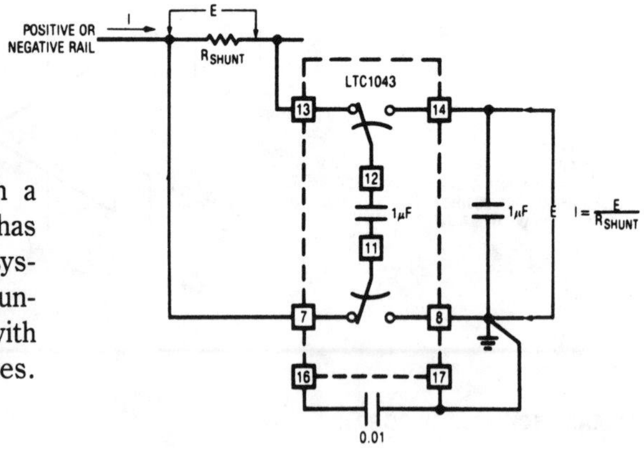

LINEAR TECHNOLOGY CORP.

Fig. 6-9

PICOAMMETER

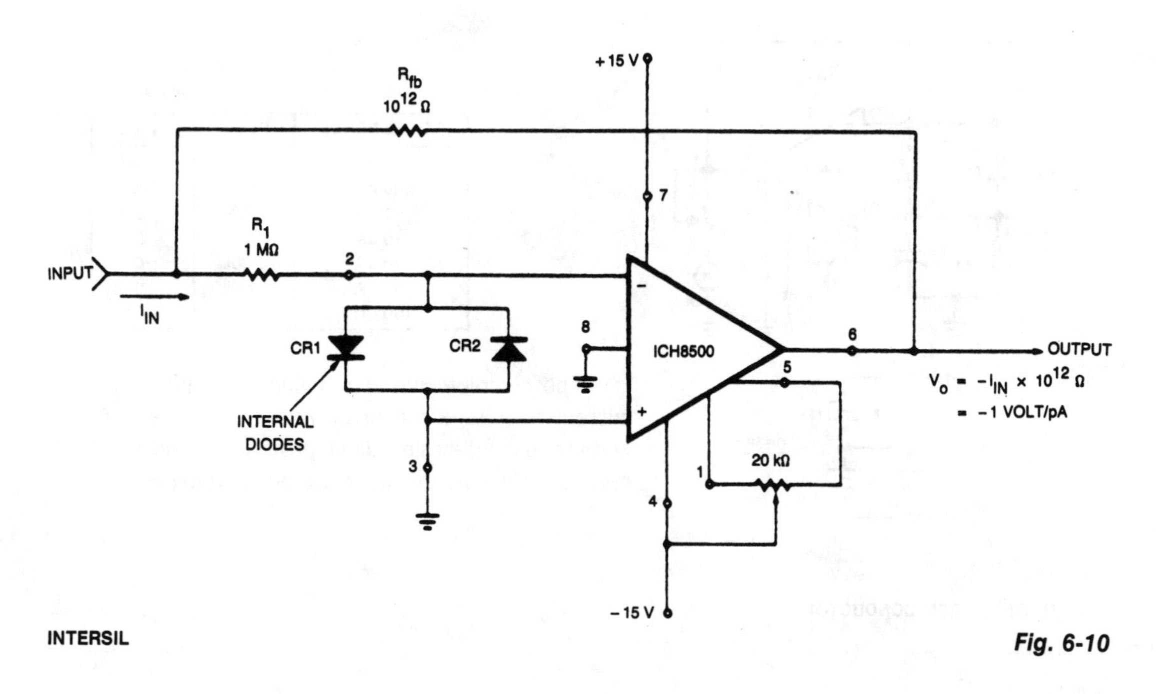

INTERSIL

Fig. 6-10

Care must be taken to eliminate any stray currents from flowing into the current summing node. This can be accomplished by forcing all points surrounding the input to the same potential as the input. In this case the potential of the input is at virtual ground, or 0 V. Therefore, the case of the device is grounded to intercept any stray leakage currents that may otherwise exist between the ±15-V input terminals and the inverting input summing junctions. Feedback capacitance should be kept to a minimum in order to maximize the response time of the circuit to step-function input currents. The time constant of the circuit is approximately the product of the feedback capacitance C_{fb} times the feedback resistor R_{fb}. For instance, the time constant of the circuit is 1 s if $C_{fb} = 1$ pF. Thus, it takes approximately 5 s (5 time constants) for the circuit to stabilize to within 1% of its final output voltage after a step function of input current has been applied. C_{fb} of less than 0.2 to 0.3 pF can be achieved with proper circuit layout.

NANOAMMETER

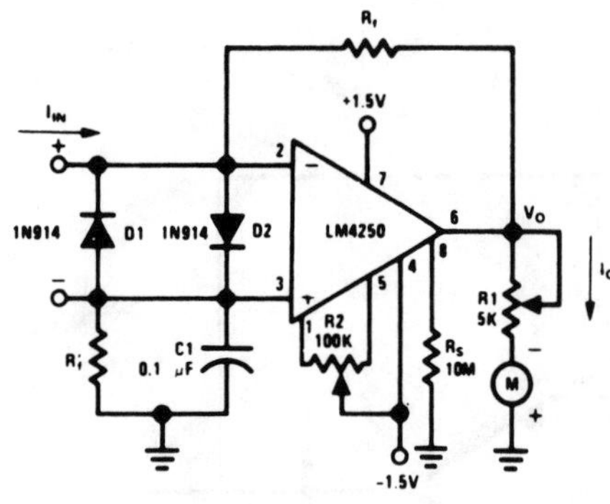

Resistance Values for DC Nano and Micro Ammeter

I FULL SCALE	R_f [Ω]	R_f' [Ω]
100 nA	1.5M	1.5M
500 nA	300k	300k
1 μA	300k	0
5 μA	60k	0
10 μA	30k	0
50 μA	6k	0
100 μA	3k	0

The complete meter amplifier is a differential current-to-voltage converter with input protection, zeroing and full-scale adjust provisions, and input resistor balancing for minimum offset voltage.

NATIONAL SEMICONDUCTOR *Fig. 6-11*

7

Dip-Meter Circuits

The sources of the following circuits are contained in the Sources section, which begins on page 217. The figure number in the box of each circuit correlates to the source entry in the Sources section.

DUAL-GATE IGFET (MOSFET) DIP METER

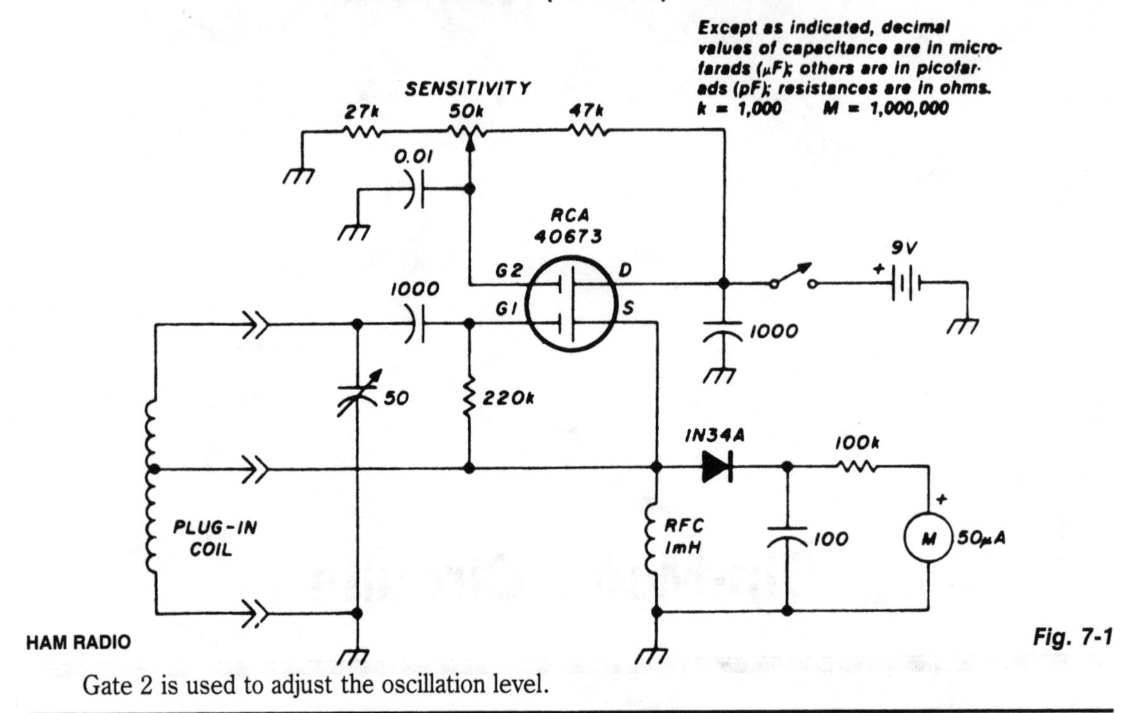

HAM RADIO

Fig. 7-1

Gate 2 is used to adjust the oscillation level.

VARICAP-TUNED FET DIP METER WITH 1-kHz MODULATION

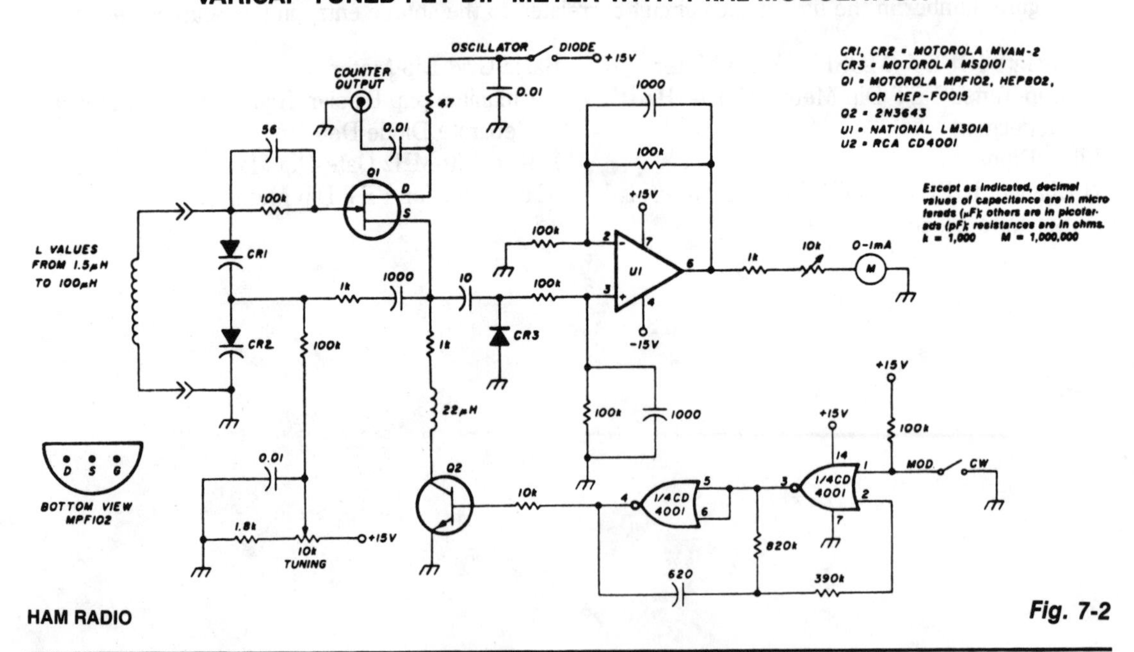

HAM RADIO

Fig. 7-2

LITTLE DIPPER

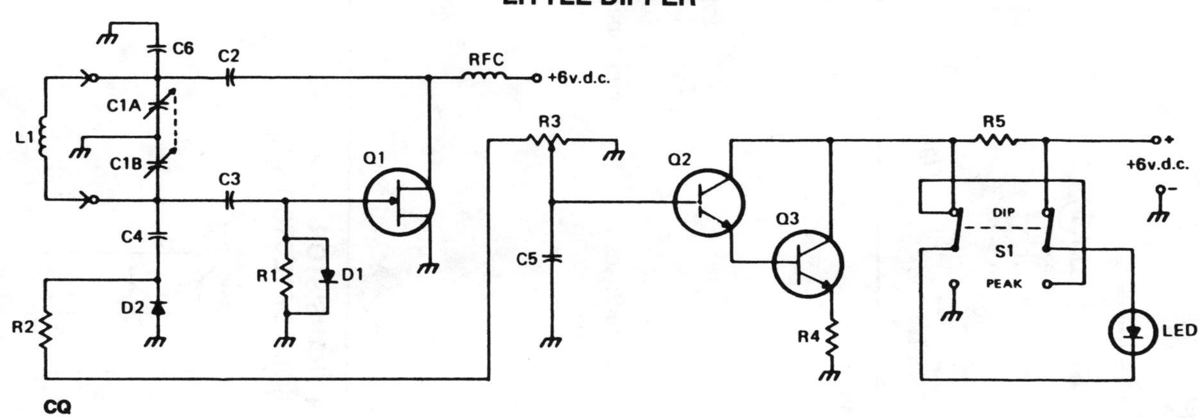

Parts List

L1—See coil data.
C1A,1B—Dual capacitor 100 pF per section (ETCO SV409 or similar)
C2,C3—100 pF mica, mylar, etc., low voltage
C4—10 pF mica, mylar, etc., low voltage
C5—.01 uF ceramic, low voltage
C6—5 pF mica, mylar, etc., low voltage
D1,D2—1N914 silicon diode or similar
R1—100 K ohms ¼ watt
R2—220 K ohms ¼ watt
R3—500 K ohms potentiometer
R4—10 ohms ¼ watt
R5—270 ohms ¼ watt
Q1—MPF 102 FET
Q2—Any general-purpose NPN transistor with a Beta (Hfe) of 40 or so (2N3904 or similar)
Q3—Any general-purpose NPN transistor capable of 20 mA collector current or more, Beta 40 or so (2N3904, 2N2222 or similar)
RFC—1 mH miniature ferrite core choke (value not critical)
LED—Panel mounting LED Radio Shack 276-068 or similar.
SW1—Sub-miniature DPDT slide switch or similar

Miscellaneous—6 volt AC adapter (Radio Shack 273-1454A)
Coaxial DC power jack (RS 274-1565)
Calibrated Dial knob (RS 274-413)
Dual phono jack (RS 274-332)

CQ

Fig. 7-3

The circuit consists of two basic circuits, the oscillator and the detector. The oscillator uses an FET in a Colpitts configuration. The energy circulating in the oscillator tank is coupled through C4 to the detector circuit, where a small diode (D2) rectifies it and feeds a dc voltage to the Darlington pair (Q2, A3) controlled by the sensitivity control (R3). Any small variations in the bias of the amplifier will cause large variations of current through the LED indicator in the DIP mode; however, in the PEAK mode the current produces a corresponding voltage drop through R5 and the action of the LED is reversed. The circuit shown will work practically on any frequency from LF to VHF, if the appropriate components are used.

N-CHANNEL IGFET (MOSFET) DIP METER AND SEPARATE DIODE DETECTOR

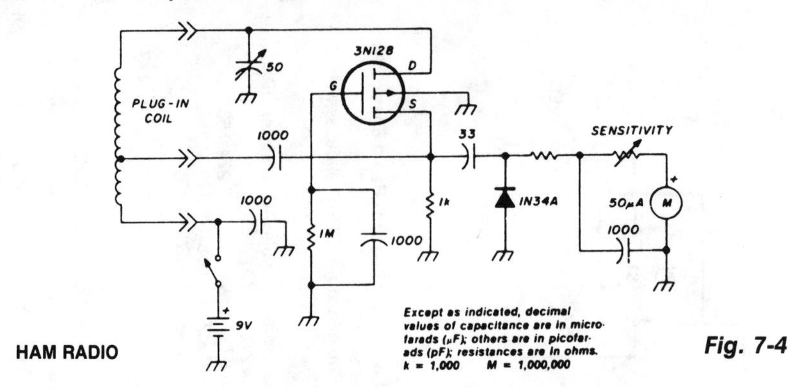

HAM RADIO

Fig. 7-4

BASIC GRID-DIP METER

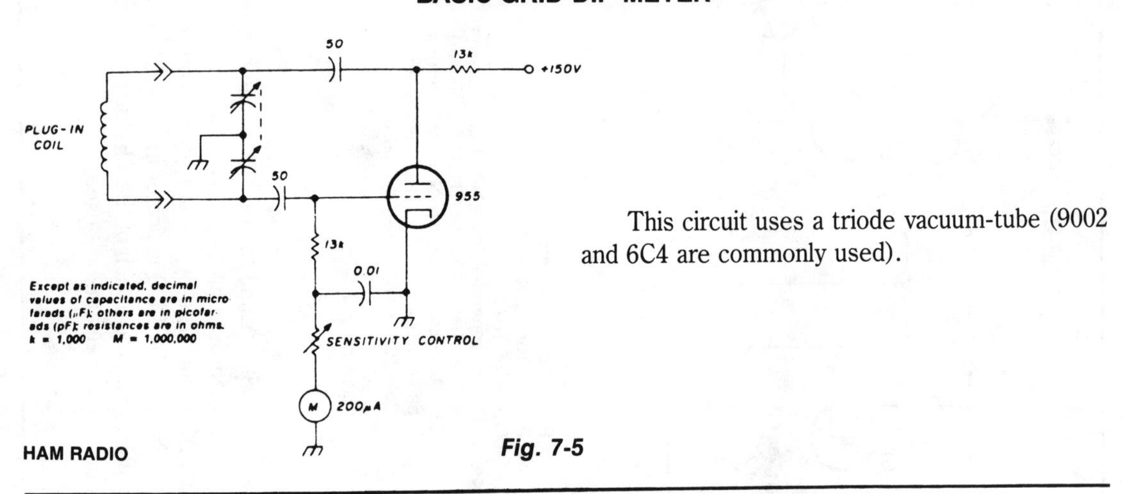

This circuit uses a triode vacuum-tube (9002 and 6C4 are commonly used).

HAM RADIO

Fig. 7-5

GERMANIUM PNP BIPOLAR TRANSISTOR DIP METER WITH SEPARATE DIODE DETECTOR

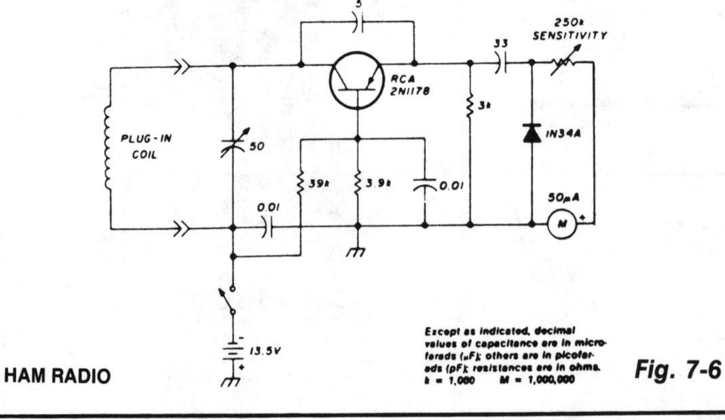

HAM RADIO

Fig. 7-6

1.8- to 150-MHz GATE-DIP METER

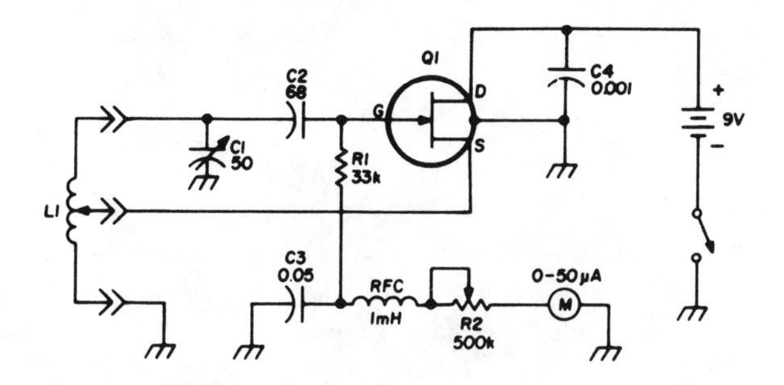

Coil data.

frequency range (MHz)	no. turns	wire size AWG	wire size (mm)	winding length inches	winding length (mm)	tap*	coil diameter inches	coil diameter (mm)
1.8 - 3.8	82	26 enamel	(0.4)	1 9/16	(40.0)	12	1¼	(32)
3.6 - 7.3	29	26 enamel	(0.4)	9/16	(14.5)	5	1¼	(32)
7.3 - 14.4	18	22 enamel	(0.6)	3/4	(19.0)	3	1	(25)
14.4 - 32	7	22 enamel	(0.6)	1/2	(12.5)	2	1	(25)
29 - 64	3½	18 tinned	(1.0)	3/4	(19.0)	3/4	1	(25)

61 - 150 Hairpin of 16 no. AWG (1.3mm) wire, 5/8 inch (16mm) spacing, 2 3/8 inches (60mm) long including coil-form pins. Tapped at 2 inches (51mm) from ground end.

PINS
5/8" (16mm)
COIL FORM SAWED OFF

*Turns from ground-end. 1 inch (25mm) forms are Millen 45004 available from Burstein-Applebee

HAM RADIO

Fig. 7-7

SILICON JUNCTION FET DIP METER

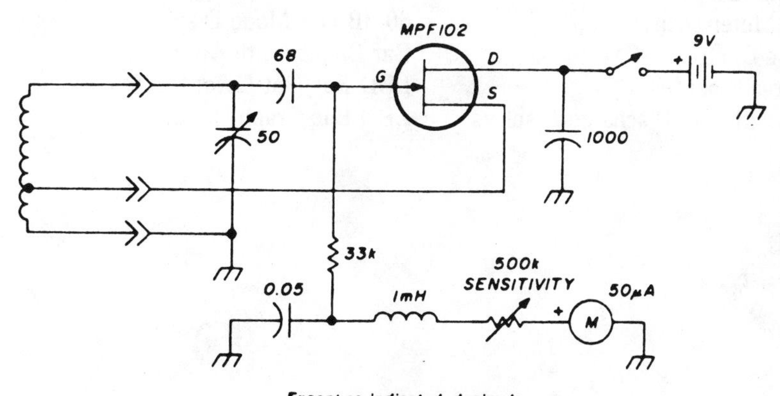

Except as indicated, decimal values of capacitance are in microfarads (μF); others are in picofarads (pF); resistances are in ohms. k = 1,000 M = 1,000,000

HAM RADIO

Fig. 7-8

8

Display Circuits

The sources of the following circuits are contained in the Sources section, which begins on page 217. The figure number in the box of each circuit correlates to the source entry in the Sources section.

VACUUM FLUORESCENT DISPLAY

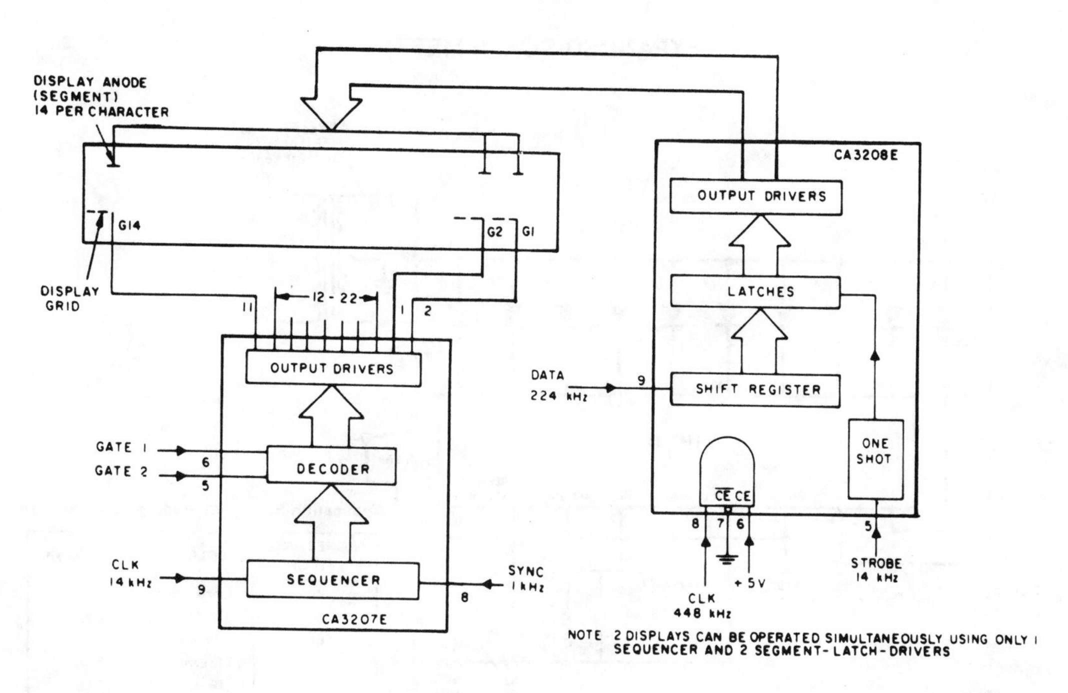

GE/RCA

Fig. 8-1

This circuit uses the CA3207 sequence driver and CA3208 segment latch-driver in combination to drive display devices of up to 14 segments with up to 14 characters of display. The CA3207 selects the digit or character to be displayed in sequence, CA3208 turns on the required alphanumeric segments.

EXPANDED-SCALE METER

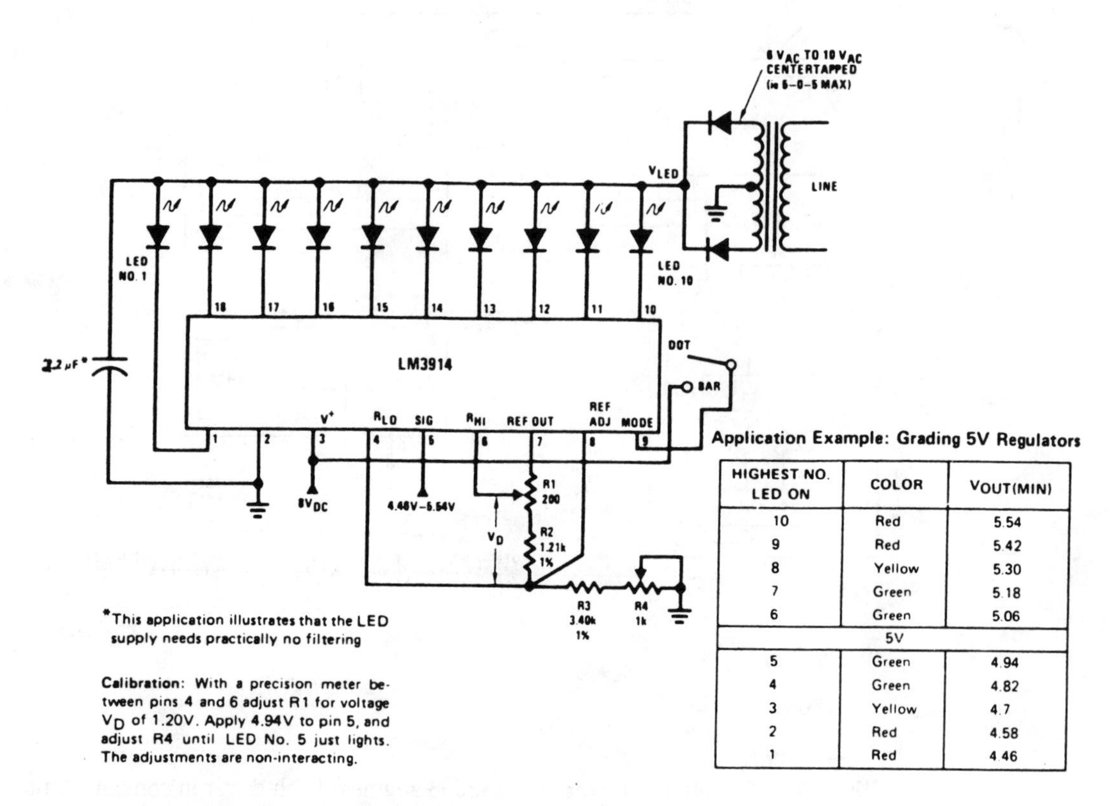

HIGHEST NO. LED ON	COLOR	VOUT(MIN)
10	Red	5.54
9	Red	5.42
8	Yellow	5.30
7	Green	5.18
6	Green	5.06
	5V	
5	Green	4.94
4	Green	4.82
3	Yellow	4.7
2	Red	4.58
1	Red	4.46

NATIONAL SEMICONDUCTOR

Fig. 8-2

A bar-graph driver IC (LM314) drives an LED display. The LEDs might be separate or in a combined (integral) bar-graph display.

EXCLAMATION POINT DISPLAY

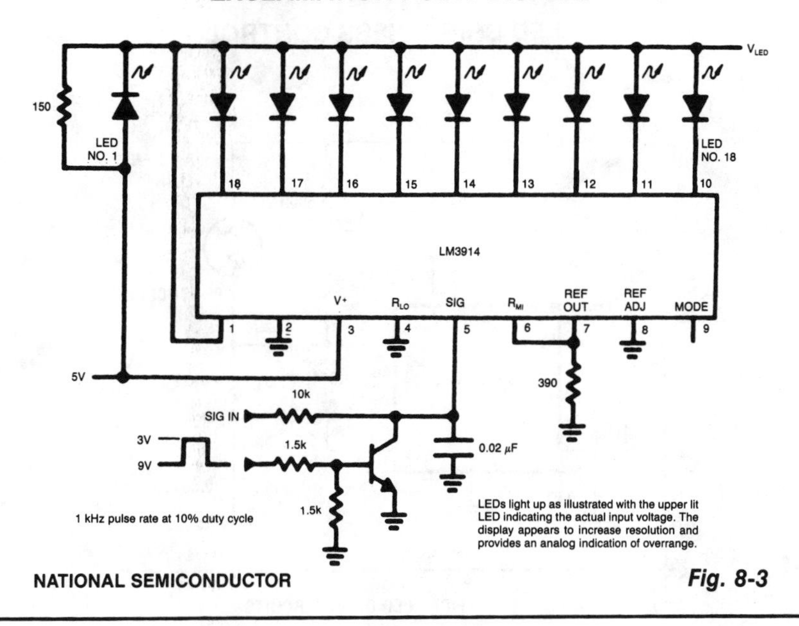

NATIONAL SEMICONDUCTOR

Fig. 8-3

LED BAR PEAK METER DISPLAY

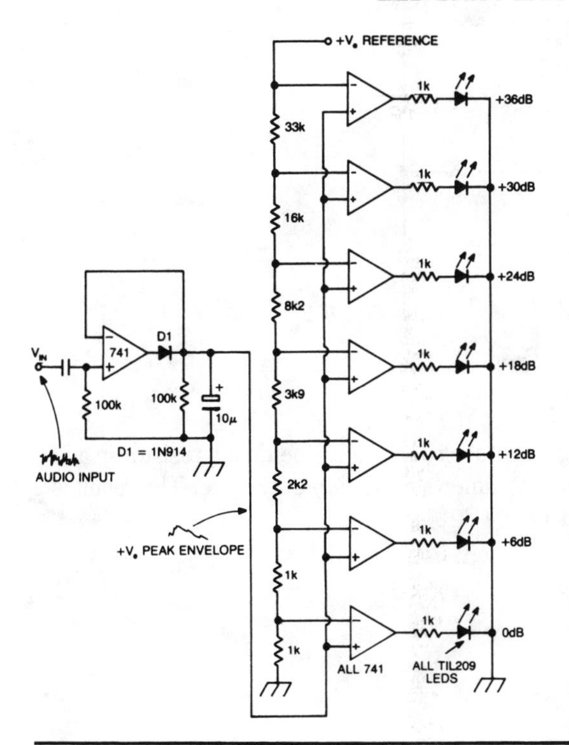

A bar column of LEDs is arranged so that as the audio signal level increases, more LEDs in the column light. The LEDs are arranged vertically in 6-dB steps. A fast response time and a 1 second decay time give an accurate response to transients and a low "flicker" decay characteristic. On each of the op amps inverting inputs is a dc reference voltage, which increases in 6-dB steps. All noninverting inputs are tied together and connected to the positive peak envelope of the audio signal. Thus, as this envelope exceeds a particular voltage reference, the op-amp output goes high and the LED lights. Also, all the LEDs below this are illuminated.

ELECTRONICS TODAY INTERNATIONAL

Fig. 8-4

LED BRIGHTNESS CONTROL

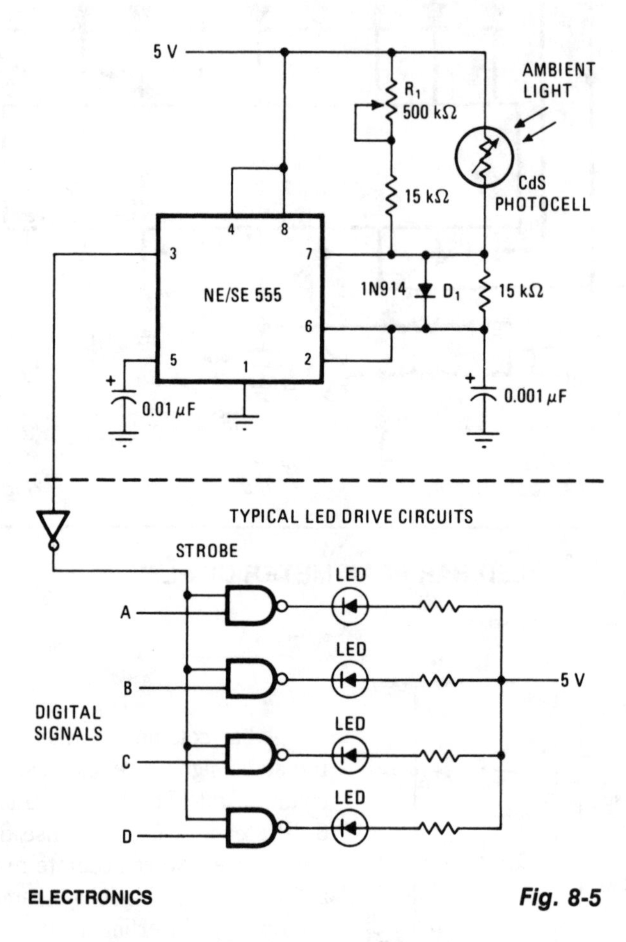

ELECTRONICS ***Fig. 8-5***

This brightness of LED display is varied by using a photocell in place of one timing resistor in a 555 timer, and bypassing the other timing resistor to boost the timer's maximum duty cycle. The result is a brighter display in sunlight and a fainter one in the dark.

10-MHz UNIVERSAL COUNTER

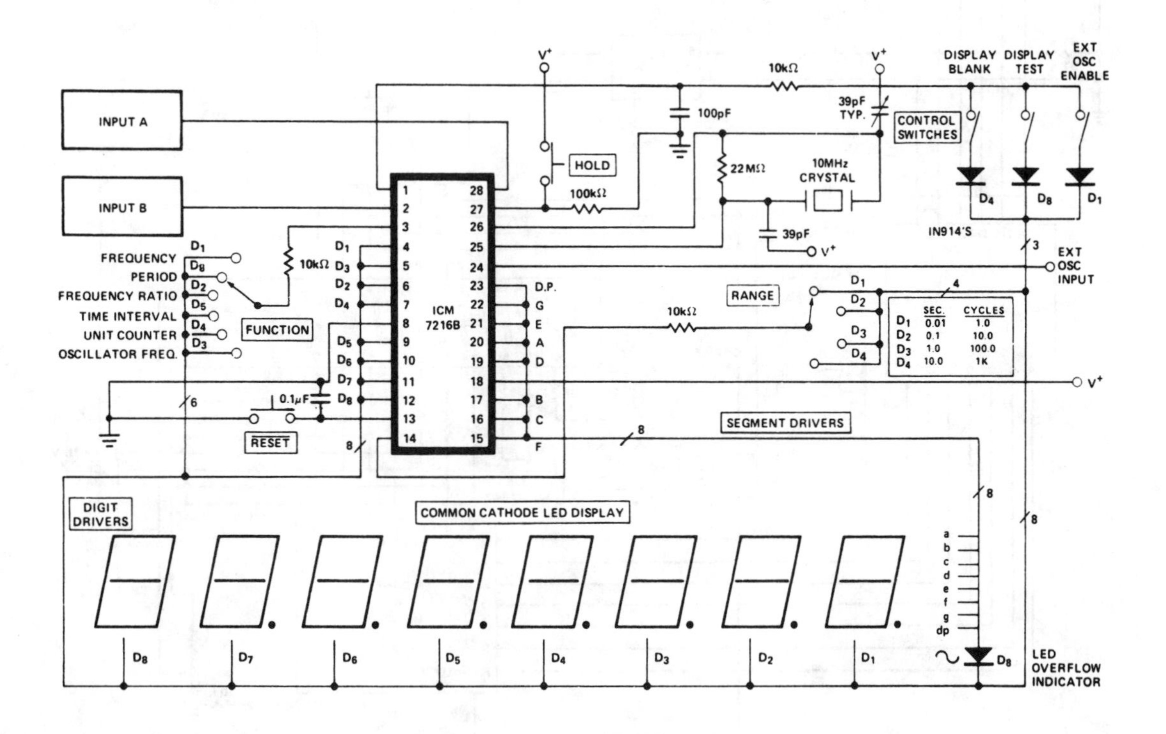

INTERSIL ***Fig. 8-6***

This is a minimum component complete Universal Counter. It can use input frequencies up to 10 MHz at input A and 2 MHz at input B. If the signal at input A has a very low duty cycle, it may be necessary to use a 74121 monostable multivibrator or similar circuit to stretch the input pulse width to be able to guarantee that it is at least 50 ns in duration.

12-HOUR CLOCK WITH GAS-DISCHARGE DISPLAYS

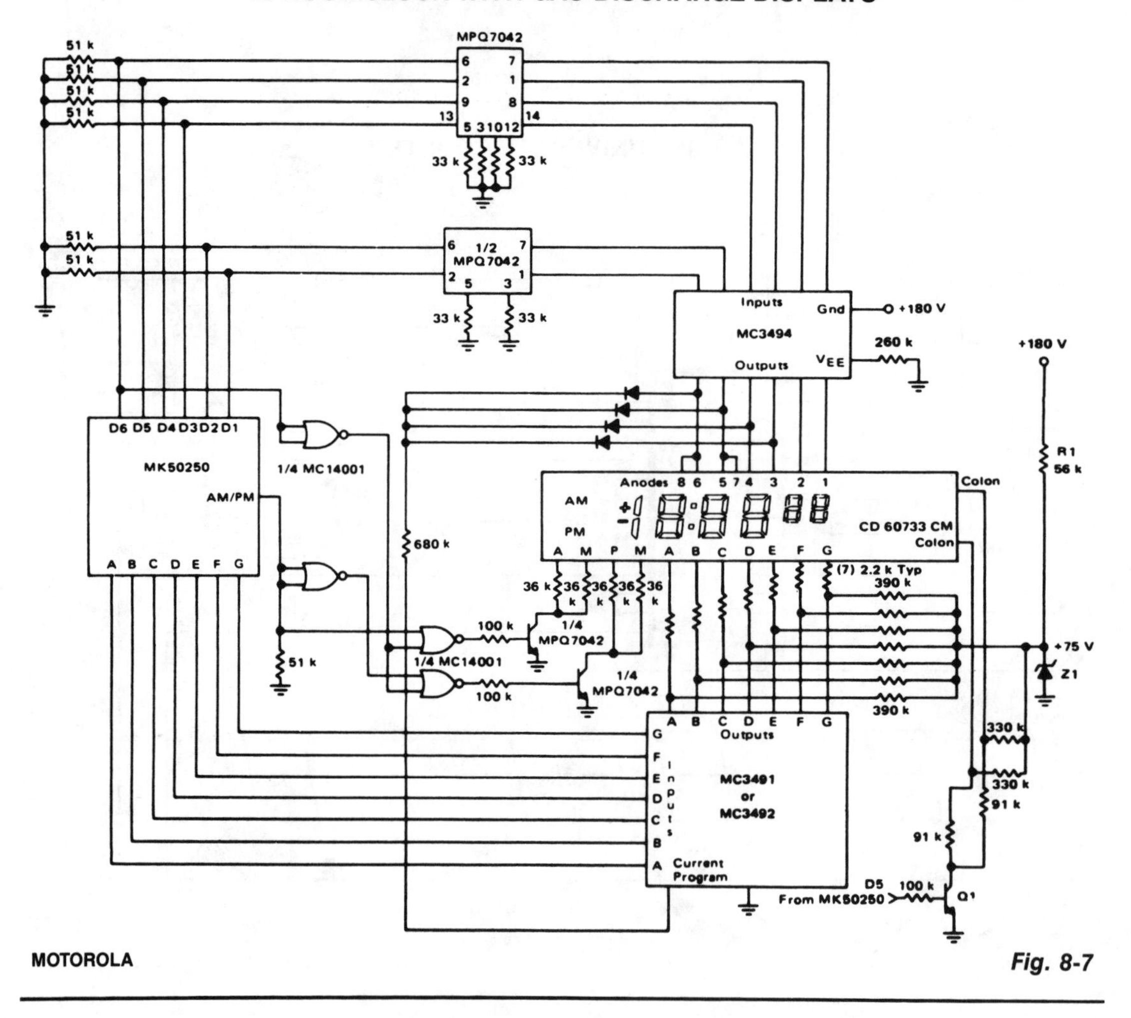

MOTOROLA

Fig. 8-7

PRECISION FREQUENCY COUNTER (~1 MHz MAXIMUM)

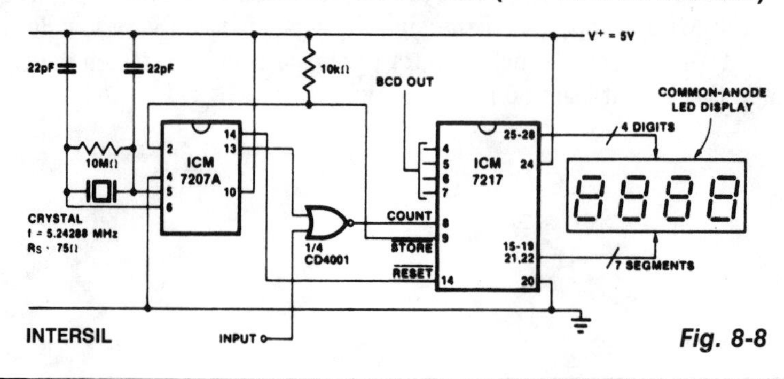

INTERSIL

Fig. 8-8

LOW-COST BAR-GRAPH INDICATOR FOR ac SIGNALS

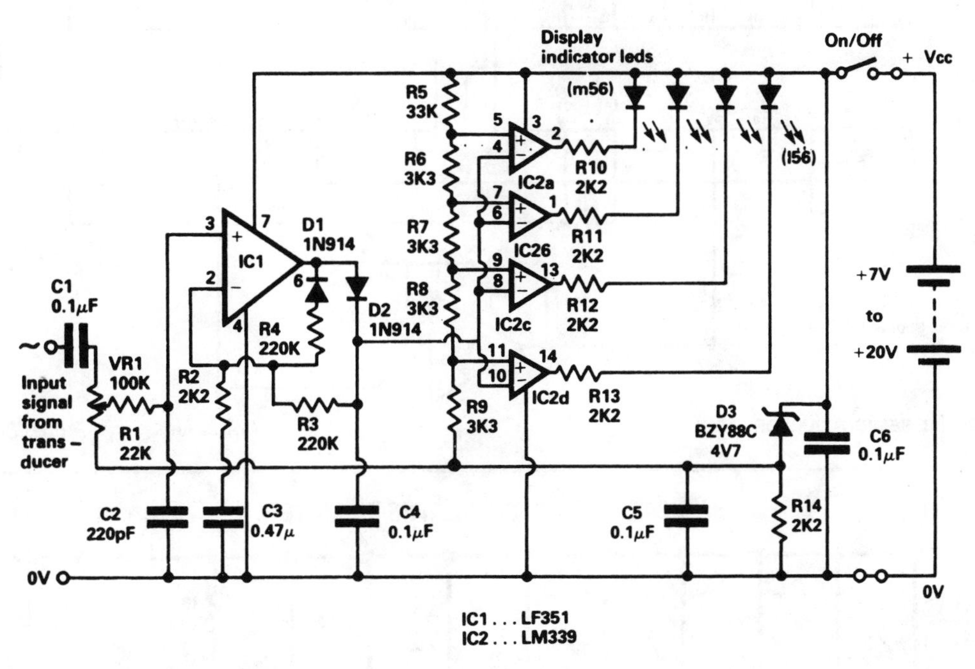

ELECTRONIC ENGINEERING *Fig. 8-9*

Indicator was designed to display the peak level of small ac signals from a variety of transducers including microphones, strain gauges, and photodiodes. The circuit responds to input signals contained within the audio frequency spectrum, i.e., 30 Hz to 20 kHz, although a reduced response extends up to 40 kHz. Maximum sensitivity, with VR1 fully clockwise, is 30 mV peak-to-peak. The indicator can be calibrated by setting VR1 when an appropriate input signal is applied.

60-dB DOT MODE DISPLAY

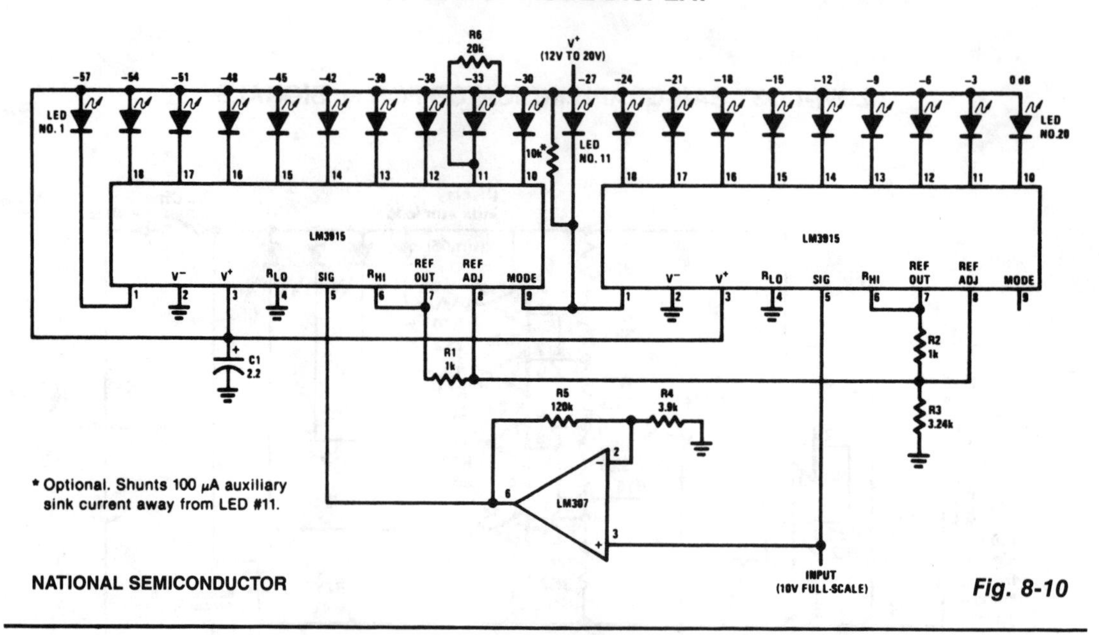

NATIONAL SEMICONDUCTOR

Fig. 8-10

BAR DISPLAY WITH ALARM FLASHER

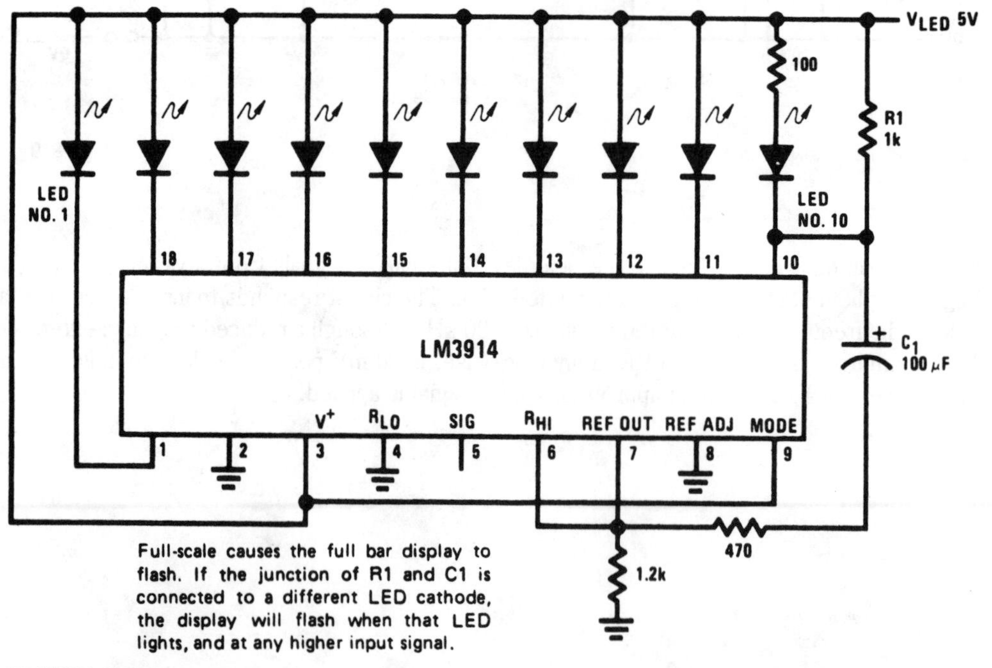

NATIONAL SEMICONDUCTOR

Fig. 8-11

LED-BAR/DOT-LEVEL METER

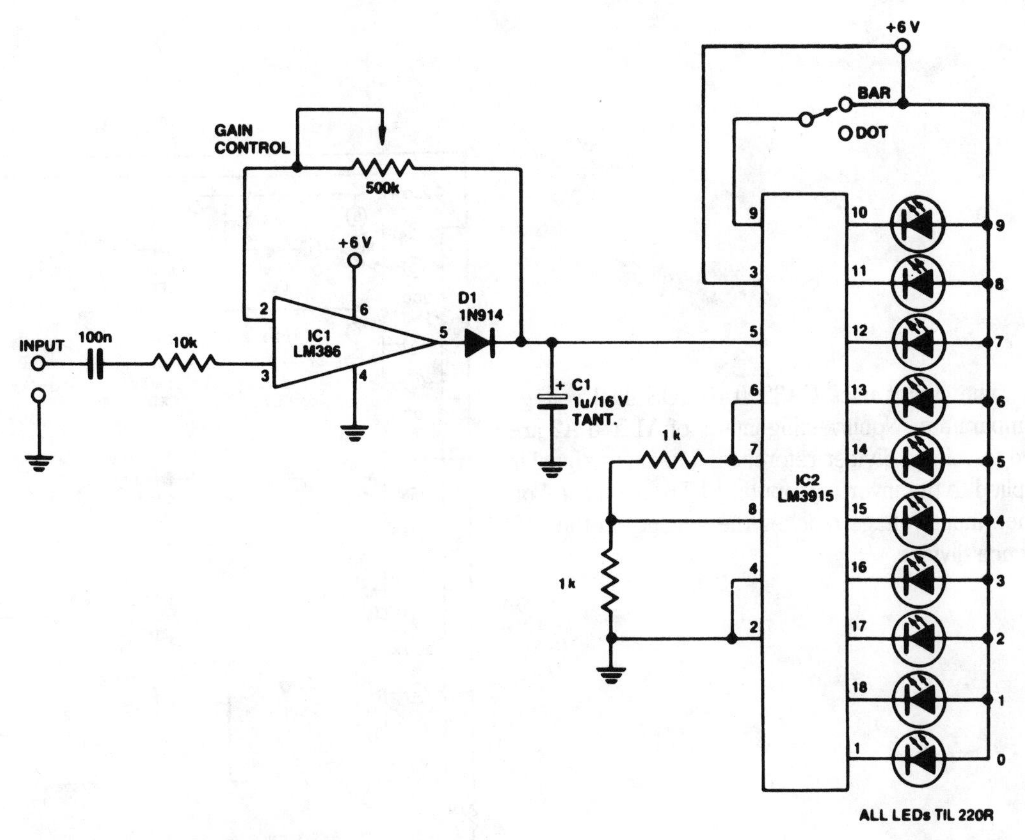

ELECTRONICS TODAY INTERNATIONAL

Fig. 8-12

A simple level of power meter can be arranged to give a bar or dot display for a hi-fi system. Use green LEDs for 0 to 7; yellow for 8 and red for 9 to indicate peak power. The gain control is provided to enable calibration on the equipment with which the unit is used. Because the unit draws some 200 mA, a power supply is advisable instead of running the unit from batteries.

LED BAR-GRAPH DRIVER

The circuit uses CA3290 BiMOS dual voltage comparators. Noninverting inputs of A1 and A2 are tied to voltage divider reference. The input signal is applied to the inverting inputs. LEDs are turned on when input voltage reaches the voltage on the reference divider.

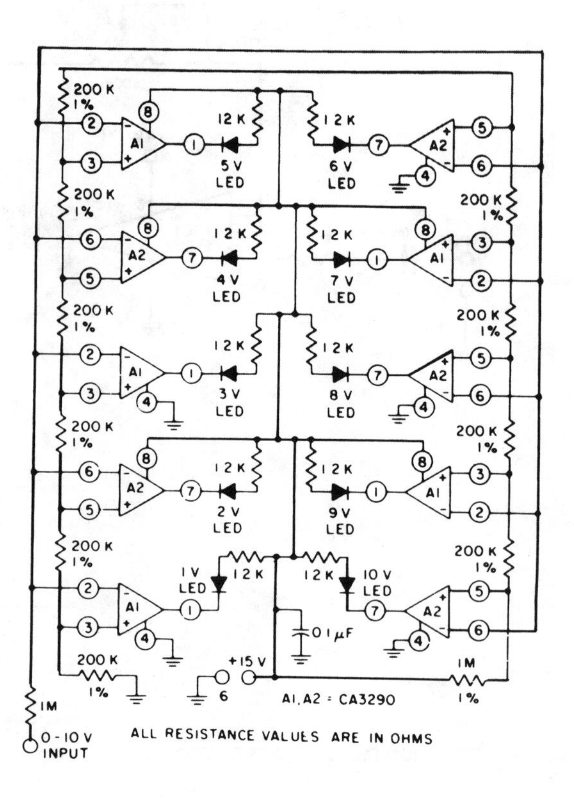

GE/RCA

Fig. 8-13

9

Field-Strength Meter Circuits

The sources of the following circuits are contained in the Sources section, which begins on page 217. The figure number in the box of each circuit correlates to the source entry in the Sources section.

LF OR HF FIELD-STRENGTH METER

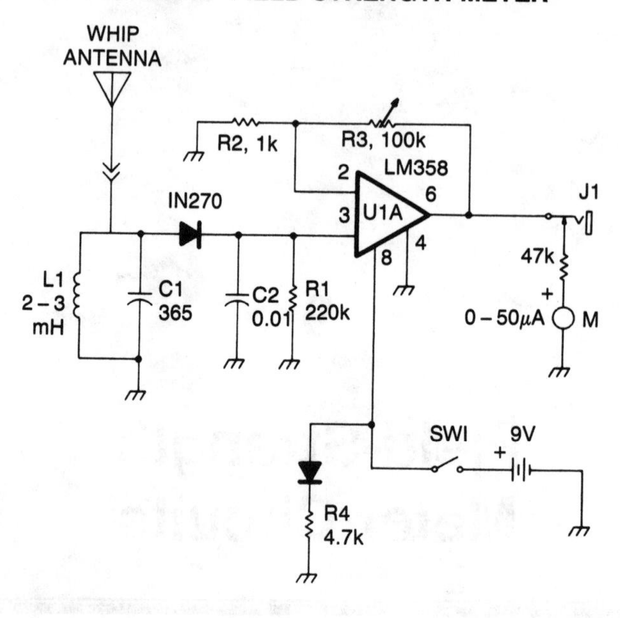

Table 1.

L1	C1 (variable)	Frequency Range	Ham Band
50 μH	30-365 pF	1- 4 MHz	160, 80 meters
3 μH	30-365 pF	5-16 MHz	40, 30, 20 meters
0.9 μH	30-365 pF	9-30 MHz	30, 20, 15, 12, 10 meters
2.5 mH	—	Broadband at reduced gain	

HAM RADIO

Fig. 9-1

C1 and L1 resonate in the 1750-meter band, with coverage from 150 to 500 kHz. L1 can be slug-tuned for 160- to 190-kHz coverage alone or a 2.5-mH choke can be used for L1, if desired, using C1 for tuning. A 1N270 germanium diode rectifies the RF signal and C2 is charged at the peak RF level. This dc level is amplified by an LM358. The gain is determined by R2 and R3, one 100-kΩ linear potentiometer that varies the dc gain from 1 to 100, driving the 50 mA meter. This field-strength meter need not be limited to LF use. The Table shows the *L*1 and *C*1 values for HF and broadband operation.

UNTUNED FIELD-STRENGTH METER

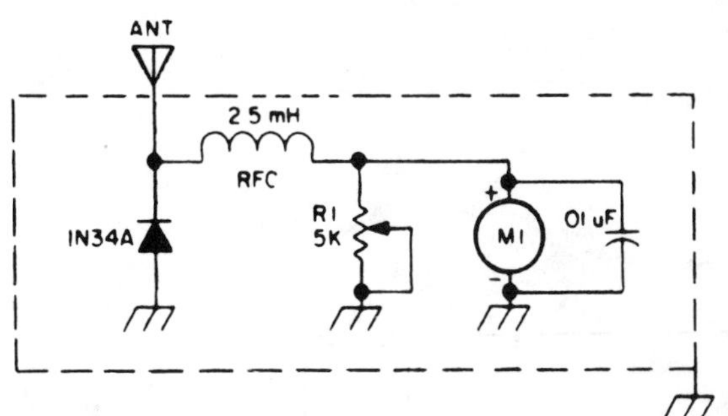

Sensitivity is controlled by R1 and by the sensitivity of meter M1.

TAB BOOKS

Fig. 9-2

TUNED FIELD-STRENGTH METER

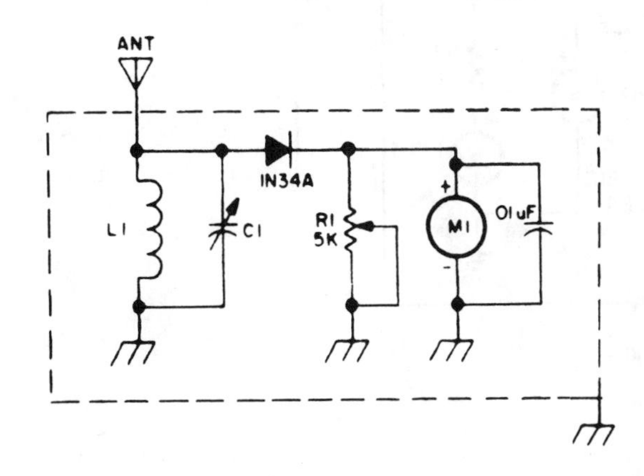

A resonant combination of L1 and C1 is selected to cover frequencies desired.

TAB BOOKS

Fig. 9-3

VOM FIELD-STRENGTH METER

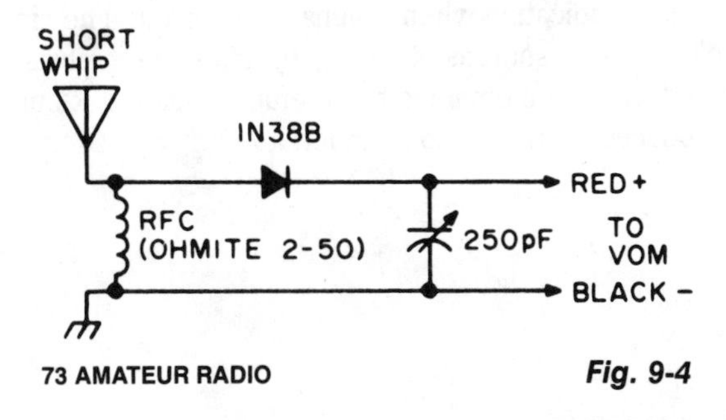

73 AMATEUR RADIO

Fig. 9-4

RF SNIFFER

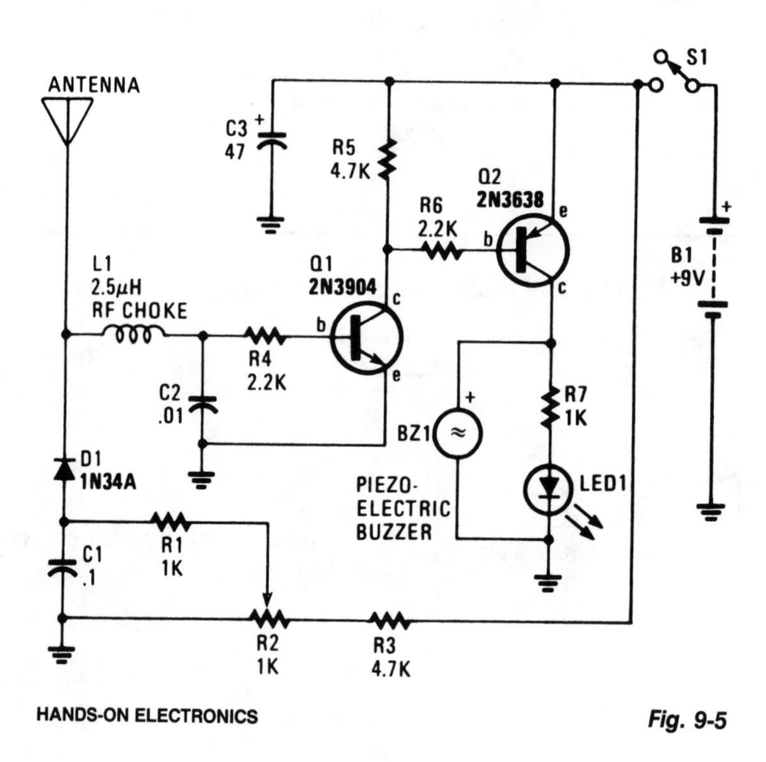

HANDS-ON ELECTRONICS

Fig. 9-5

This circuit responds to RF signals from below the standard broadcast band to well over 500 MHz, and provides a visual and audible indication when a signal is received. The circuit is designed to receive low-powered signals as well as strong sources of energy by adjusting the bias on the pick-up diode, D1, with R2. A very sensitive setting can be obtained by carefully adjusting R2 until the LED just begins to light and a faint sound is produced by the piezo sounder.

SENSITIVE FIELD-STRENGTH METER

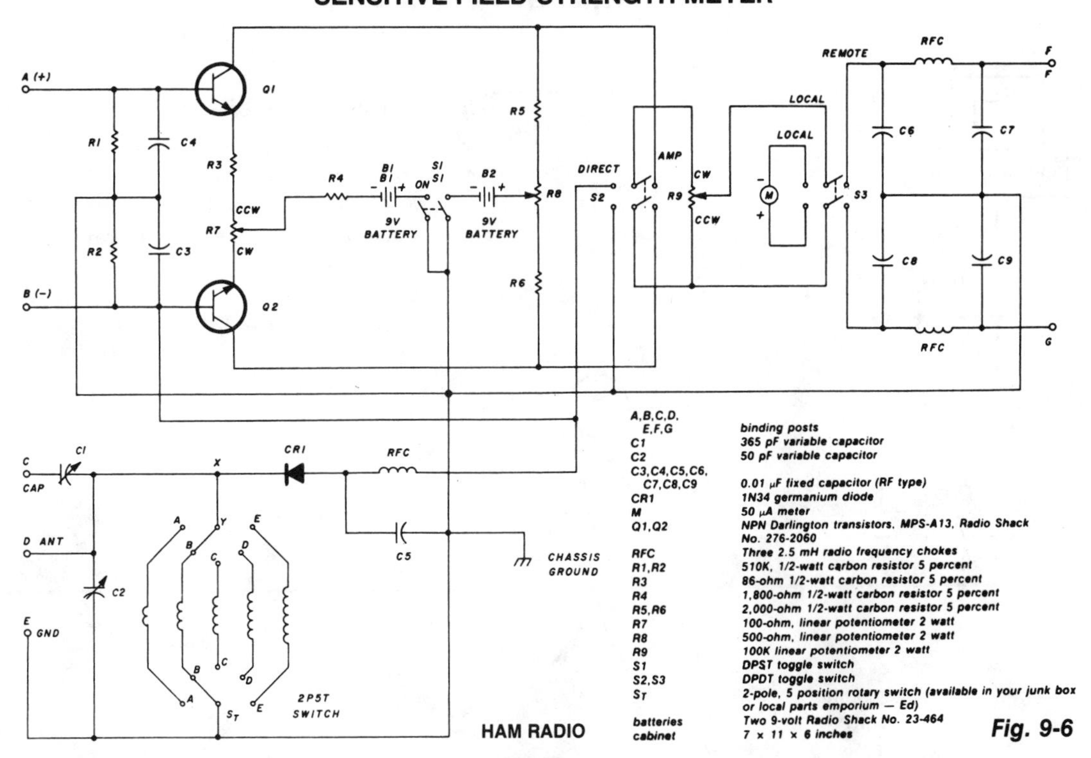

Fig. 9-6

The two-pole, five-position switch, coils, and 365-pF variable capacitor cover a range from 1.5 to 30 MHz. The amplifier uses Darlington npn transistors whose high beta, 5000, provides high sensitivity with S1 used as the amplifier on/off switch. Switch S2 in the left position allows the output of the 1N34 diode to be fed directly into the 50-µA meter (M) for direct reading. When S2 is in the right position, the amplifier is switched into the circuit. Switch S3 is for local or remote monitoring. At full gain setting, the input signal is adjusted to give a full-scale reading of 50 mA on the meter. Then, with the amplifier switched out of the circuit, the meter reading drops down to about 0.5 mA. A 2.5-mH RF choke and capacitors C3, C4, and C5 effectively keep RF out of the amplifier circuit.

FIELD-STRENGTH METER I

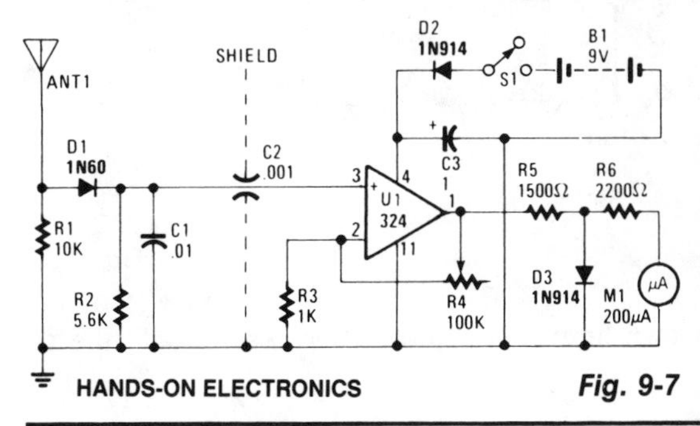

Fig. 9-7

The untuned, but amplified, FSM can almost sense that mythical flea's whisper—from 3 through 148 MHz no less—and yet, is so immune to overload that the meter pointer won't pin. The key to the circuit is the amplifier, a 324 quad op amp, of which only one section is used. It's designed for a single-ended power supply, will provide at least 20-dB dc gain, and the output current is self-limiting. The pointer can't be pinned.

FIELD-STRENGTH METER II

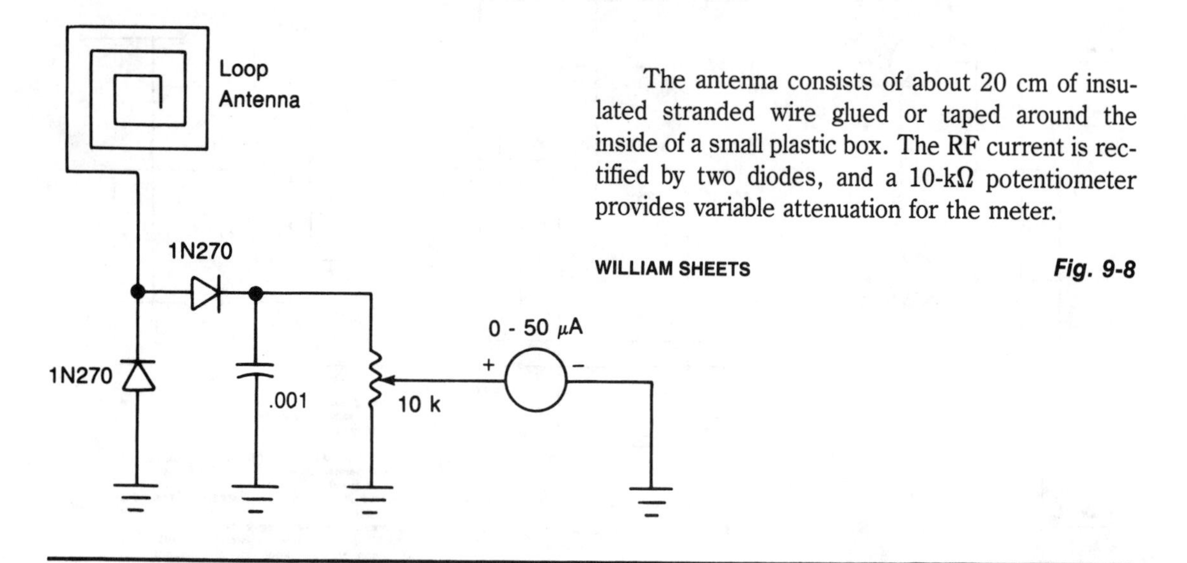

The antenna consists of about 20 cm of insulated stranded wire glued or taped around the inside of a small plastic box. The RF current is rectified by two diodes, and a 10-kΩ potentiometer provides variable attenuation for the meter.

WILLIAM SHEETS *Fig. 9-8*

FIELD-STRENGTH METER III

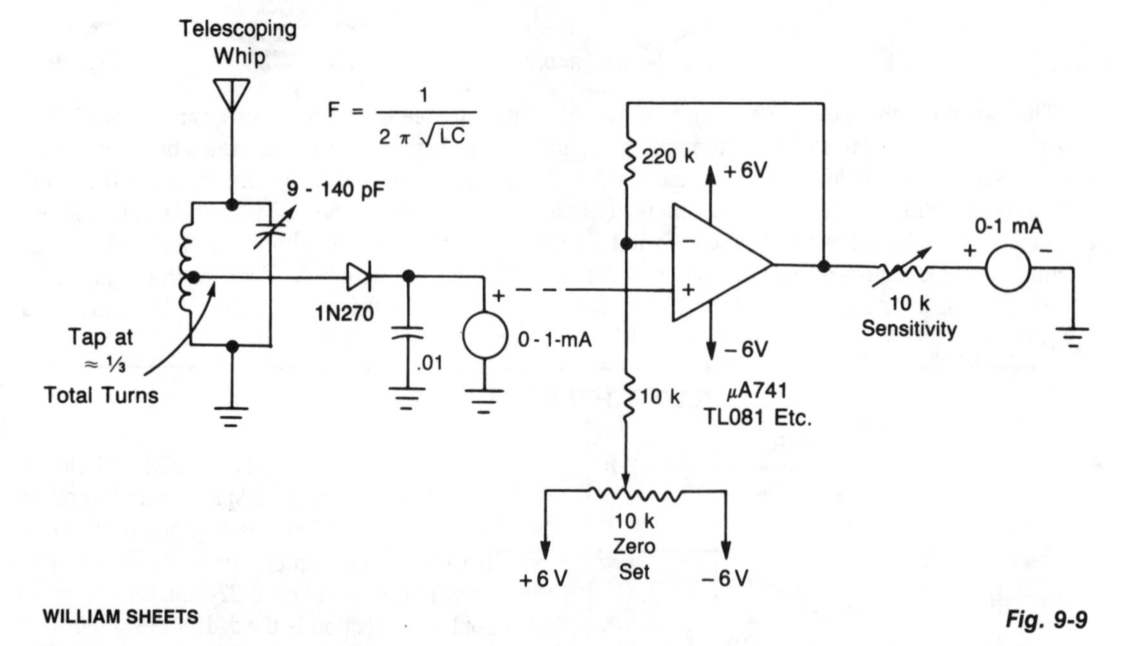

WILLIAM SHEETS *Fig. 9-9*

A "minimum-parts" field-strength meter is shown here. For more distant testing, add the dc amplifier.

RECEIVER SIGNAL ALARM

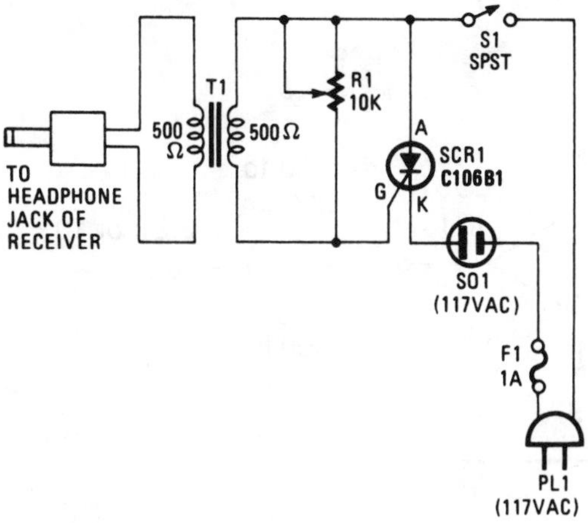

HANDS-ON ELECTRONICS

Fig. 9-10

RF-ACTUATED RELAY

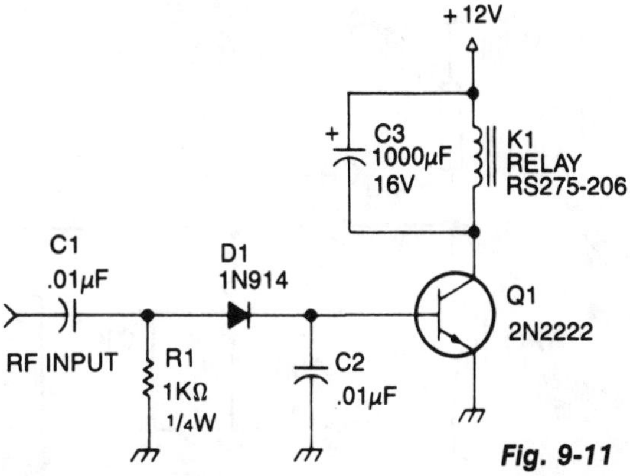

73 AMATEUR RADIO

Fig. 9-11

Automatic antenna switching or RF power indication can be achieved with this circuit. Relay will key with less than 150 mW drive on 2 m.

ON-THE-AIR INDICATOR

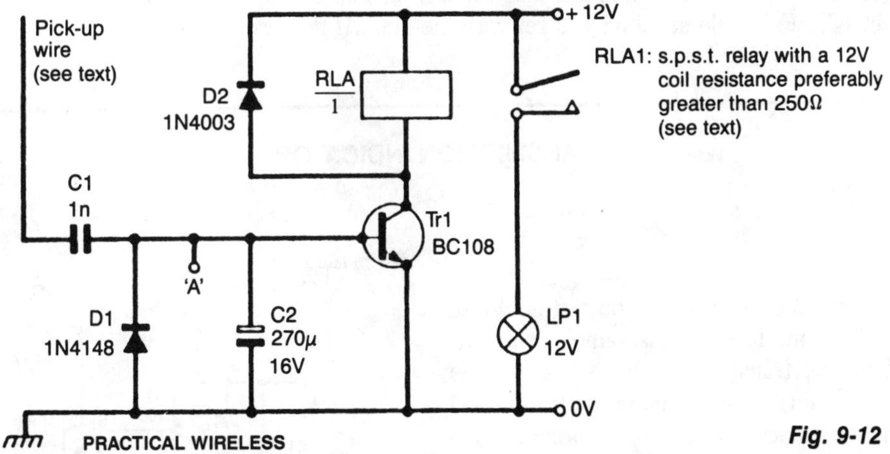

PRACTICAL WIRELESS

Fig. 9-12

The circuit is a simple RF-actuated switch which will respond to any strong field in the region of the pickup wire. The length of the wire will depend on how much coupling is needed, but a 250-mm length wrapped around the outside of the coaxial cable feeding the antenna should suffice for most power levels. If only one band is used, the wire can be made a resonant length—495 mm for 144-MHz band operation, for example. When RF energy is picked up by the device, diode D1 will conduct on the negative half-cycles, but will be cut off on the positive half-cycles. The result will be a net positive voltage at the base of transistor Tr1, forward biasing it into conduction. On SSB and cw transmissions, where the transmission is not continuous, that bias would be constantly varying and the relay RLA would chatter. However, capacitor C2 holds the bias voltage steady until a long gap in transmissions occurs.

HIGH-SENSITIVITY FIELD-STRENGTH METER

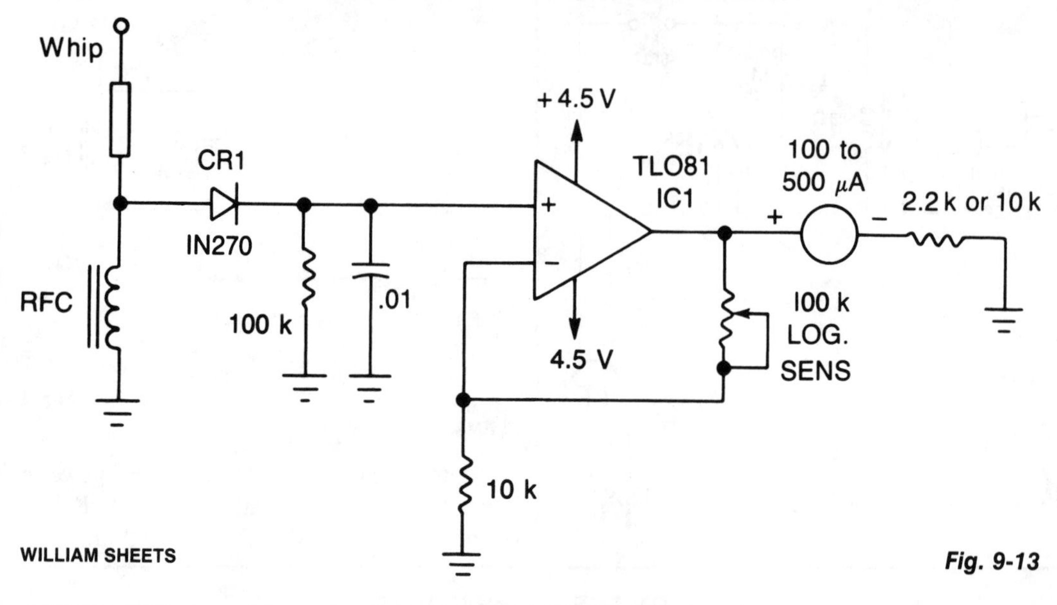

WILLIAM SHEETS

Fig. 9-13

A TL081 (IC1 op amp) is used to increase sensitivity. The RF signal is detected by CR1 and is then amplified by IC1. Full-scale sensitivity is set with the 100-kΩ potentiometer.

TRANSMISSION INDICATOR

Every time the push-to-talk button is closed the light will go on. The antenna samples the output RF from the transmitter. That signal is then rectified (detected) by germanium diode D1, and used to charge capacitor C2. The dc output is used to trigger a small silicon-controlled rectifier (SCR1), which permits the current to flow through the small pilot lamp. For lower-power applications, such as CB radio, the indicator antenna will have to be close-coupled to the transmitter antenna.

HANDS-ON ELECTRONICS

Fig. 9-14

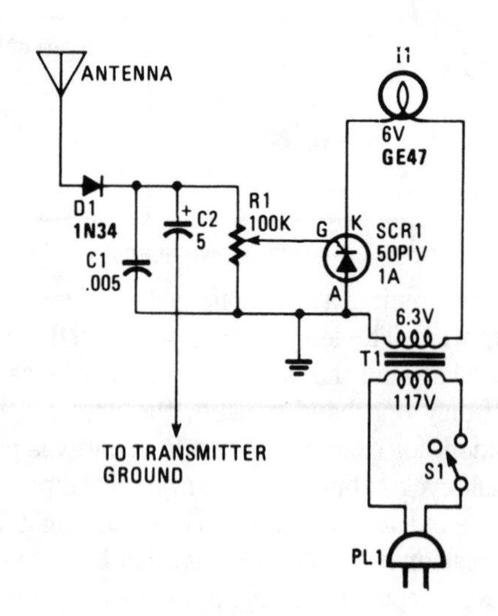

ADJUSTABLE-SENSITIVITY FIELD-STRENGTH INDICATOR

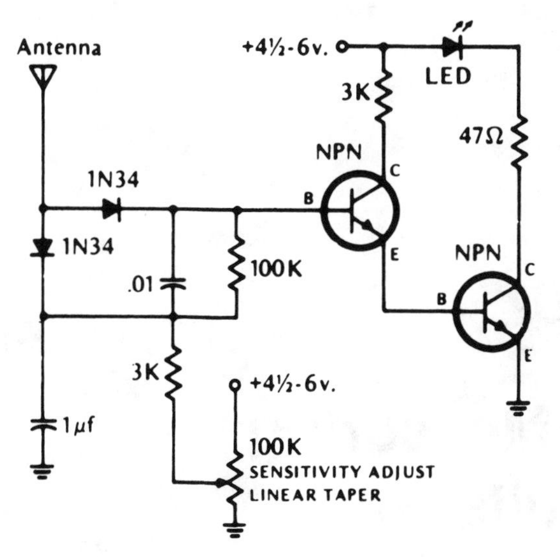

The LED lights if the RF field is higher than the preset field strength level. Diodes should be germanium. Transistors (npn)-2N2222, 2N3393, 2N3904 or equivalent.

MODERN ELECTRONICS

Fig. 9-15

SIMPLE FIELD-STRENGTH METER

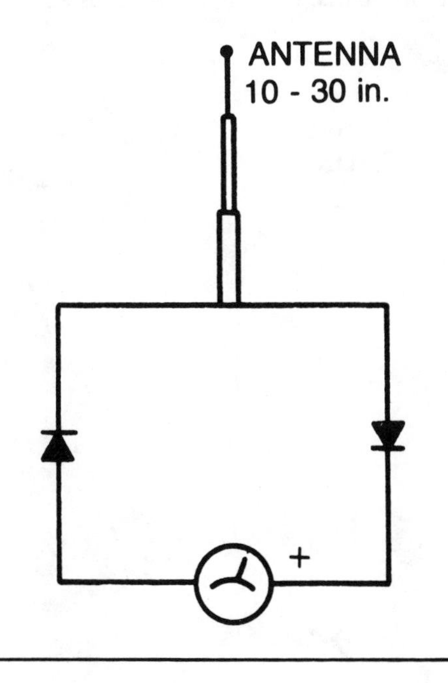

The circuit is not frequency selective. It has been used from 2 meters through 160 meters. The telescoping antenna may be adjusted to its shortest length when working at 2 meters to keep the needle on the scale. Meter should be a 100- to 500-μA movement. The diodes are germanium, such as the 1N34, etc. Silicon diodes will also work, but they are a bit less sensitive.

TAB BOOKS

Fig. 9-16

10

Frequency-Measuring Circuits

The sources of the following circuits are contained in the Sources section, which begins on page 217. The figure number in the box of each circuit correlates to the source entry in the Sources section.

POWER-LINE FREQUENCY METER

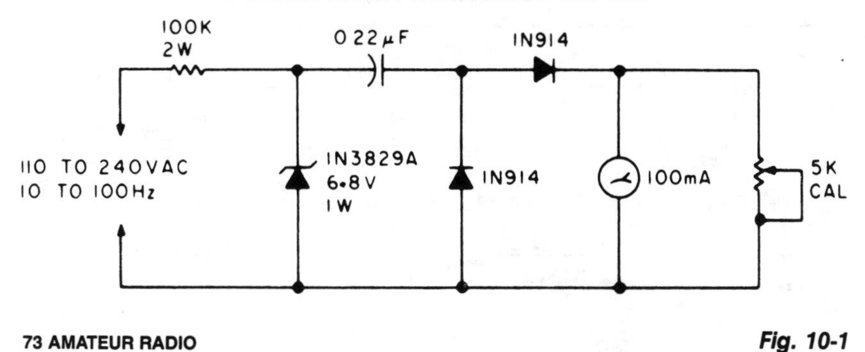

73 AMATEUR RADIO

Fig. 10-1

The meter will indicate the frequency from a power generator. Incoming sine waves are converted to square waves by the 100 kΩ resistor and by the 6.8-V zener. The square wave is differentiated by the capacitor and the current is averaged by the diodes. The average current is almost exactly proportional to the frequency and can be read directly on a 100-mA meter. To calibrate, hook the circuit up to a 60-Hz powerline and adjust the 5-KΩ pot to read 60mA.

AUDIO FREQUENCY METER

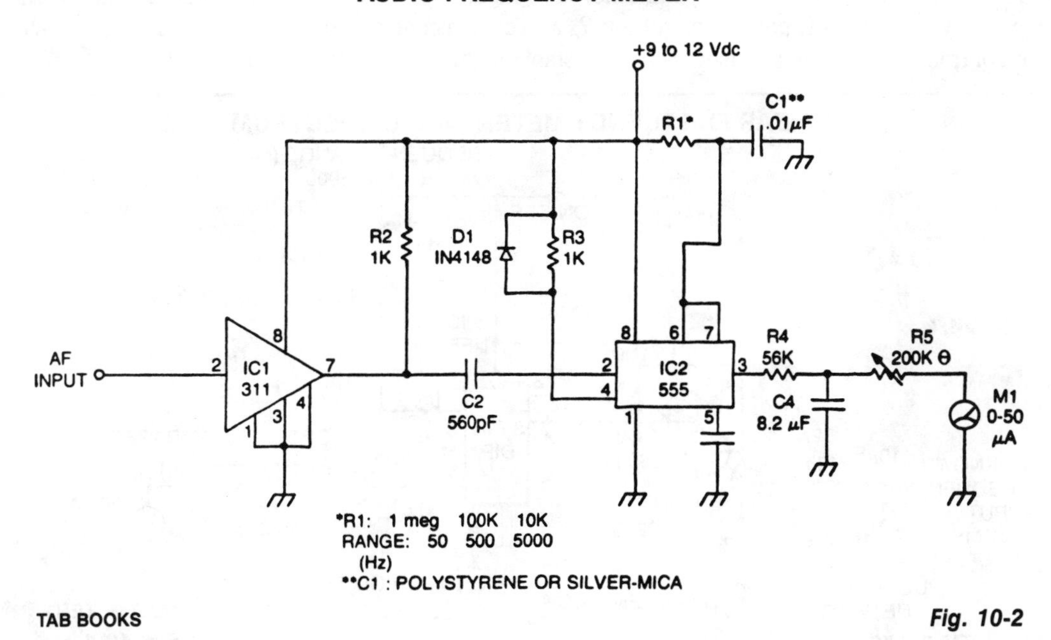

TAB BOOKS

Fig. 10-2

The meter uses time averaging to produce a direct current that is proportional to the frequency of the input signal.

INEXPENSIVE FREQUENCY COUNTER/TACHOMETER

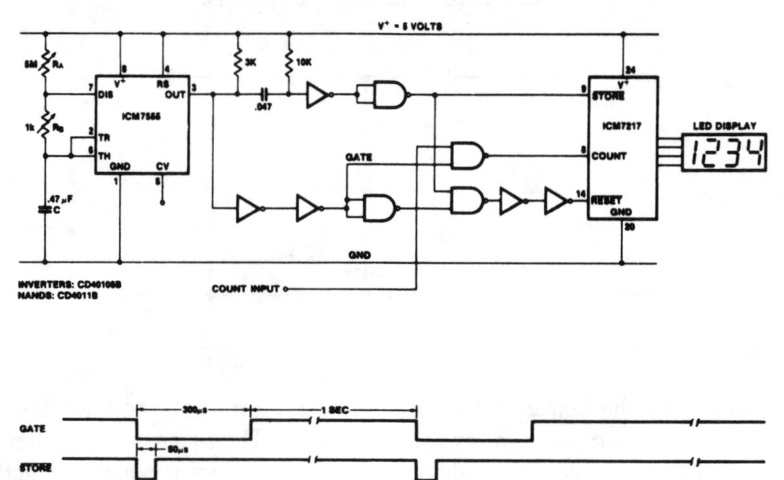

INTERSIL **_Fig. 10-3_**

This circuit uses the low power ICM7555 (CMOS 555) to generate the gating, $\overline{\text{STORE}}$ and $\overline{\text{RESET}}$ signals. To provide the gating signal, the timer is configured as an astable multivibrator. The system is calibrated by using a 5-MΩ potentiometer for R_A as a coarse control and a 1-kΩ potentiometer for R_B as a fine control. CD40106Bs are used as a monostable multivibrator and reset time delay.

LINEAR FREQUENCY METER (AUDIO SPECTRUM)

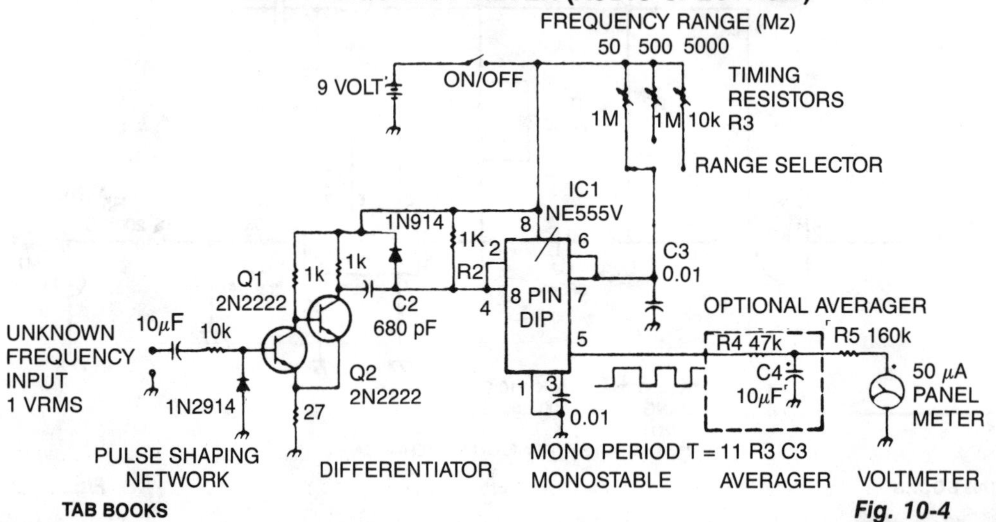

TAB BOOKS **_Fig. 10-4_**

The 555 is used in a monostable multivibrator circuit that puts out a fixed timewidth pulse, which is triggered by the unknown input frequency.

LOW-COST FREQUENCY INDICATOR

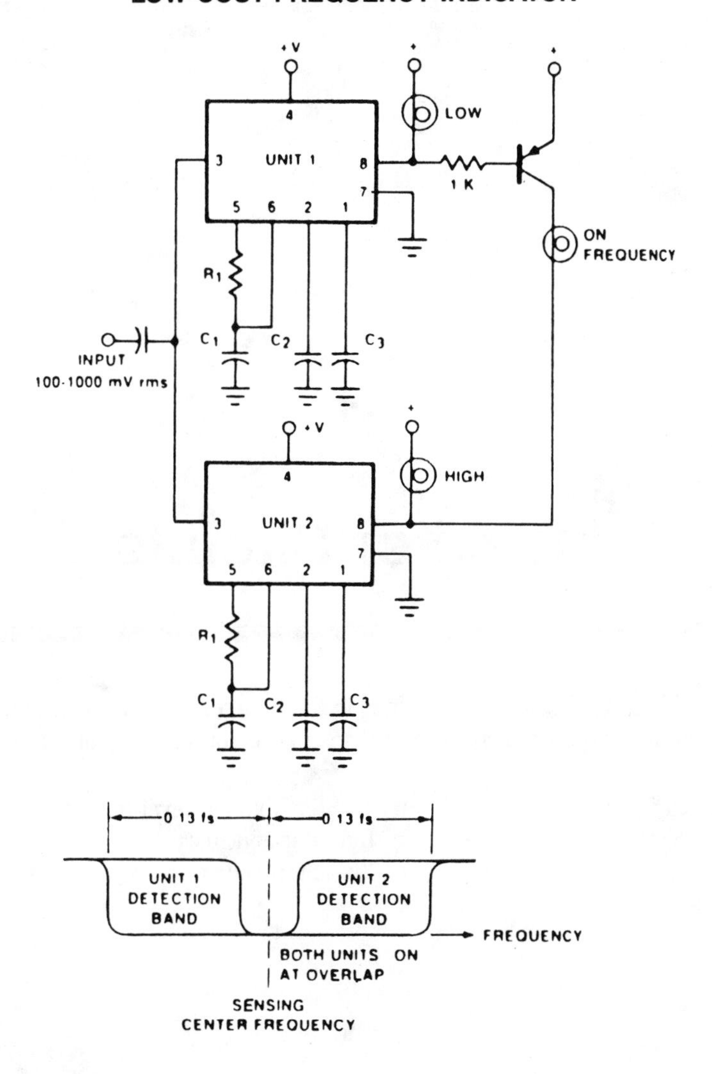

EXAR *Fig. 10-5*

The circuit shows how two tone decoders set up with overlapping detection bands can be used for a go/no-go frequency meter. Unit 1 is set 6% above the desired sensing frequency and Unit 2 is set 6% below the desired frequency. Now, if the incoming frequency is within 13% of the desired frequency, either Unit 1 or Unit 2 will give an output. If both units are on, the incoming frequency is within 1% of the desired frequency. Three light bulbs and a transistor allow low-cost read-out. The IC is an EXAR 567.

11

Indicator Circuits

The sources of the following circuits are contained in the Sources section, which begins on page 217. The figure number in the box of each circuit correlates to the source entry in the Sources section.

BEAT-FREQUENCY INDICATOR

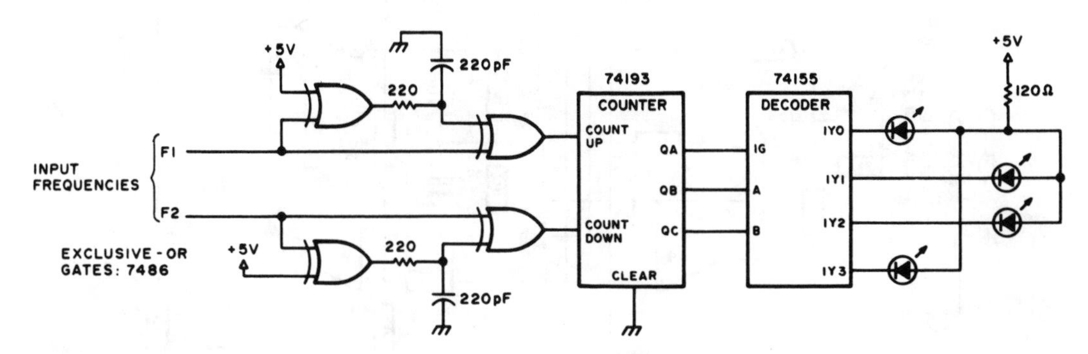

73 AMATEUR RADIO ***Fig. 11-1***

This circuit uses LEDs to display the beat frequency of two-tone oscillators. Only one LED is on at a time, and the apparent rotation of the dot is an exact indication of the best frequency. When f_1 is greater than f_2, a dot of light rotates clockwise; when f_1 is less than f_2, the dot rotates counterclockwise; and when f_1 equals f_2, there is no rotation.

THREE-STEP LEVEL INDICATOR

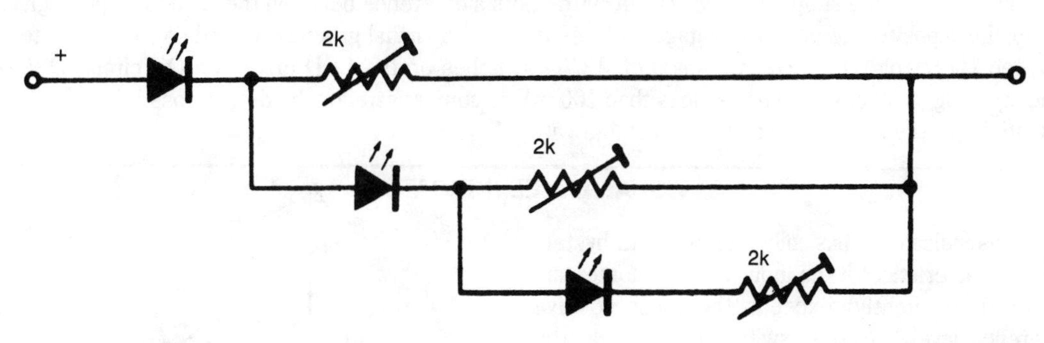

ELECTRONICS TODAY INTERNATIONAL ***Fig. 11-2***

This circuit makes a very compact level indicator where a meter would be impractical or not justified because of cost. Resistor values will depend on type of LED used. For MV50 LEDs the resistors are 2 kΩ for steps of approx 2 V and current drain with all three LEDs on of 5 mA. The chain can be extended, but current drain increases rapidly and the first LED carries all the current drawn from the supply.

STEREO INDICATOR

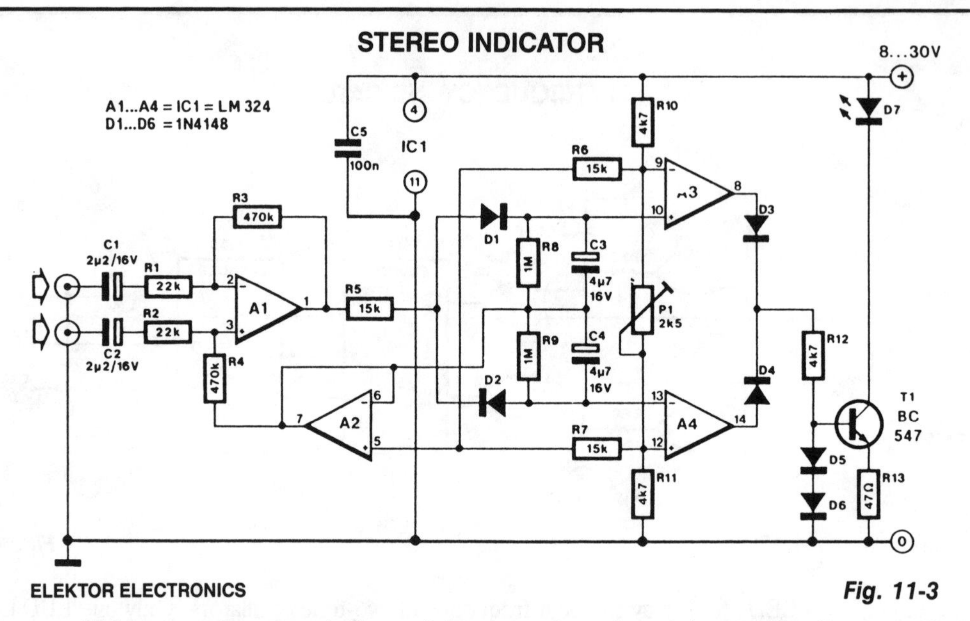

ELEKTOR ELECTRONICS

Fig. 11-3

On most FM tuners, the stereo indicator lights upon detection of the 19-kHz pilot tone. However, this doesn't mean that the program is actually stereophonic, since the pilot tone is often transmitted with mono programs also. A similar situation exists on stereo amplifiers, where the stereo LED is simply controlled from the mono/stereo switch.

The LED-based stereo indicator described here lights only when a true stereo signal is fed to the inputs. Differential amplifier A1 raises the difference between the L and R input signals. When these are equal, the output of A1 remains at the same potential as the output of A2, which forms a virtual ground rail at half the supply voltage. When A1 detects a difference between the L and R input signals, it supplies a positive or negative voltage with respect to the virtual ground rail, and so causes C3 to be charged via D1 or C4 via D2. Comparator A3/A4 switches on the LED driver via OR circuit D3/D4. The input signal level should not be less than 100 mV to compensate for the drop across D1 or D2. The sensitivity of the stereo indicator is adjustable with P1.

VISUAL LEVEL INDICATOR

This indicator is basically a switch with hysteresis characteristics. If the input voltage momentarily (or permanently) exceeds the most positive reference level, LED1 is switched on. If, on the other hand, the voltage falls below the negative, or least positive, reference level, LED1 will be switched off and LED2 switched on. The output voltage, V_O is clamped either to the diode voltage V_{D1}, or V_{D2} depending on which LED is conducting. For V_O to be positive, V_B has to be positive, with respect to the reference voltage V_R; for V_O to be negative, V_B has to be negative, with respect to V_R.

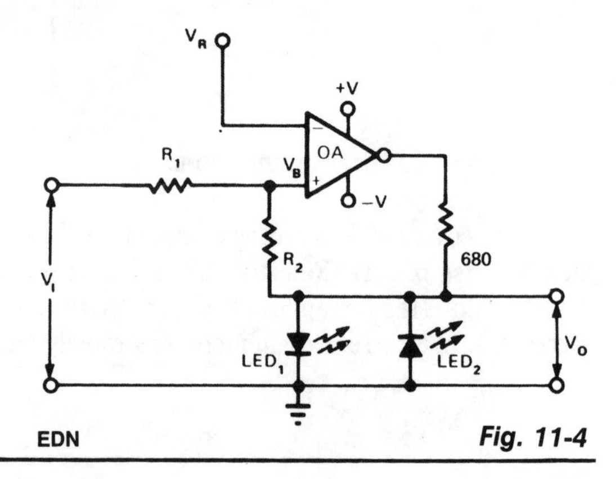

EDN

Fig. 11-4

10-STEP VOLTAGE-LEVEL INDICATOR

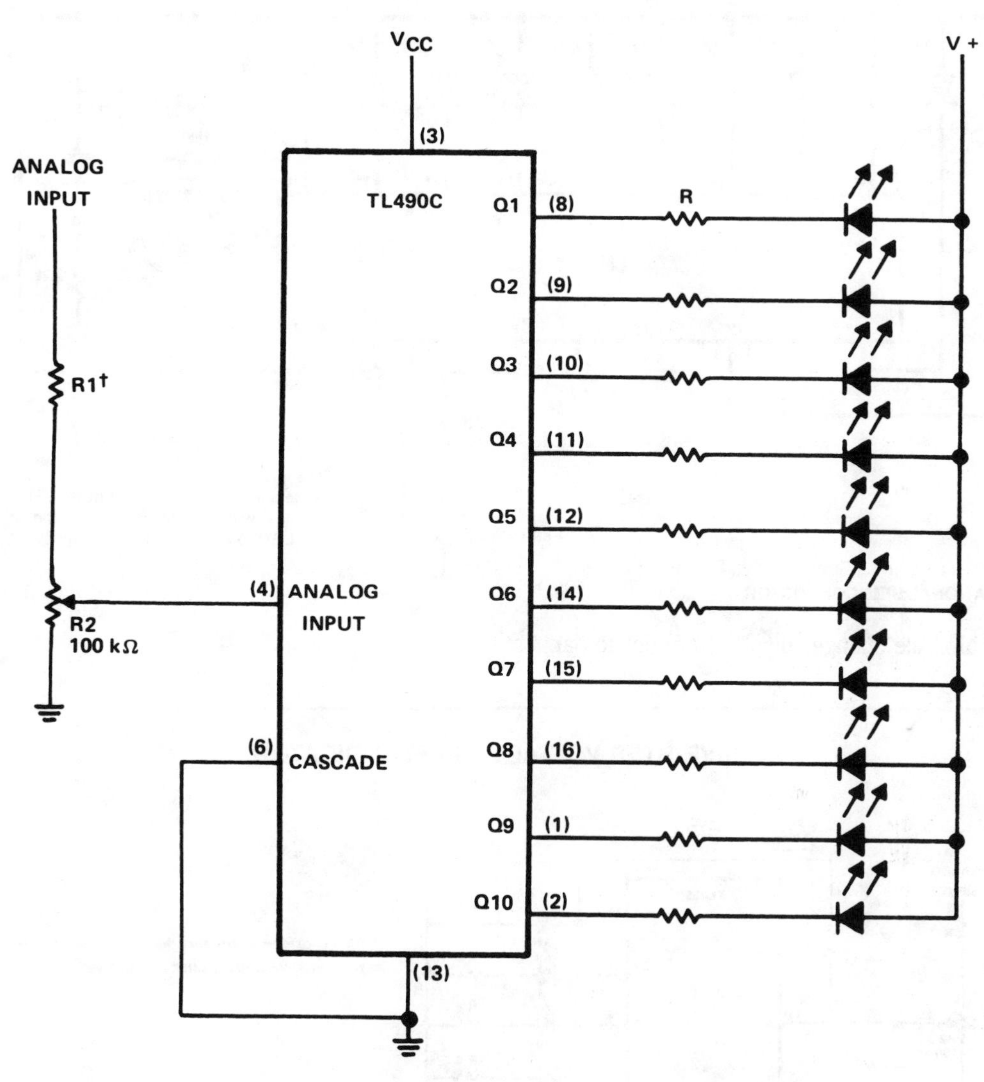

TEXAS INSTRUMENTS

Fig. 11-5

This 10-step adjustable analog level detector is capable of sinking up to 40 mA at each output. The voltage range at the input pin should range from 0 to 2 V. Circuits of this type are useful as liquid-level indicators, pressure indicators, and temperature indicators. They can also be used with a set of active filters to provide a visual indication of harmonic content of audio signals.

INDICATOR AND ALARM

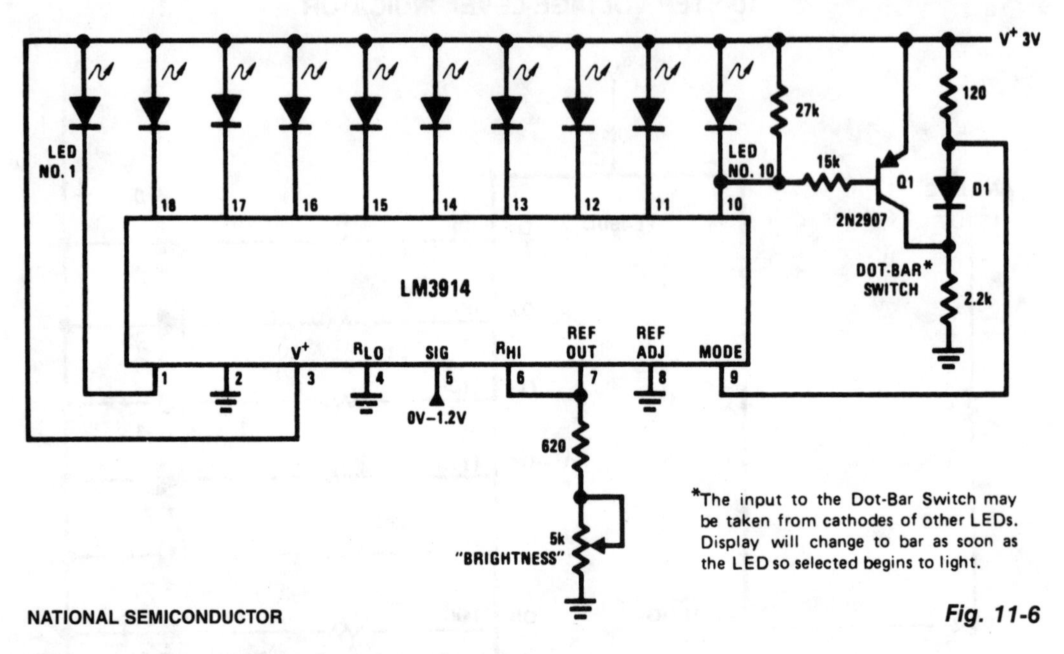

NATIONAL SEMICONDUCTOR

Fig. 11-6

Full-scale changes display from dot to bar.

FIVE-STEP VOLTAGE-LEVEL INDICATOR

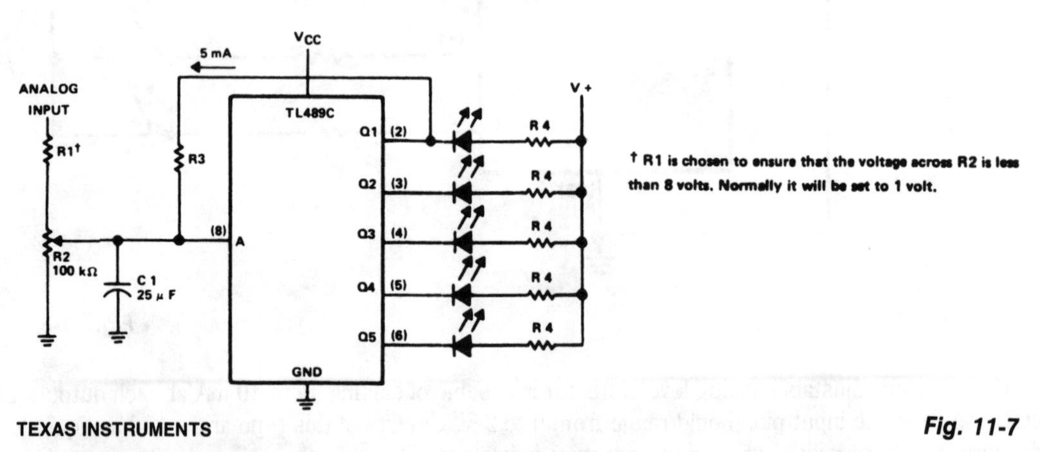

TEXAS INSTRUMENTS

Fig. 11-7

This circuit provides a visual indication of the input analog voltage level. It has a high input impedance at pin 8 and open-collector outputs capable of sinking up to 40 mA. It is suitable for driving a linear array of five LEDs to indicate the level is five steps. The voltage at the analog input should be in the range of 0 to approximately 1 V and should never exceed 8 V.

12

Light-Measuring Circuits

The sources of the following circuits are contained in the Sources section, which begins on page 217. The figure number in the box of each circuit correlates to the source entry in the Sources section.

LINEAR LIGHT-METER CIRCUIT

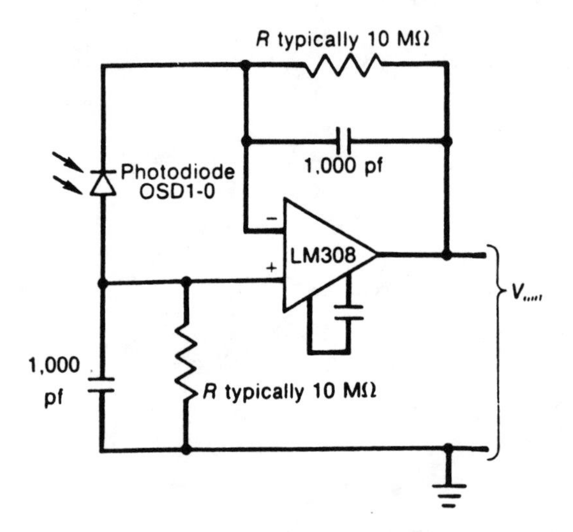

This circuit uses a low-input-bias op amp to give a steady dc indication of light level. To reduce circuit sensitivity to light, R1 can be reduced, but should not be less than 100 K. The capacitor values in the circuit are chosen to provide a time constant sufficient to filter high-frequency light variations that might arise, for example, from fluorescent lights.

MACHINE DESIGN

Fig. 12-1

LOGARITHMIC LIGHT-METER CIRCUIT

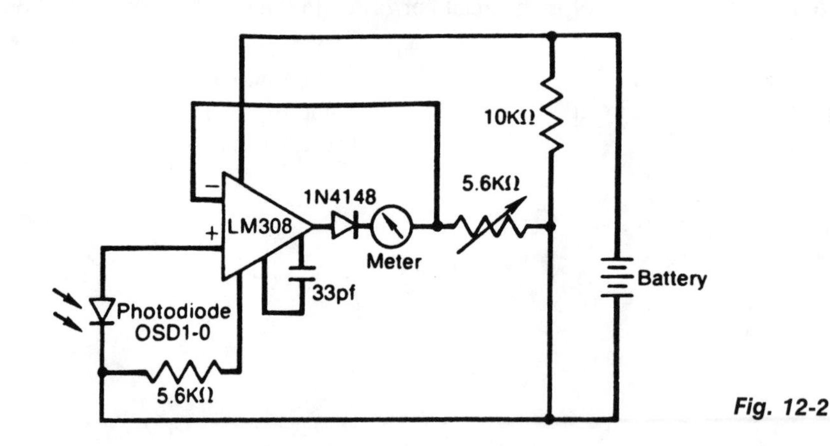

MACHINE DESIGN

Fig. 12-2

The meter reading is directly proportional to the logarithm of the input light power. The logarithmic circuit behavior arises from the nonlinear diode pn junction current/voltage relationship. The diode in the amplifier output prevents output voltage from becoming negative (thereby pegging the meter), which may happen at low lightlevels due to amplifier bias currents. R1 adjusts the meter full-scale deflection, enabling the meter to be calibrated.

LIGHT METER I

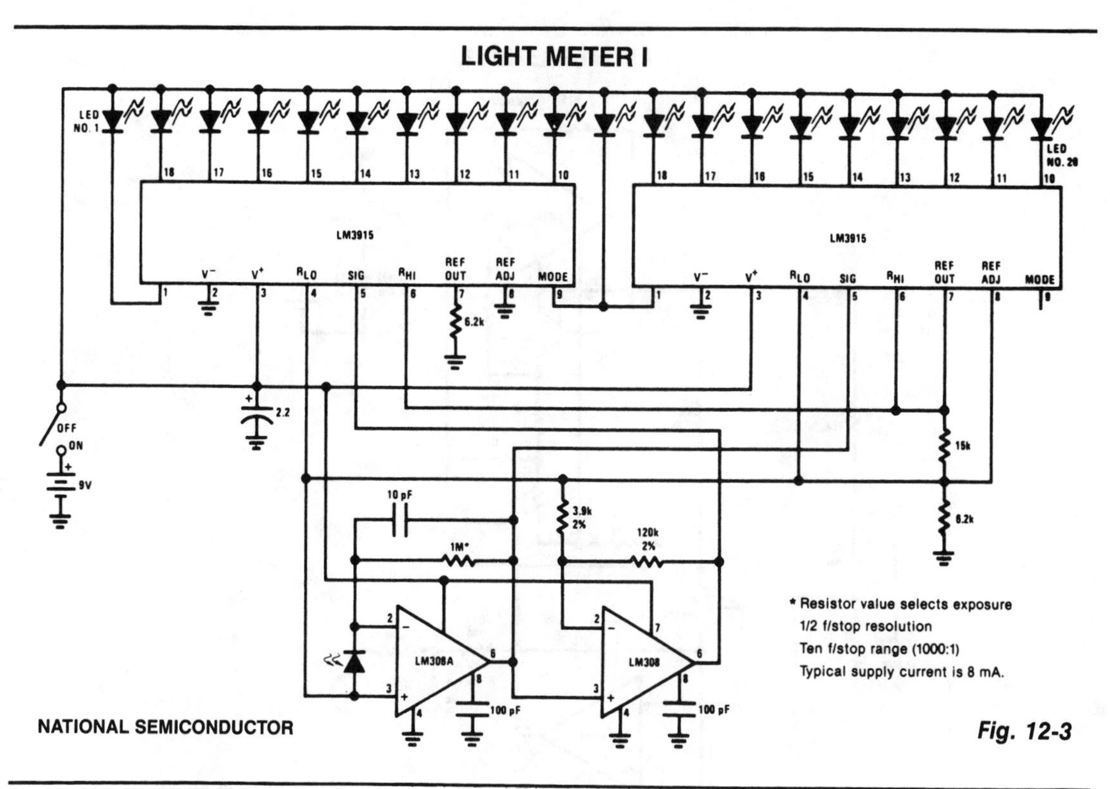

NATIONAL SEMICONDUCTOR

Fig. 12-3

LIGHT METER II

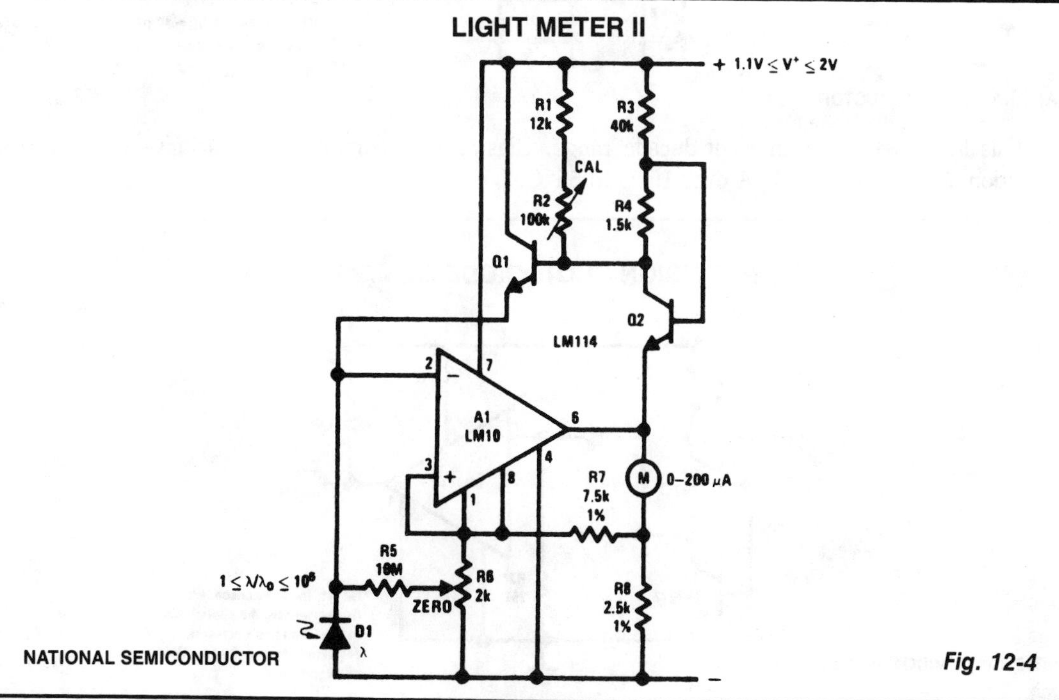

NATIONAL SEMICONDUCTOR

Fig. 12-4

LIGHT METER III

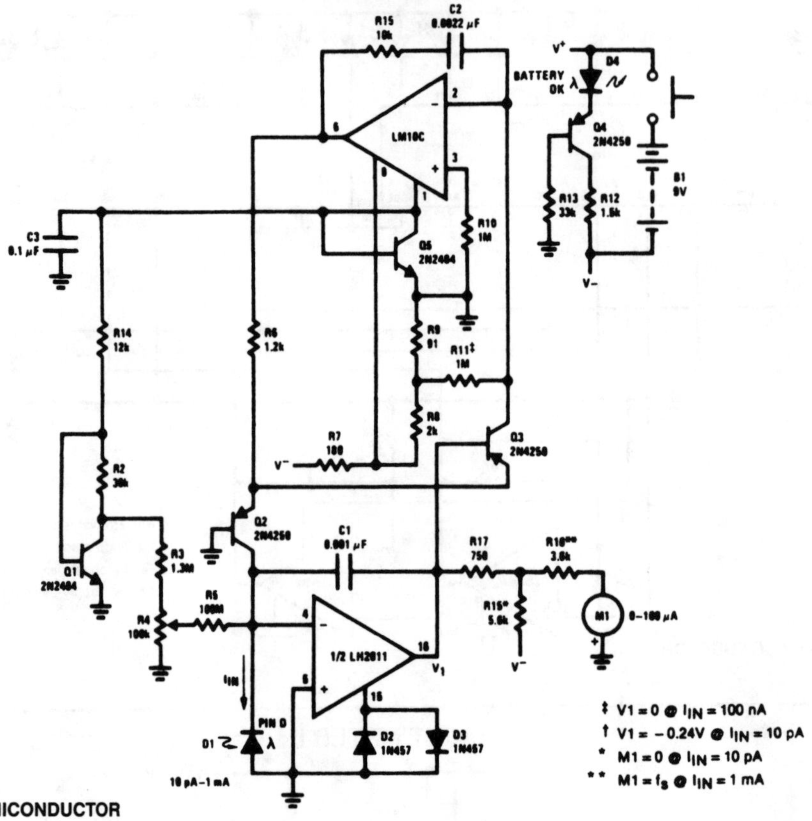

NATIONAL SEMICONDUCTOR

Fig. 12-5

This light meter has an eight-decade range. Bias-current compensation can give input-current resolution of better than ±2 pA over 15°C to 55°C.

PRECISION PHOTODIODE COMPARATOR

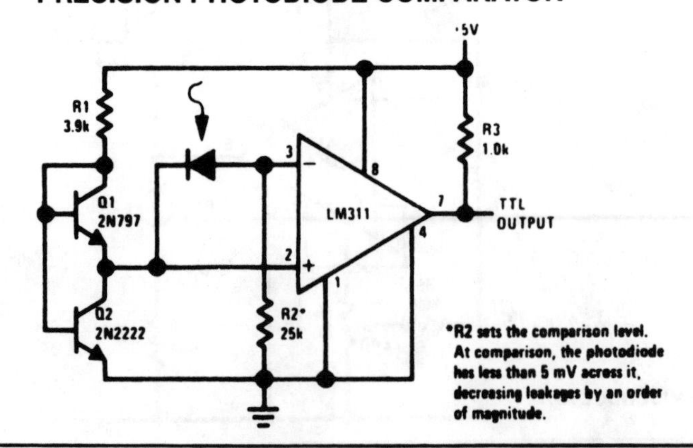

NATIONAL SEMICONDUCTOR

Fig. 12-6

13

Measuring and Test Circuits

The sources of the following circuits are contained in the Sources section, which begins on page 217. The figure number in the box of each circuit correlates to the source entry in the Sources section.

PHASE METER

$$\phi = \frac{2\pi V_{AVG}}{V_{OUT,\,PEAK}} - \pi$$

FAIRCHILD CAMERA & INSTRUMENT *Fig. 13-1*

PRECISION CALIBRATION STANDARD

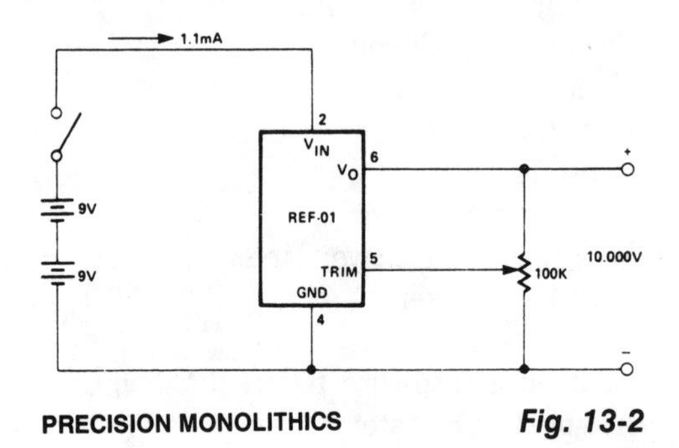

PRECISION MONOLITHICS *Fig. 13-2*

ZENER DIODE CHECKER

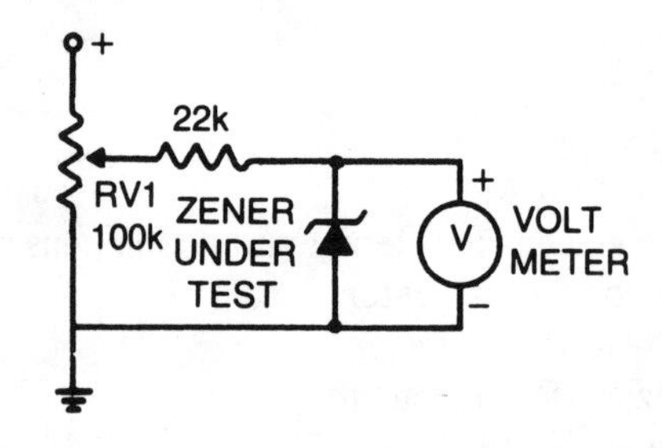

ELECTRONICS TODAY INTERNATIONAL *Fig. 13-3*

This circuit requires an external power supply that produces a voltage that is higher than the highest expected rating of the zener diodes to be tested. Potentiometer RV1 is adjusted until the meter reading stabilizes. This reading is the zener diode's breakdown voltage.

SOUND-LEVEL MONITOR

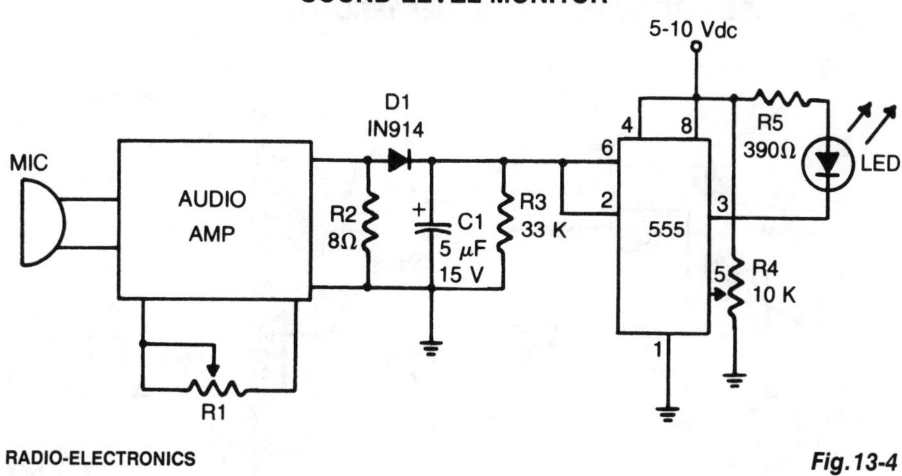

RADIO-ELECTRONICS

Fig. 13-4

This loudness detector consists of a 555 IC wired as a Schmitt trigger. The output changes state—from high to low—whenever the input crosses a certain voltage. That threshold voltage is established by the setting of R4.

LINEAR VARIABLE DIFFERENTIAL TRANSFORMER (LVDT) DRIVER DEMODULATOR

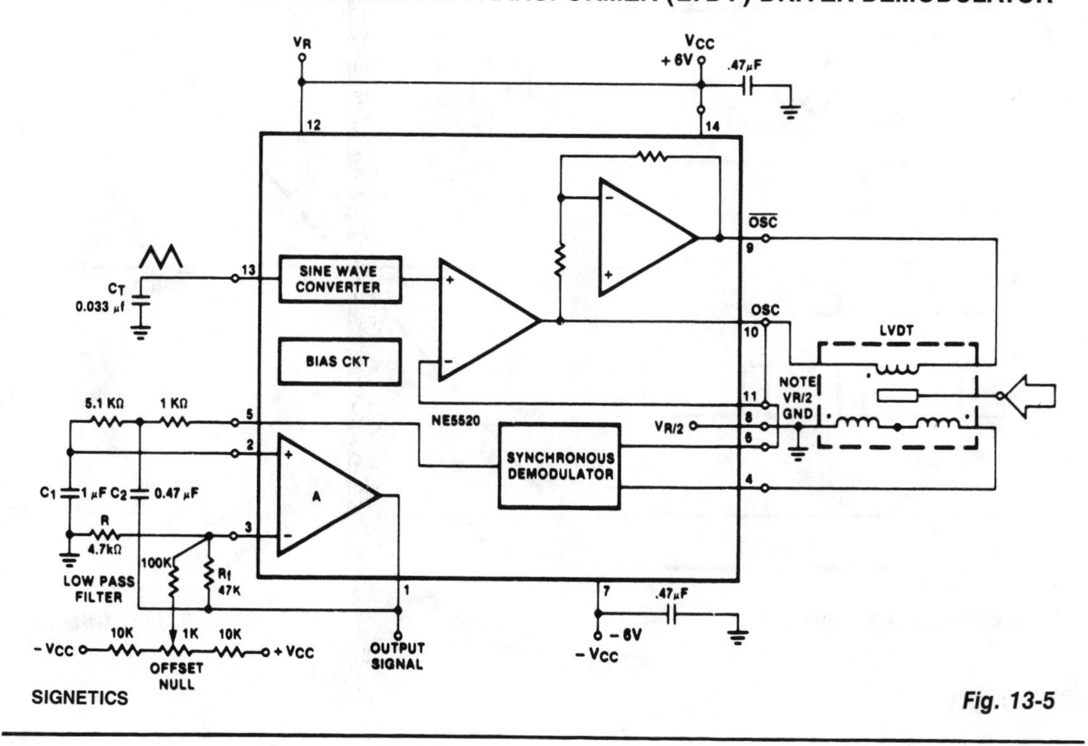

SIGNETICS

Fig. 13-5

PHASE DIFFERENCE MEASURER (FROM 0° TO ±180°)

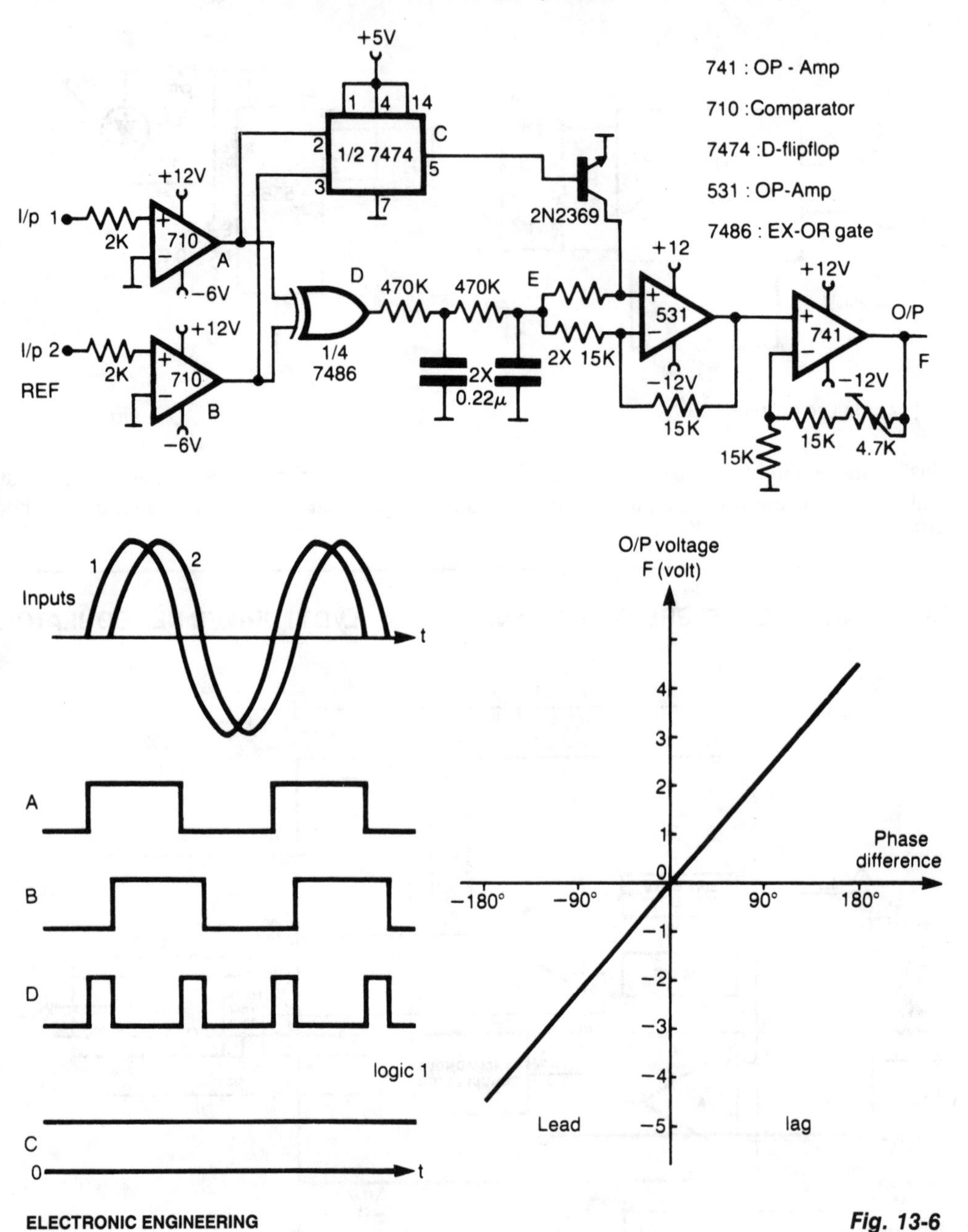

ELECTRONIC ENGINEERING

Fig. 13-6

PHASE DIFFERENCE MEASURER (FROM 0° TO ±180°) *(Cont.)*

This method is capable of measuring phase between 0 to ±180°. The generated square waves (A and B) are fed to a (D) flip-flop, which gives an output (C) equal to logic 1 when input 1 leads input 2 and equal to logic 0 in case of lagging. When C = logic 0, the output of the amplifier (F) will be positive, proportional to the average value (E) of the output of the EX-OR. When C = logic 1, F will be negative and also proportional to E by the same factor. Hence, the output of the meter is positive in case of lagging and negative for leading. The circuit is tested for sinusoidal inputs and indicates a linearity within 1%. Measurements are unaffected by the frequency of the inputs up to 75 kHz.

GROUND TESTER

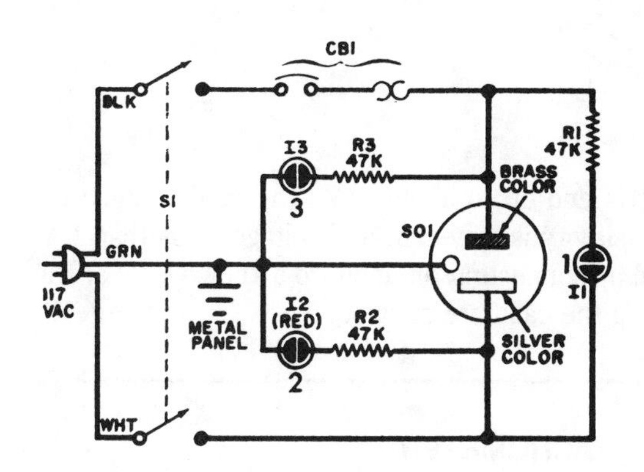

The circuit is simple and foolproof if wired correctly. Under normal conditions, only lamps 1 and 3 should be lit. If lamp 2 comes on, the cold lead is 117 V above ground.

POPULAR ELECTRONICS *Fig. 13-7*

MAKING SLOW LOGIC PULSES AUDIBLE

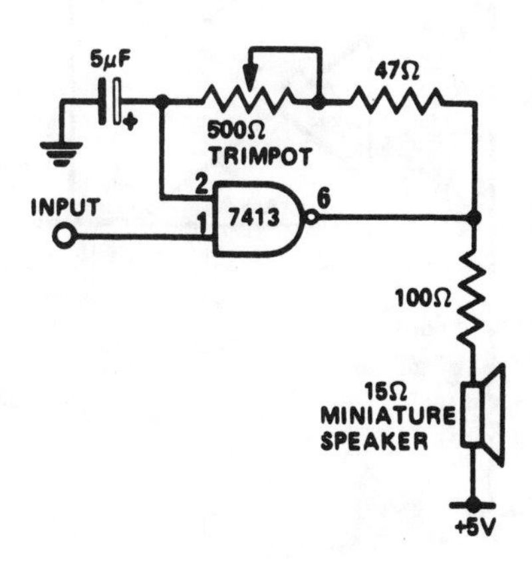

This circuit is useful for monitoring slow logic pulses as a keying monitor or digital clock alarm. The Schmitt trigger is connected as an oscillator. The trimpot controls the pitch of the output. When the input goes high, the circuit will oscillate.

ELECTRONICS TODAY INTERNATIONAL *Fig. 13-8*

SENSITIVE RF VOLTMETER

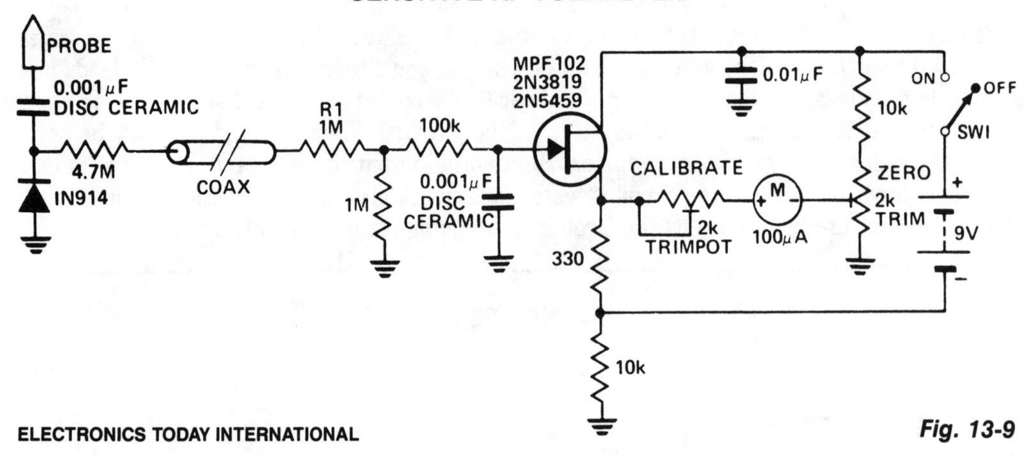

ELECTRONICS TODAY INTERNATIONAL

Fig. 13-9

This circuit measures RF voltages beyond 200 MHz and up to about 5 V. The diode should be mounted in a remote probe, close to the probe tip. Sensitivity is excellent and voltages less than 1 V peak can be easily measured. The unit can be calibrated by connecting the input to a known level of RF voltage, such as a calibrated signal generator, and setting the calibrate control.

MINIMUM COMPONENT TACHOMETER

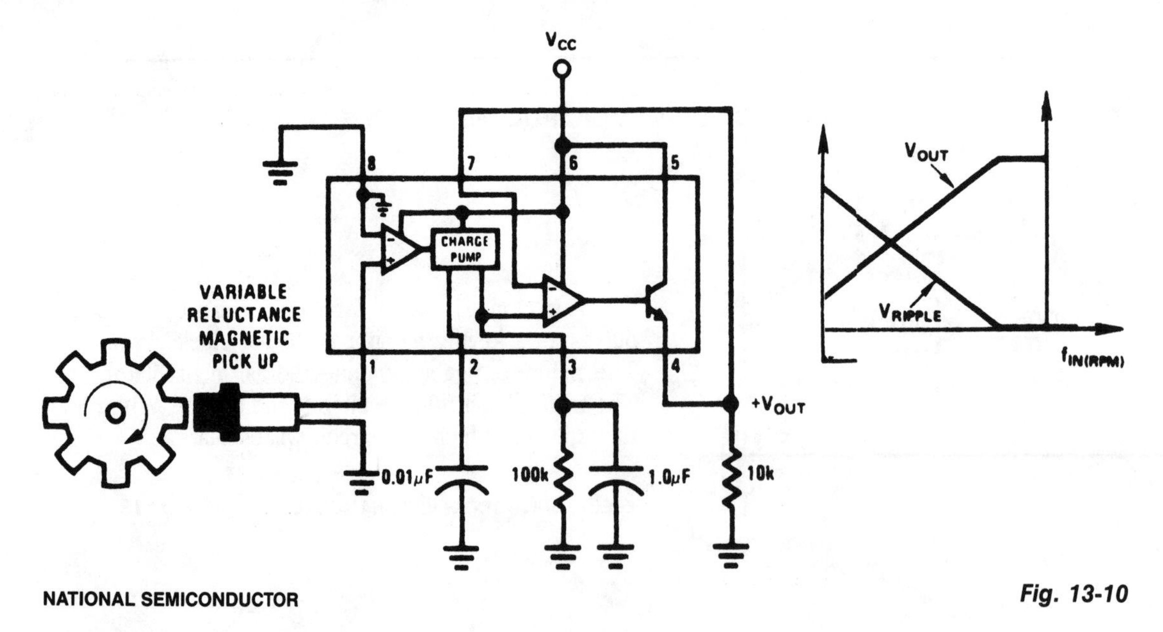

NATIONAL SEMICONDUCTOR

Fig. 13-10

LINEAR VARIABLE DIFFERENTIAL TRANSFORMER (LVDT) MEASURING GAUGE

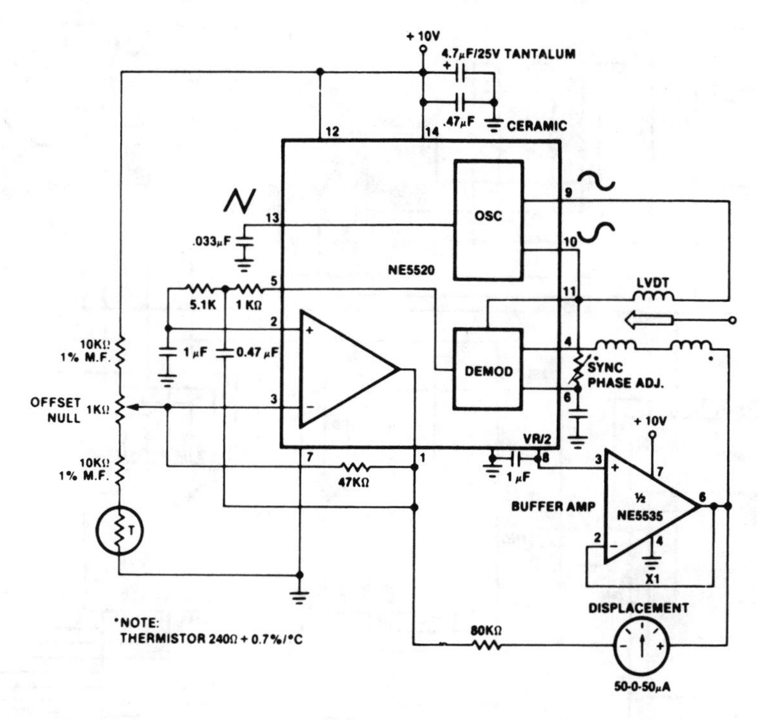

SIGNETICS

Fig. 13-11

VIBRATION METER

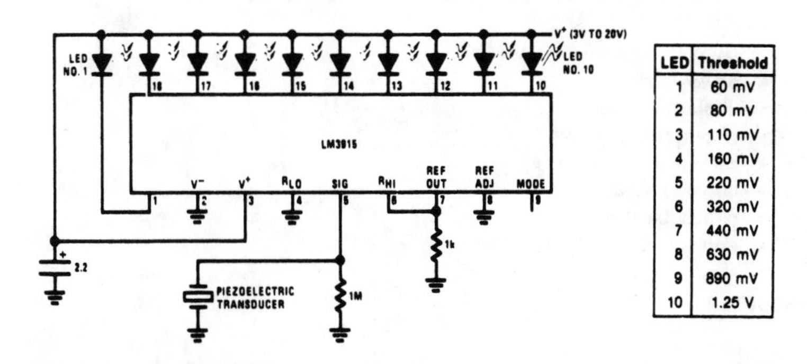

LED	Threshold
1	60 mV
2	80 mV
3	110 mV
4	160 mV
5	220 mV
6	320 mV
7	440 mV
8	630 mV
9	890 mV
10	1.25 V

NATIONAL SEMICONDUCTOR

Fig. 13-12

DIGITAL WEIGHT SCALE

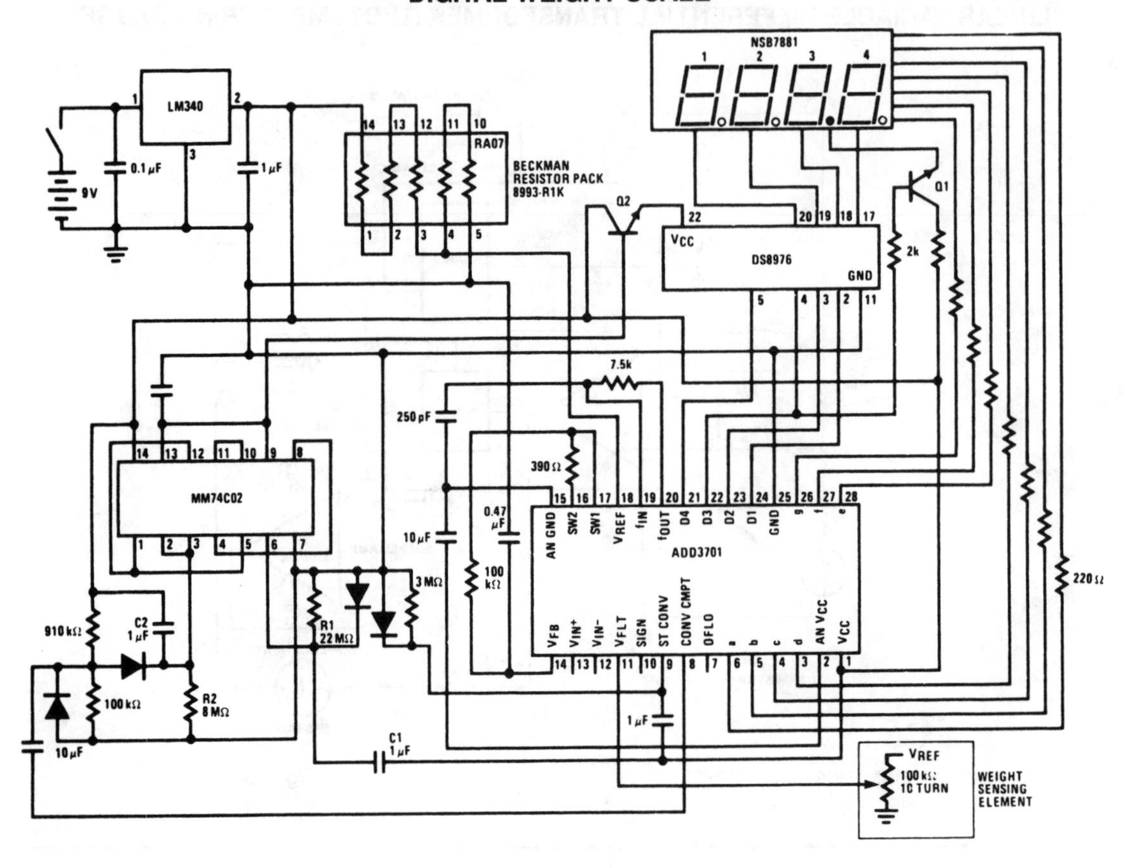

Notes:

1. R1, C1 defines POWER ON display blanking interval. R2, C2 defines display ON time.
2. All V_{CC} connections should use a single V_{CC} point and all ground/analog ground connections should use a single ground/analog ground point.
3. Display sequence for Rev A ckt implementation:

t = 0 sec	• power ON
t = 0 → 5 sec	• display blanked
	• system converging
t = 5 → 10 sec	• conversion complete
	• display ENABLE
t ⩾ 10 sec	• display blanked
	• wait for new POWER UP cycle

NATIONAL SEMICONDUCTOR ***Fig. 13-13***

This circuit uses a potentiometer as the weight-sensing element. An object placed on the scale displaces the potentiometer wiper, an amount proportional to its weight. Conversion of the wiper voltage to digital information is performed, decoded, and interfaced to the numeric display.

LOW-COST pH METER

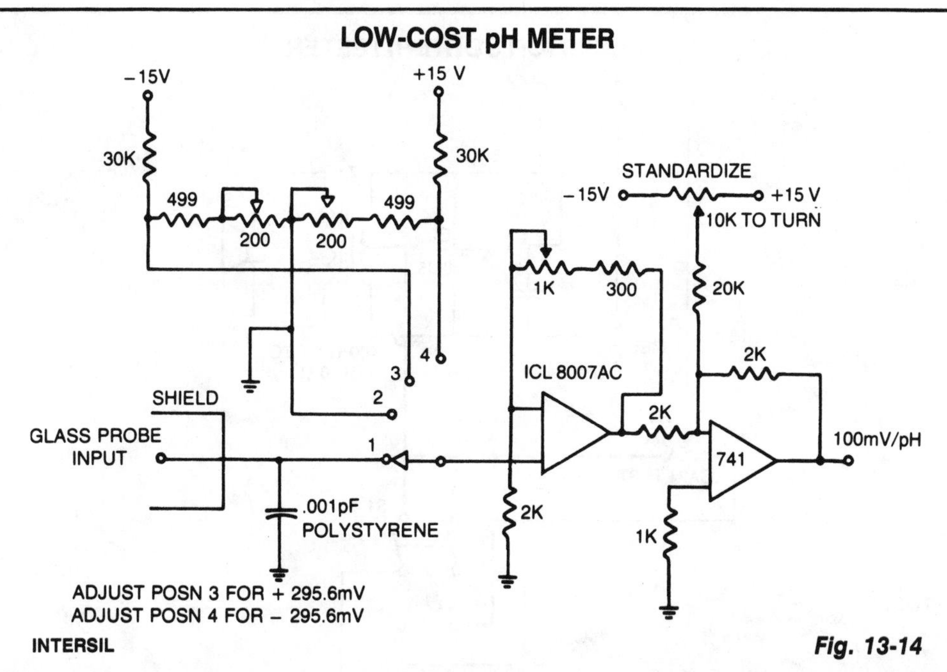

INTERSIL

Fig. 13-14

With guaranteed 1 pA input bias, the ICL 8007A is ideal as a pH meter as a long-term sample and hold.

PH PROBE AMPLIFIER/TEMPERATURE COMPENSATOR

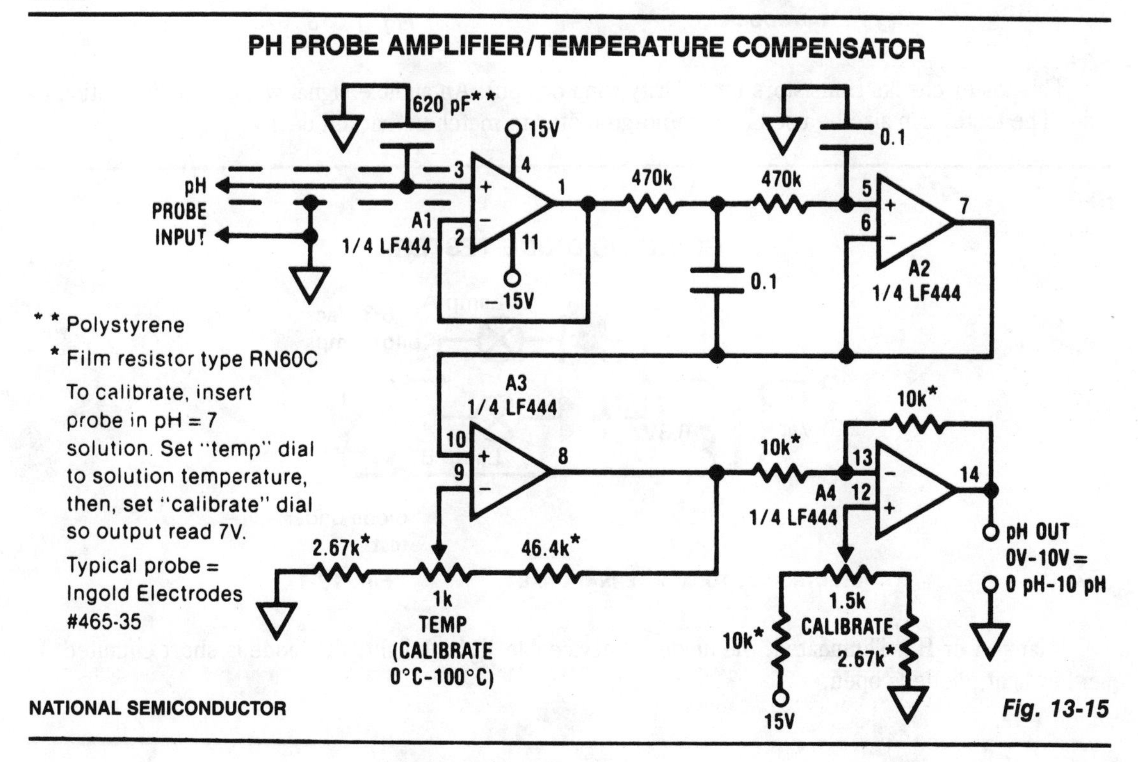

NATIONAL SEMICONDUCTOR

Fig. 13-15

TRANSISTOR SORTER/TESTER

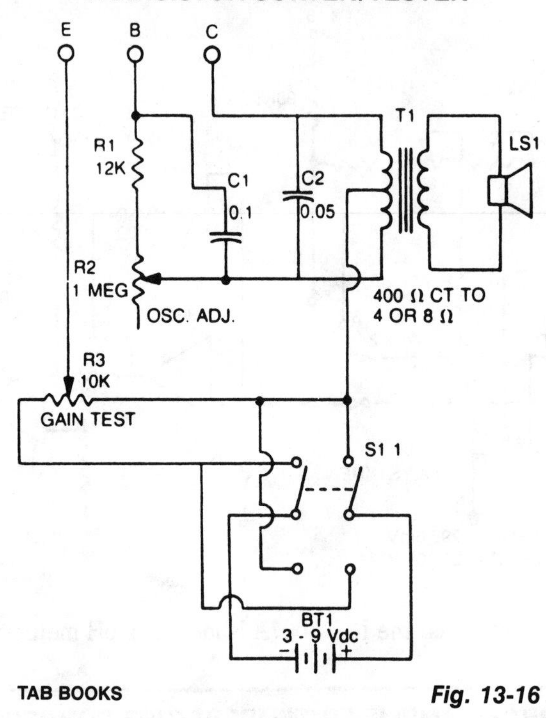

TAB BOOKS *Fig. 13-16*

This tester checks transistors for polarity (pnp or npn). An audible signal will give an indication of gain. The tester can also be used as a go/no-go tester to match unmarked devices.

GO/NO-GO DIODE TESTER

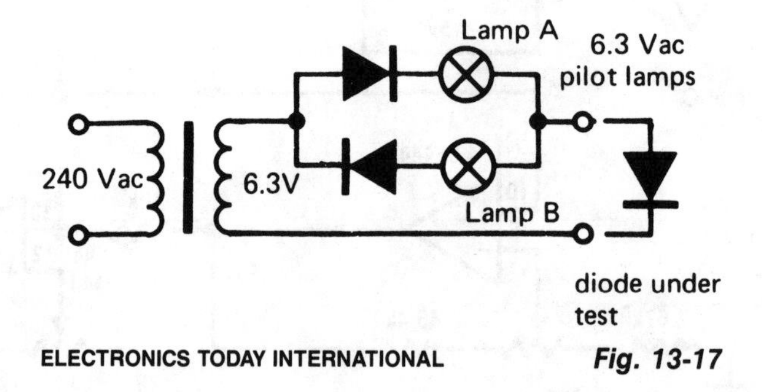

ELECTRONICS TODAY INTERNATIONAL *Fig. 13-17*

If lamp A or B is illuminated, the diode is serviceable. If both light, the diode is short circuited. If neither light, diode is open.

RESISTANCE-RATIO DETECTOR

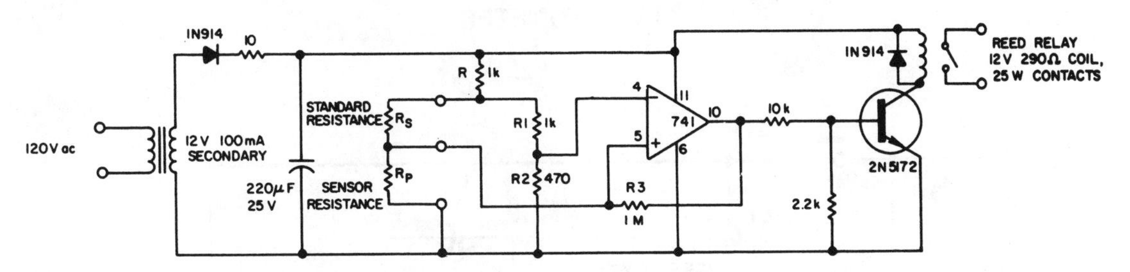

ELETRONIC DESIGN *Fig. 13-18*

Some applications (such as photoelectric control, temperature detection, and moisture sensing) require a circuit that can accurately detect a given resistance ratio. A simple technique that uses an op amp as a sensing element can provide 0.5% accuracy with low parts cost. The reed-relay contacts close when the resistance (R_p) of the sensor equals 47% of the standard (R_s). Adjusting either R1 or R2 provides a variable threshold; the threshold is controlled by varying R3. For the most part, the type of resistors used for R1 or R2 determines the accuracy and stability of the circuit. With metal-film resistors, less than 0.5% change in ratio sensing occurs over the commercial temperature range (0 to 70°C) with ac input variations from 105 to 135 V.

CONTINUITY TESTER FOR PCBs

This continuity tester is for tracing wiring on printed circuit boards. It only consumes any appreciable power when the test leads are shorted,so no on/off switch is used or required. The applied voltage at the test terminals is insufficient to turn on diodes or other semiconductors. Resistors below 50 Ω act as short circuit; above 100 Ω as open circuit. The circuit is a simple multivibrator—T1 and T2, are switched on by transistor T3. The components in the base of T3 are D1, R1, R2, and the test resistance. With a 1.5-V-supply, there is insufficient voltage to turn on a semiconductor connected to the test terminals. The phone is a telephone earpiece, but a 30 Ω speaker would work equally as well.

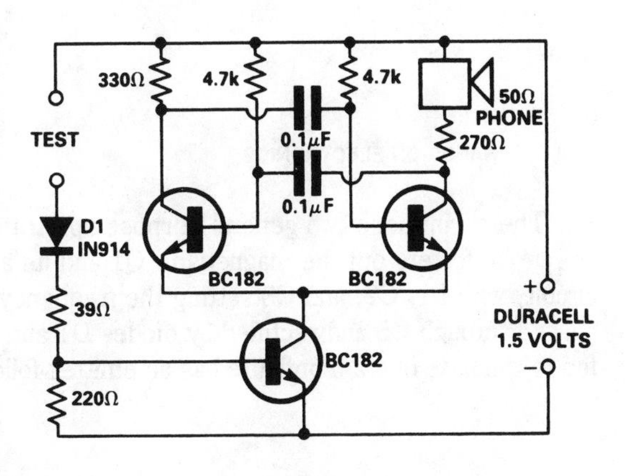

ELECTRONIC ENGINEERING *Fig. 13-19*

MAGNETOMETER

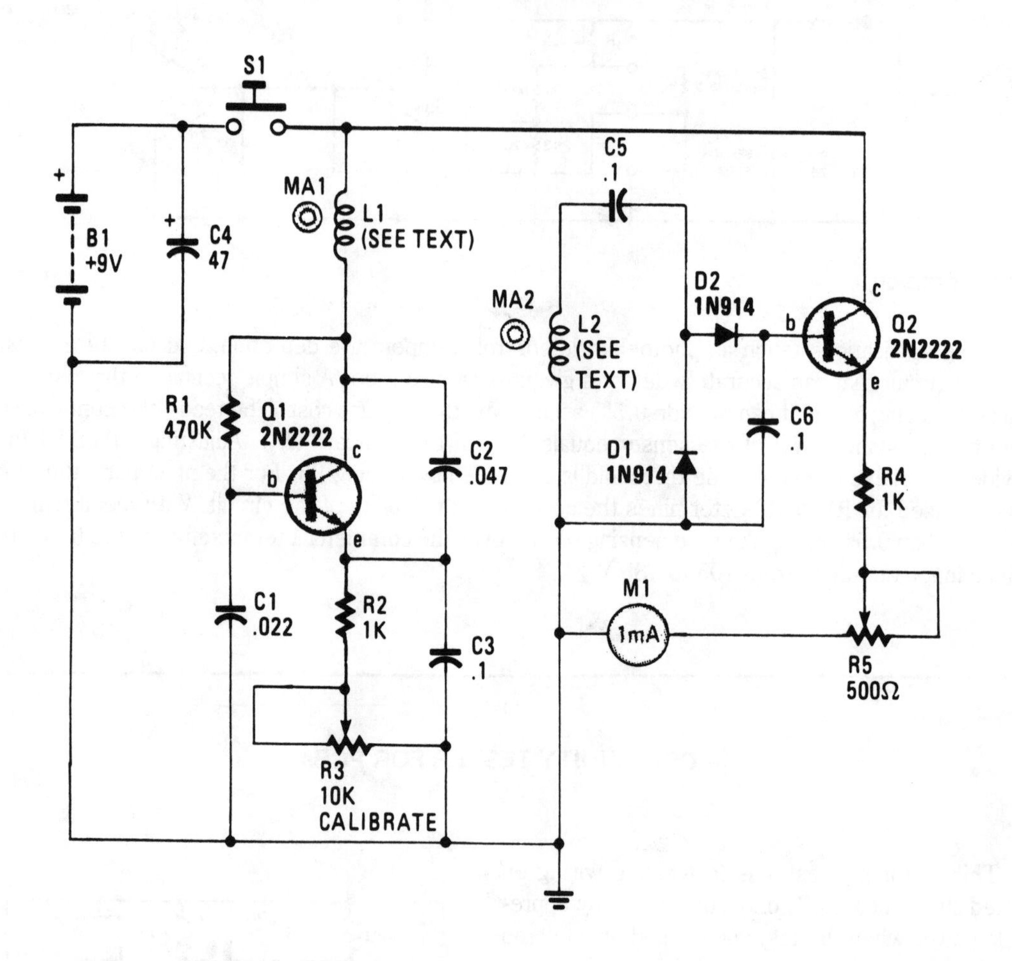

HANDS-ON ELECTRONICS ***Fig. 13-20***

The circuit uses two general-purpose npn transistors, Q1 and Q2, and a special handwound, dual-coil probe ferrets out the magnetism. Q1 and its associated components form a simple VLF oscillator circuit, with L1, C2, and C3 setting the frequency. The VLF signal received by the pickup coil, L2 is passed through C5 and rectified by diodes D1 and D2. The small dc signal output from the rectifier is fed to the base of Q2 (configured as an emitter-follower), which is then fed to a 0- to 1-mA meter, M1.

DIODE TESTER

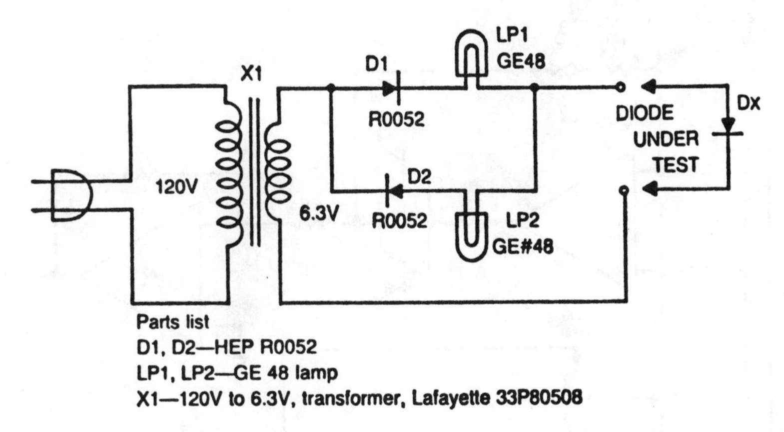

TAB BOOKS *Fig. 13-21*

The circuit tests whether or not a diode is open, shorted, or functioning correctly. If lamp A lights, the diode under test is functional. When lamp B is lit, the diode is good, but is connected backwards. When both lamps are lit, the diode is shorted, and it is open if neither lamp is lit.

PEAK LEVEL INDICATOR

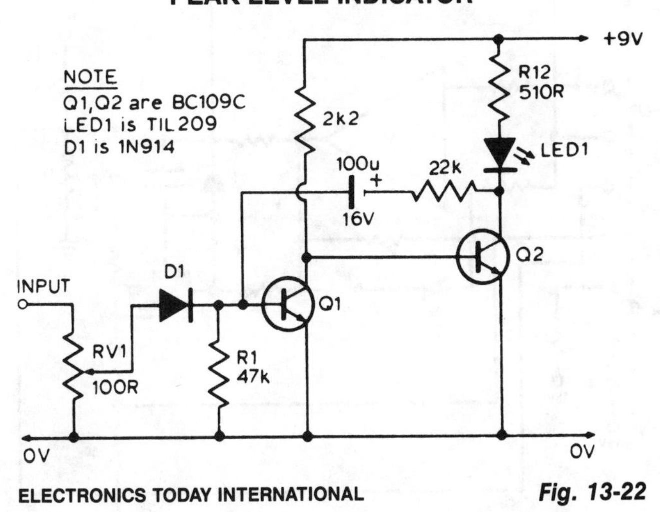

ELECTRONICS TODAY INTERNATIONAL *Fig. 13-22*

The LED is normally lit, but it will be briefly extinguished if the input exceeds a preset (by RV1) level. A possible application is to monitor the output voltage across a loudspeaker; the LED will flicker with large signals.

FET CURVE TRACER

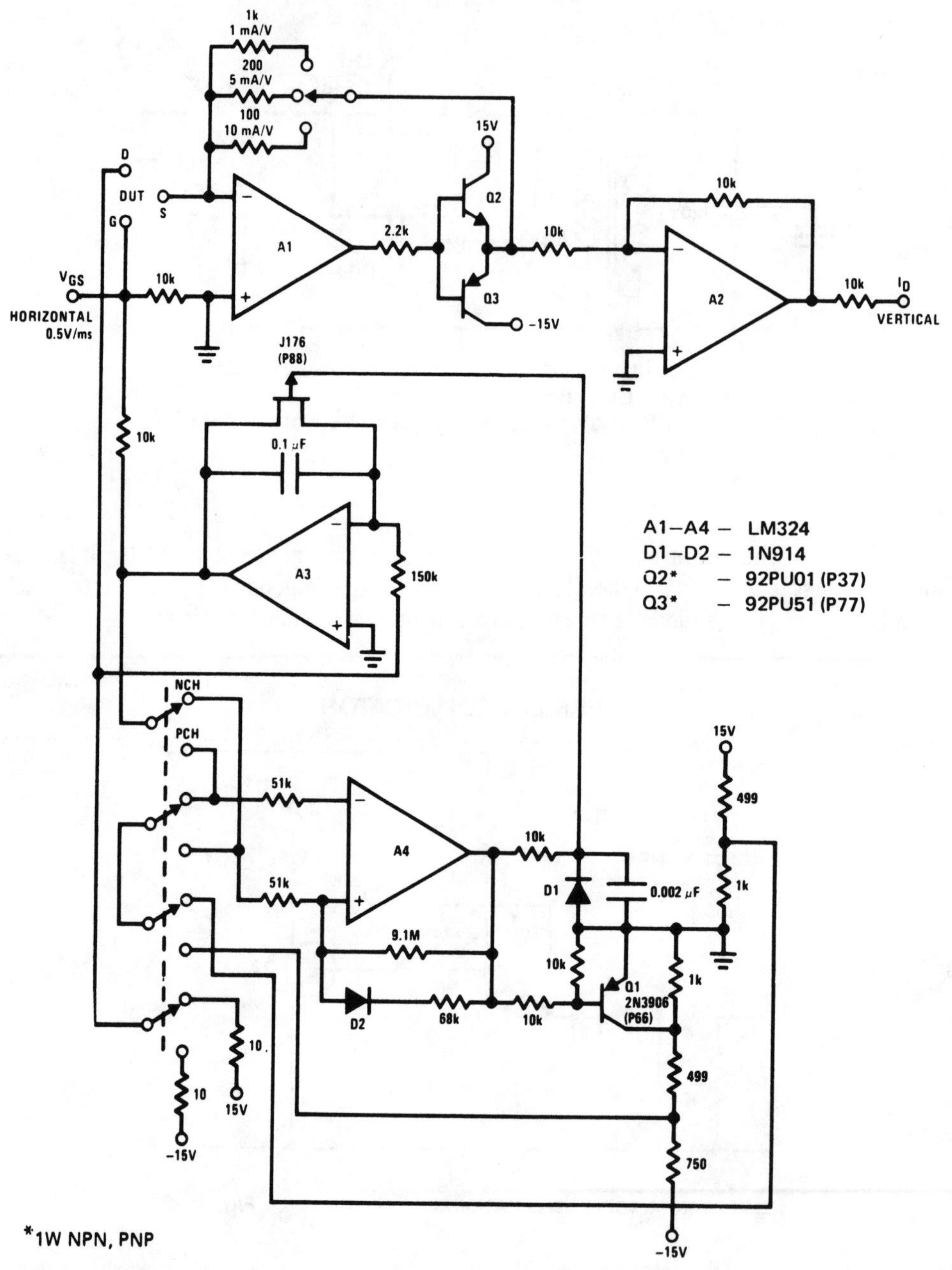

NATIONAL SEMICONDUCTOR ***Fig. 13-23***

The circuit displays drain current versus gate voltage for both p and n-channel JFETS at a constant drain voltage.

WIRE TRACER

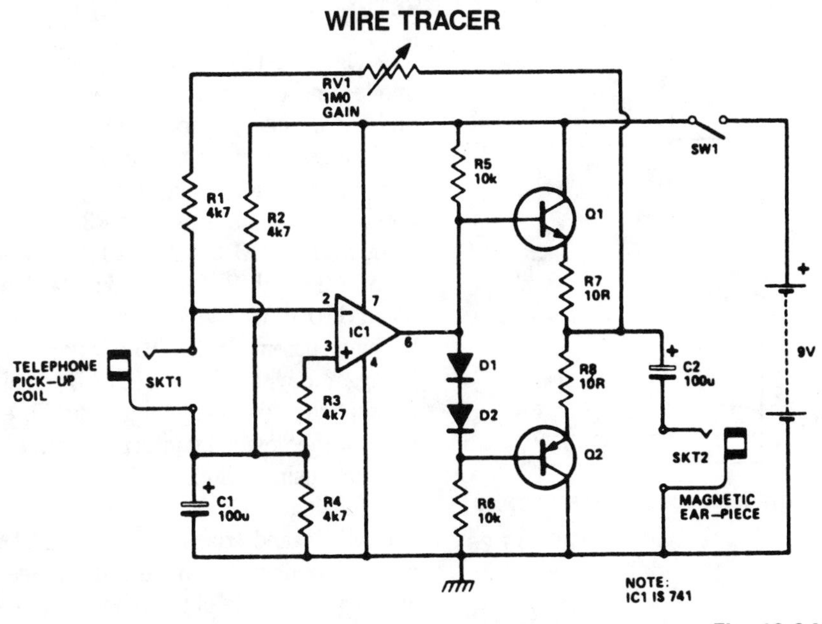

ELECTRONICS TODAY INTERNATIONAL *Fig. 13-24*

The tracer detects the weak magnetic field of any current-carring house wiring and amplifies this signal to a level that is adequate for driving a magnetic earpiece. The unit uses a telephone pickup coil to detect the magnetic field.

DIODE TESTER

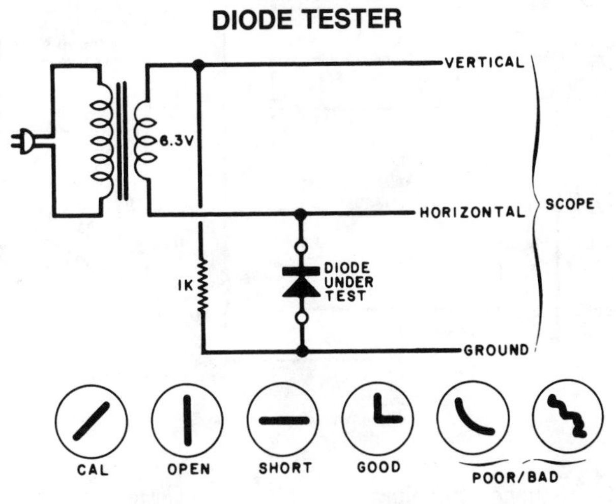

POPULAR ELECTRONICS *Fig. 13-25*

The circuit will display curves on a scope, contingent on the state of the diode. To "calibrate" substitute a 1000-Ω resistor for the diode and adjust the scope gains for a 45-degree line. The drawings show some expected results.

SCR TESTER

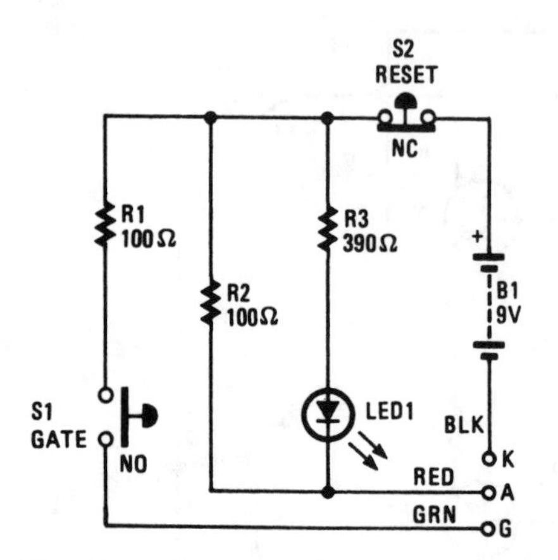

HANDS-ON ELECTRONICS/POPULAR ELECTRONICS

Fig. 13-26

The DUT's (device under test) cathode, anode, and gate are connected to the units K, A, and G terminals, respectively. Pressing switch S1 feeds a gate current to the DUT, which triggers it on. Resistor R1 limits the gate current to the appropriate level. Resistor R3 limits the current through the LED to about 20 mA, which, with the current through R2, results in a latching current of about 110 mA. The LED is used to monitor the latching current. If the DUT is good, once the gate is triggered with S1, the LED will remain lit, indicating that the device is conducting. To end the test, turn off the device by interrupting the latching current flow using switch S2. The LED should turn off and remain off. The preceeding procedure will work with SCRs and triacs. To check LEDs and other diodes, connect the anode and cathode leads to the anode and cathode of the diode; LED1 should light. When the leads are reversed the LED should remain off.

DIGITAL FREQUENCY METER

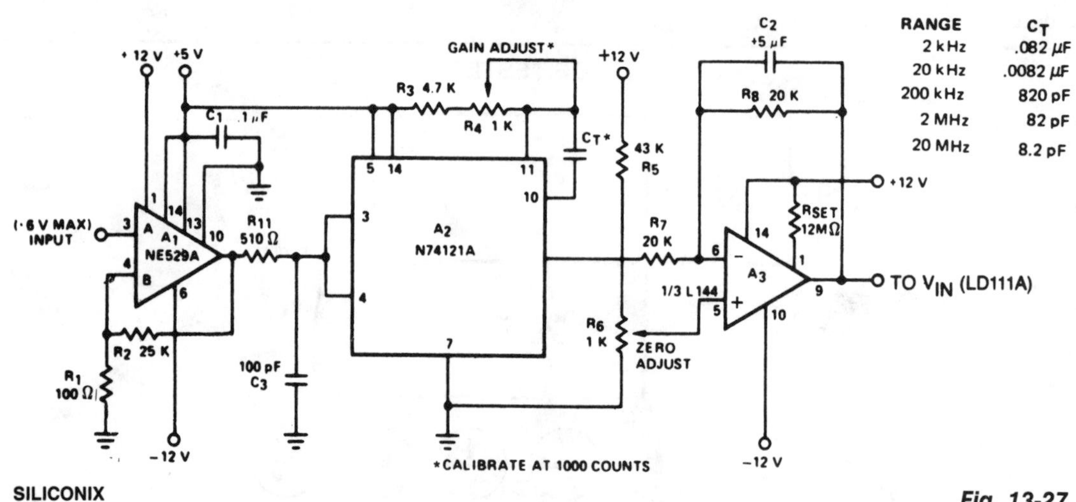

RANGE	C_T
2 kHz	.082 μF
20 kHz	.0082 μF
200 kHz	820 pF
2 MHz	82 pF
20 MHz	8.2 pF

SILICONIX

Fig. 13-27

The circuit converts frequency to voltage by taking the average dc value of the pulses from the 74121 monostable multivibrator. The one shot is triggered by the positive-going ac signal at the input of the 529 comparator. The amplifier acts as a dc filter, and also provides zeroing. This circuit will maintain an accuracy of 2% over five decades of range. The input signal to the comparator should be greater than 0.1 V pk-pk, and less than 12 V pk-pk for proper operation.

S METER

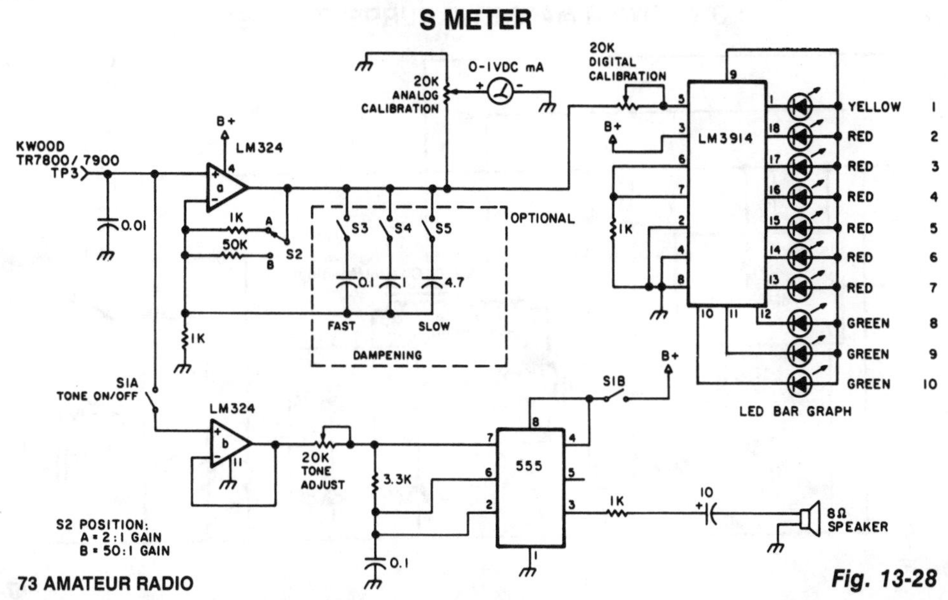

73 AMATEUR RADIO

Fig. 13-28

This design is for an external signal strength meter that is analog, digital, and audible for mobile transmitter hunters. The S meter also incorporates a gain circuit. The digital LED bar graph display has a very fast response time. The 3.3 kΩ resistor near LM3914 can be replaced with a 5-kΩ pot to control LED brightness. The S2A position gives a 2:1 gain and the S2B position gives about a 50:1 gain. The calibration pots control the amount of meter action relative to the gain. The optional dampening circuit is used for the averaging of a transmitted signal that has modulated power or when a dip on the voice peaks occur. The capacitors can be switched one by one or switched into a very slow response using 5.8 μF total capacitance.

HOT-WIRE ANEMOMETER

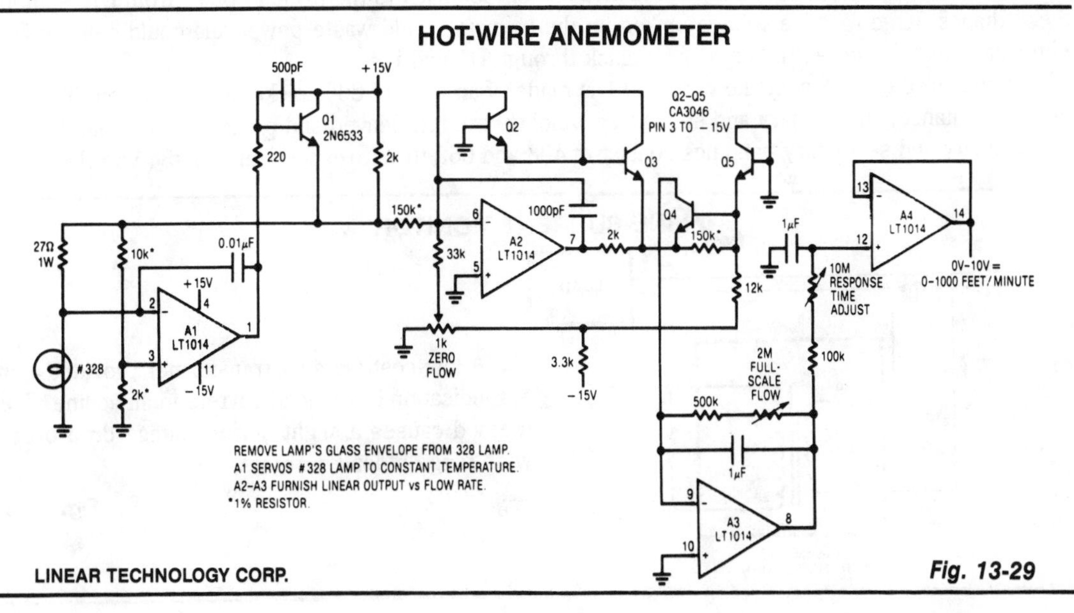

LINEAR TECHNOLOGY CORP.

Fig. 13-29

LOW-POWER MAGNETIC CURRENT SENSOR

D_1 Through D_6	1N5619
C_1	4.7 μF
C_2	100 μF
C_3	0.22 μF
C_4	15 μF
C_5	15 μF
R_1	20 Ω
R_2	10 kΩ
R_3	196 kΩ
R_4	100 kΩ

NASA

Fig. 13-30

A transducer senses a direct current magnetically, providing isolation between the input and the output. The detecting-and-isolating element is a saturable reactor, in which the input current, to be measured, passes through a one-turn control coil. The transducer provides an output of 0 to 3 Vdc, an input current of 0 to 15 Adc, and consumes 22 mW at 10 Adc input.

Line driver U1 excites the saturable reactor L1 by feeding a 2.3-kHz square wave through transformer T1. The output of L1 is rectified by the bridge rectifier composed of diodes D3 through D6, then amplified by op amp U2, which has a gain of 20.

Diodes D1 and D2 commutate the reactive current fed back to the primary of T1 from L1. Without these diodes, large reactive voltage spikes on the primary would waste power and could destroy U1. Filter capacitor C1 stores the energy fed back through D1 and D2.

To minimize core losses, the core of T1 is made of an alloy of 80% nickel and 20% iron. To minimize capacitance, the primary and secondary windings are interleaved and progressively wound 350°. The primary and secondary windings consist of 408 and 660 turns, respectively, of #34 wire.

LINE-CURRENT MONITOR

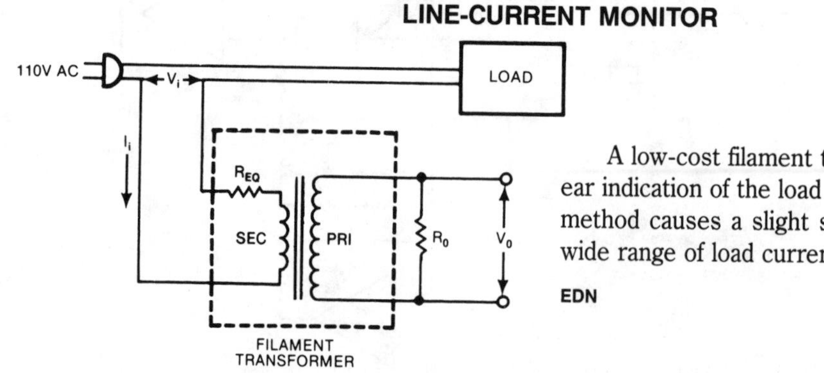

A low-cost filament transformer provides a linear indication of the load current in an ac line. This method causes a slight series voltage drop over a wide range of load currents.

EDN

Fig. 13-31

PRECISION FREQUENCY COUNTER/TACHOMETER

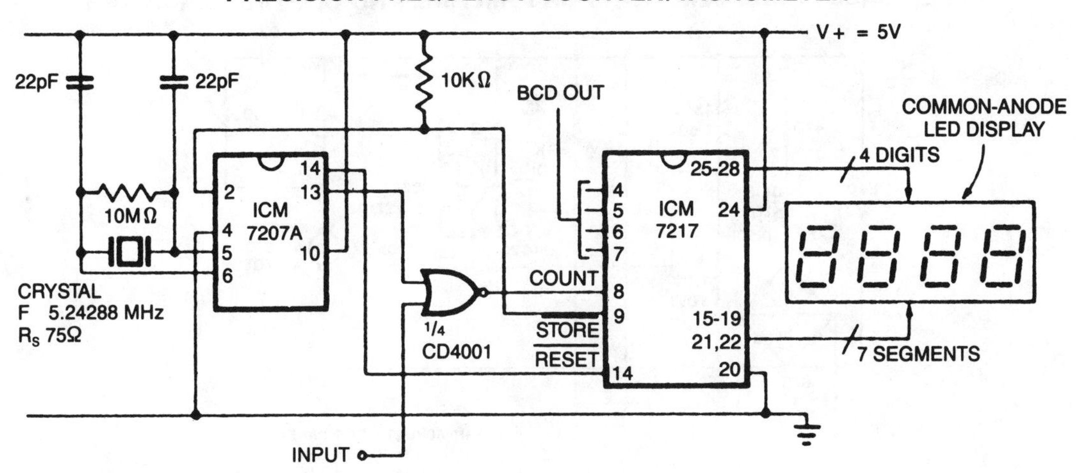

INTERSIL *Fig. 13-32*

In this configuration, the display reads hertz directly. With pin 11 of the ICM7027A connected to V_{DD},the gating time will be 0.1 second; this will display tens of hertz as the least significant digit. For shorter gating times, an ICM7207 can be used with a 6.5536-MHz crystal, giving a 0.01 second gating with pin 11 connected to V_{DD} and a 0.1 second gating with pin 11 open.

MOTOR HOUR METER

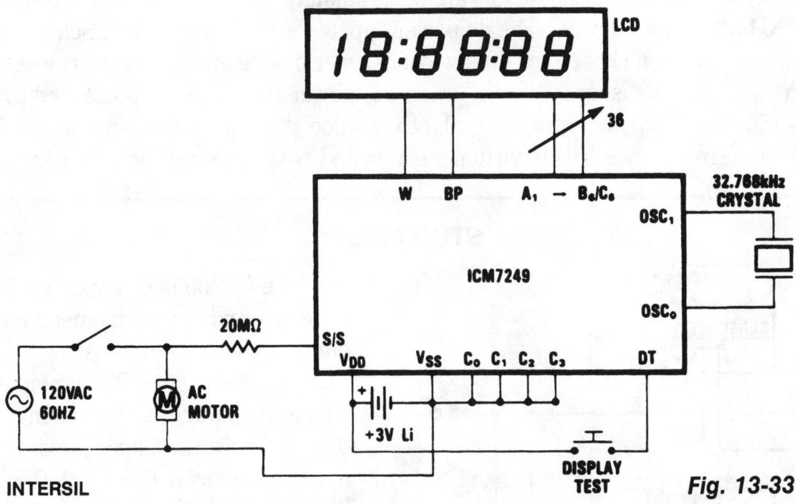

INTERSIL *Fig. 13-33*

In this application, the ICM7249 is configured as an hours-in-use meter and shows how many whole hours of line voltage have been applied. The 20-MΩ resistor and high-pass filtering allow ac line activation of the S/S input. This configuration, which is powered by a 3-V lithium cell, will operate continuously for $2^1/_2$ years. Without the display, which only needs to be connected when a reading is required, the span of operation is extended to 10 years.

PAPER SHEET DISCRIMINATOR FOR PRINTING AND COPYING MACHINES

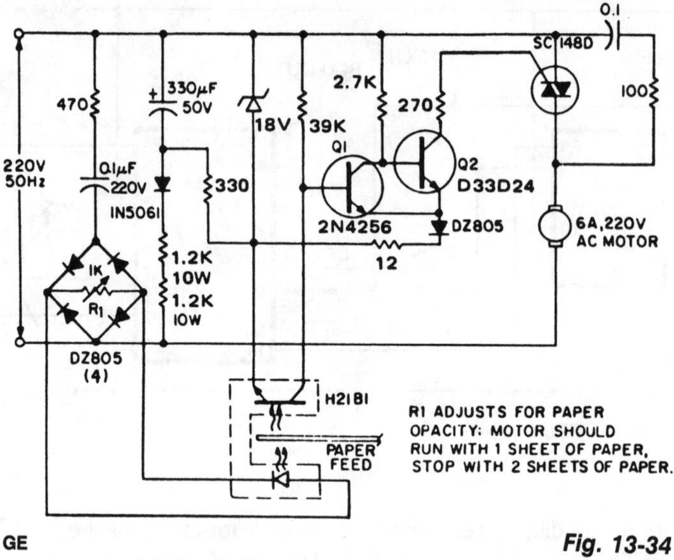

GE *Fig. 13-34*

The circuit outputs power to the drive motor when one or no sheets are being fed, but interrupts motor power when two or more superimposed sheets pass through the optodetector slot. The optodetector can be either an H2aB Darlington interruptor module or an H23B matched emitter-detector pair. The output from the optodarlington is coupled to a Schmitt trigger, comprising transistors Q1 and Q2 for noise immunity and minor paper opacity variation immunity. When the Schmitt is on, gate current is applied to the SC148D output device. The dc power supply for the detector and Schmitt is a simple RC diode half-wave configuration chosen for its low cost (fewer diodes and no transformers) and minimum bulk. Although such a supply is directly coupled to the power triac, This is precluded by current drain considerations (50 mA dc for the gate drive alone). Notice that direct coupling of the Schmitt to the output triacs is perferred, since RFI is virtually eliminated with the quasi-dc gate drive.

STUD FINDER

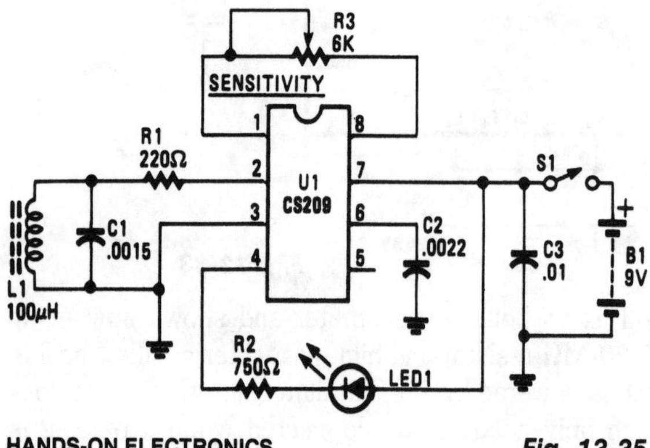

HANDS-ON ELECTRONICS *Fig. 13-35*

The CS209 is designed to detect the presence or proximity of magnetic metals. It has an internal oscillator that, along with its external LC resonant circuit, provides oscillations whose amplitude is dependent upon the Q of the LC network. Close proximity to magnetic material reduces the Q of the tuned circuit, thus the oscillations tend to decrease in amplitude. The decrease in amplitude is detected and used in turn on LED1, indicating the presence of a magnetic material (i.e., nail or screw).

CURRENT MONITOR AND ALARM

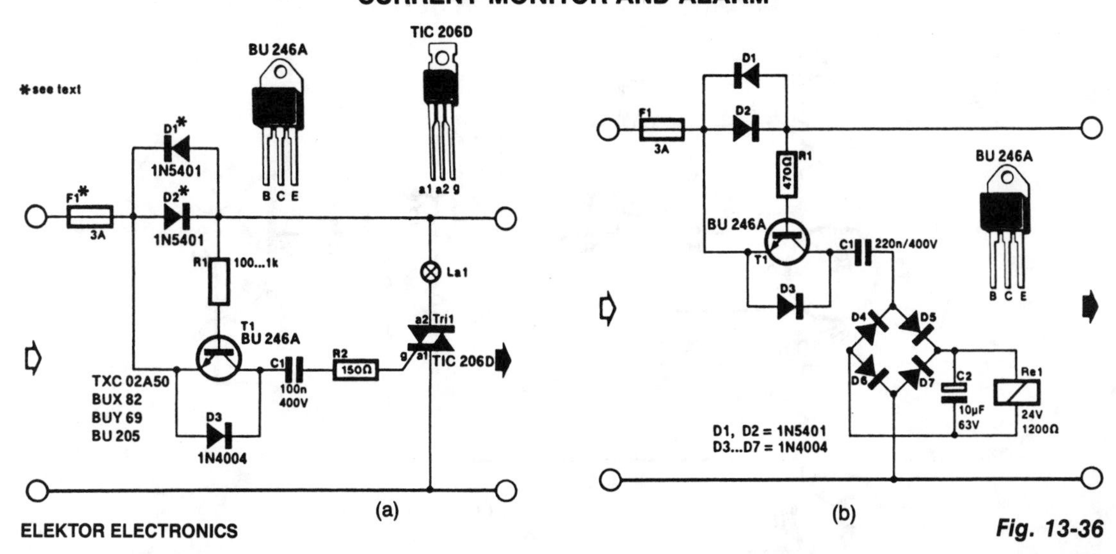

ELEKTOR ELECTRONICS

Fig. 13-36

The circuit in Fig. 13-36a lights the signal lamp upon detecting a line current consumption of more than 5 mA, and handles currents of several amperes with appropriate diodes fitted in the D1 and D2 positions. Transistor T1 is switched on when the drop across D1/ D2 exceeds a certain level. Diodes from the well-known 1N400x series can be used for currents of up to 1 A, while 1N540x types are rated for up to 3 A. Fuse F1 should suit the particular application.

The circuit in Fig. 13-36b is a current-triggered alarm. Rectifier bridge D4 through D7 can only provide the coil voltage for Rel when the current through D1/D2 exceeds a certain level, because then series capacitor C1 passes the alternating main current. Capacitor C1 needs to suit the sensitivity of the relay coil. This is readily affected by connecting capacitors in parallel until the coil voltage is high enough for the relay to operate reliably.

PICOAMMETER CIRCUIT

This circuit uses the exceptionally low input current 0.1 pA of the CA3420 BiMOS op amp. With only a single 10-MΩ resistor, the circuit covers the range from ± 50 pA to a maximum full-scale sensitivity of ± 1.5 pA.

GE/RCA

Fig. 13-37

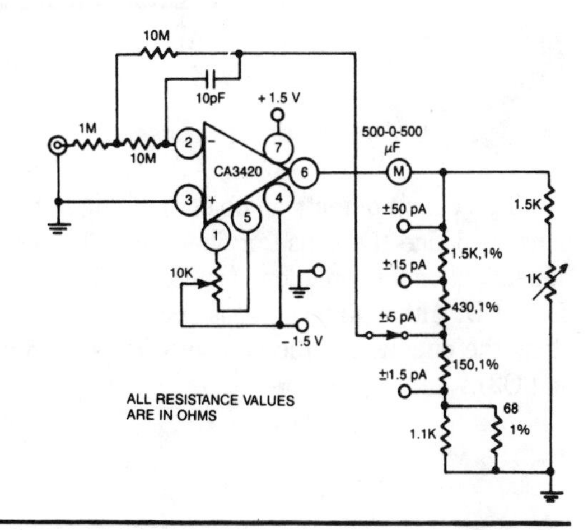

UNIDIRECTIONAL MOTION SENSOR

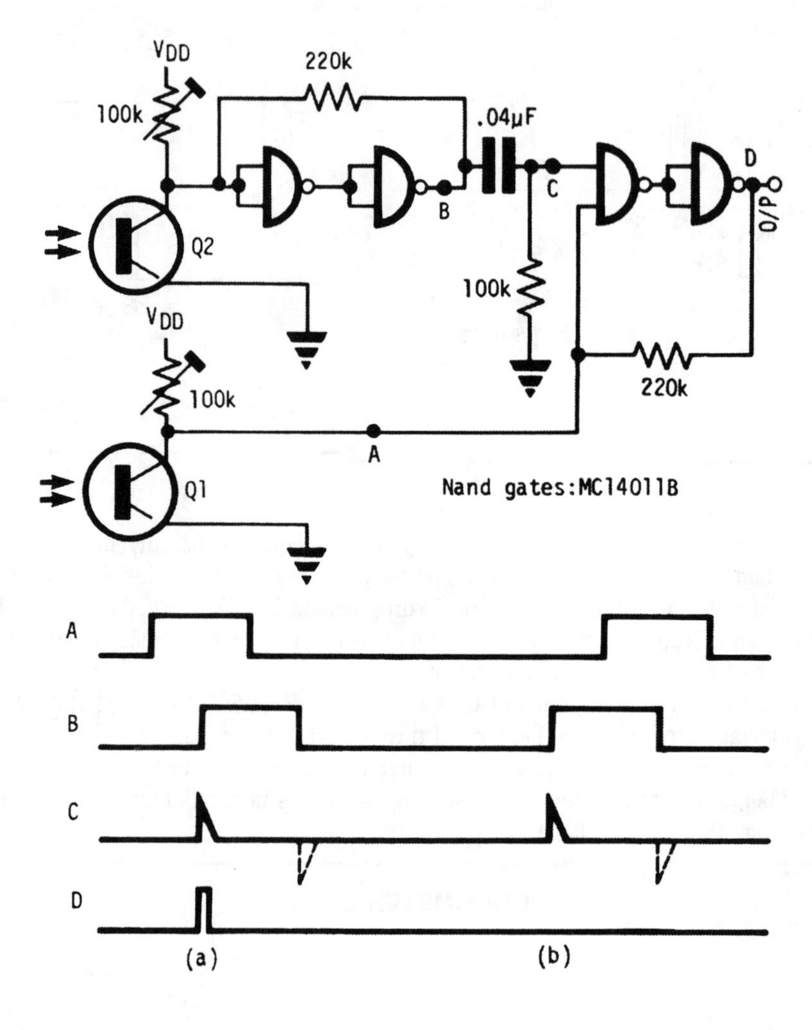

ELECTRONIC ENGINEERING *Fig. 13-38*

This circuit detects an object passing in one direction, but ignores it going the opposite way. Two sensors define the sense of direction. The object blocks the light to phototransistor Q1 or Q2, first dependent on the direction of approach. When the object passes Q1, then Q2, an output pulse is generated at D. Although no pulse is seen at D as the object passes Q2, then Q1. Object length (measured along the direction of the two sensors) should be greater than the separation of the two sensors (Q1 and Q2).

DUTY CYCLE MONITOR

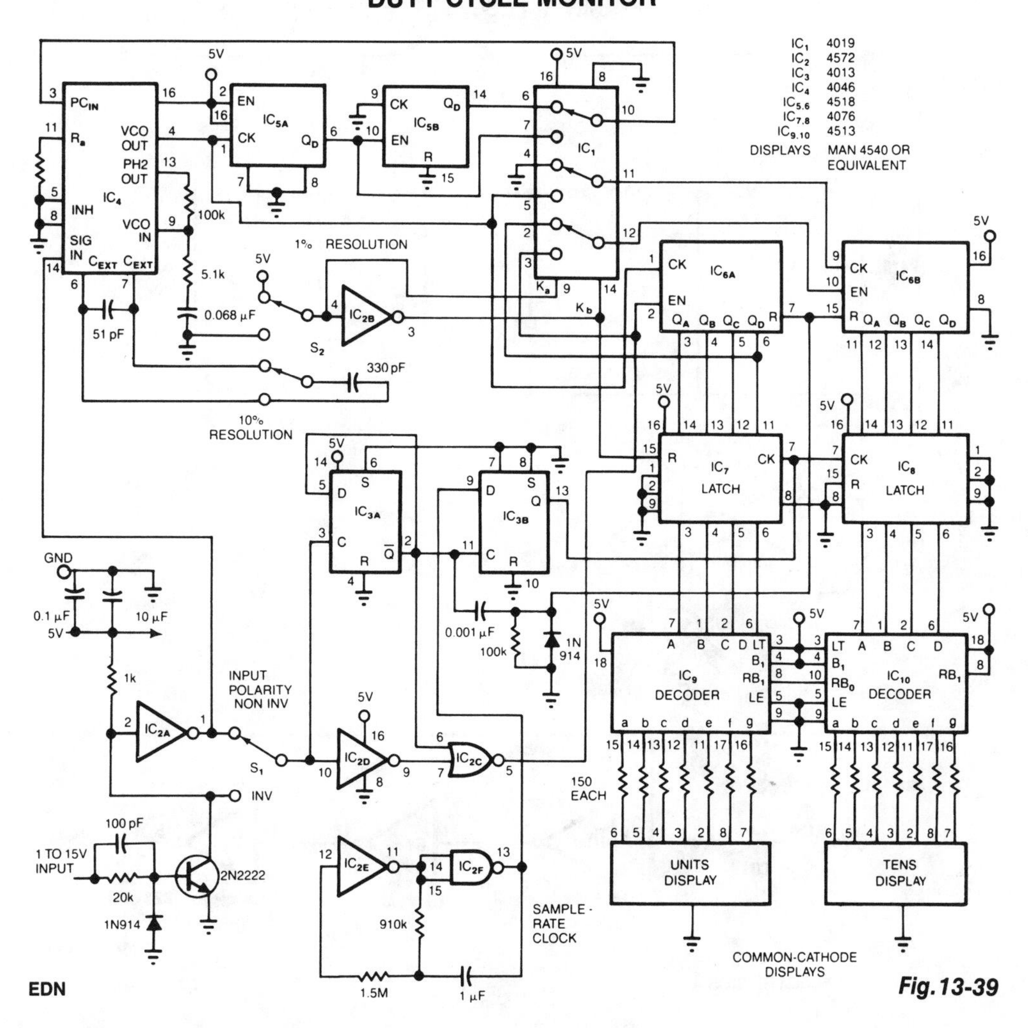

Fig. 13-39

The circuit monitors and displays a digital signal's duty cycle and provides accuracy as high as ± 1%. Using switch S2, you can choose a frequency range of either 250 Hz to 2.5 kHz at ± 1% accuracy or 2 kHz to 50 kHz at ± 10% accuracy. The common-cathode display gives the signal's duty-cycle percentage. Phase-locked loop IC4 and counters IC5A and IC5B multiply the input frequency by a factor of either 10 or 100, depending on switch S2's setting. IC6A and IC6B count this multiplied frequency during the incoming signal's mark interval. IC7 and IC8 then latch this count and display it at the clock's sample rate. For example, if you select a 1% resolution, when the signal's mark period is 40% of the total period, the circuit will enable the counter comprising IC6A and IC6B for 40 counts. To obtain space-interval sampling, you can reverse the input polarity using switch S1. IC2A samples the input signal's period, enables gate IC2C, and resets the counter. IC2E and IC2F form the sample-rate clock; IC3B synchronizes the clock's output with the input, so that the circuit can update latches IC7 and IC8.

3-IN-1 TEST SET

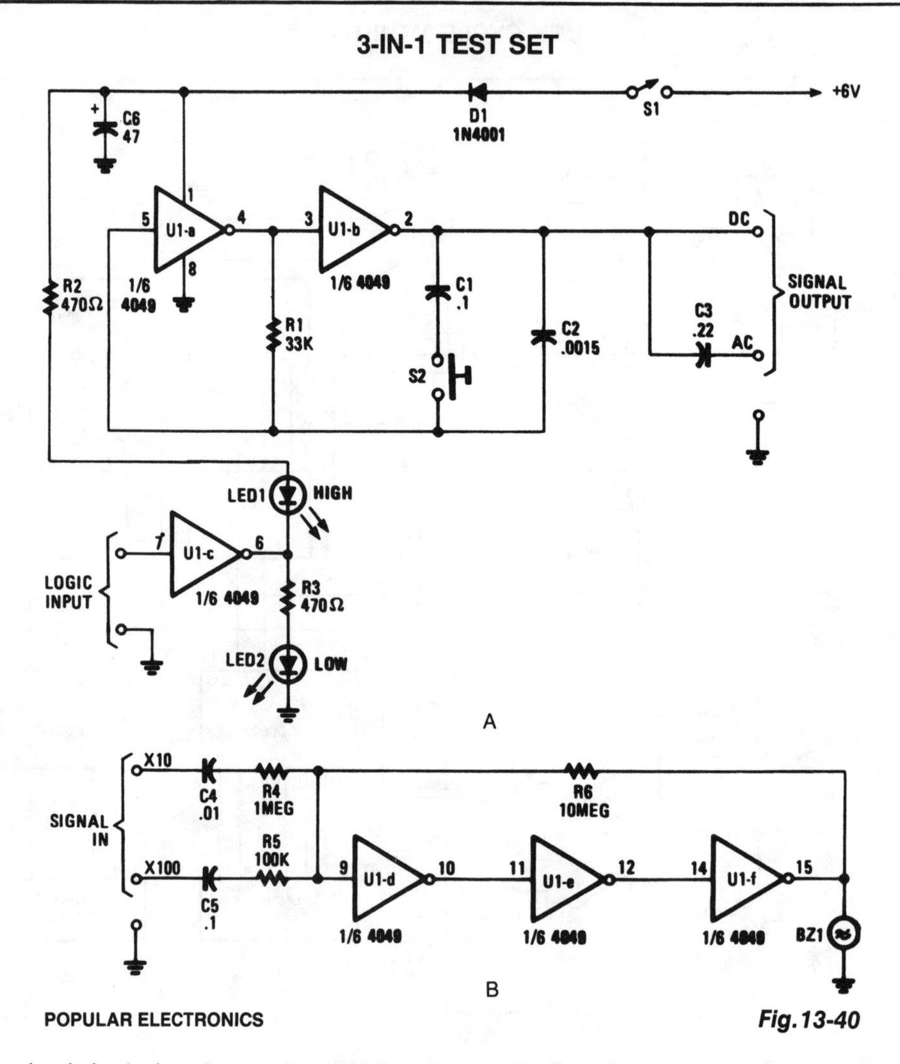

POPULAR ELECTRONICS *Fig. 13-40*

This circuit is designed around a 4049 hex inverter/buffer. Two inverters are used in a dual-frequency signal-injector circuit, another inverter is used as a logic probe, and the remaining three inverters are used as a sensitive dual-input, audio-signal tracer.

The signal-injector portion gates are configured as a two-frequency, pulse-generator circuit. Under normal conditions, the generator's output frequency is around 10 kHz, but when S2 is closed, the output frequency drops to about 100 Hz. The logic-probe portion is made up of U1c, the output of the inverter decreases. The low output of U1c reverse biases LED2, so it remains off. That low output also forward biases LED1, causing it to light. But when a logic low is present U1c's input, the situation is reversed, so LED2 lights and LED1 darkens.

The audio-signal tracer portion is made up of the three remaining inverters, which are configured as a linear audio amplifier to increase the input signal level by a factor of 10 or 100. The amplified output signal feeds a miniature piezo element for audible detection.

STEREO POWER METER

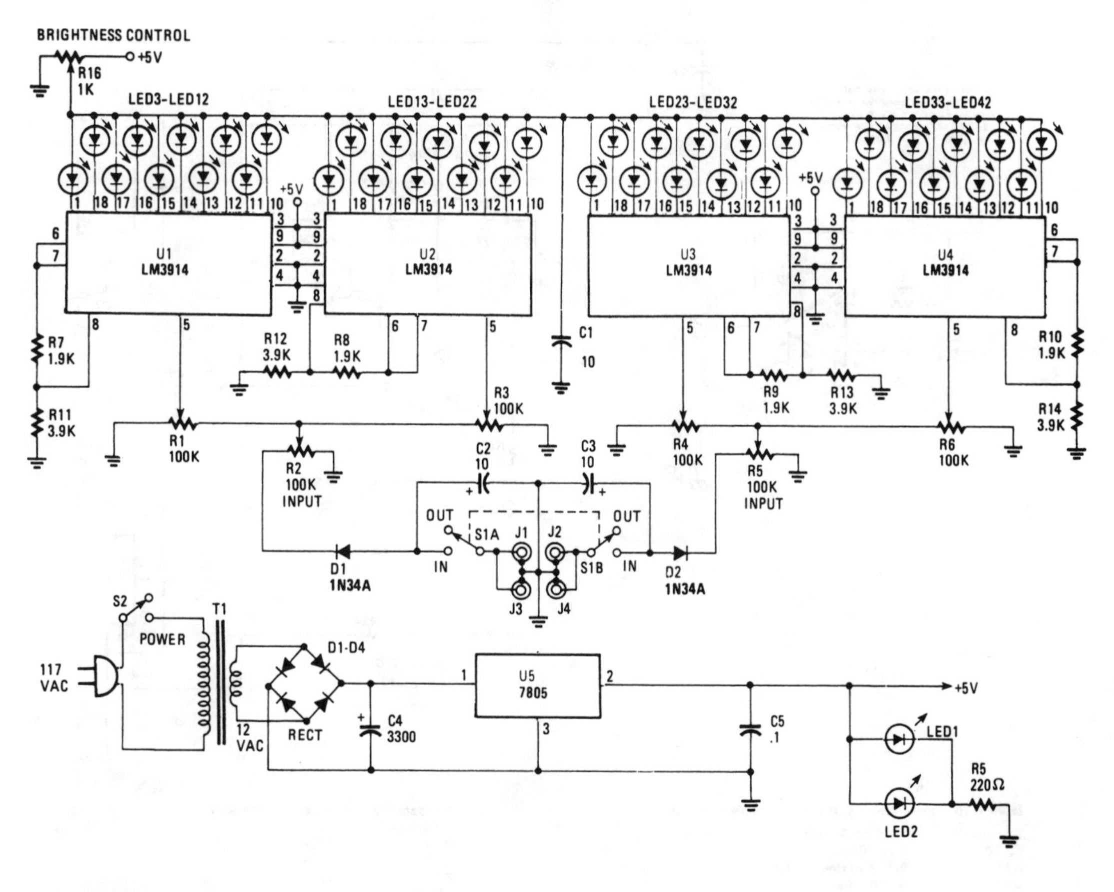

HANDS-ON ELECTRONICS *Fig. 13-41*

This stereo power meter is made up of two identical circuits and a power supply. Each circuit contains two LM3914 display chips that contain 10 voltage comparators, a 10-step voltage divider, a reference voltage source, and a mode-select circuit that selects a bar or dot display via a logic input at pin 9. The brightness of the LEDs is controlled by the 1900-Ω resistors and the reference voltage is controlled by the 3900-Ω resistors. The 10-step voltage divider within the chips is connected between the reference voltage and ground. Because each step of the voltage divider is separated by a 1-kΩ resistor, each comparator senses a voltage 10% greater than the preceding comparator. The signal is applied to pin 5, which is buffered through a resistor-diode network and then amplified as it passes to each of the 10 comparators. Each LED is grounded through the comparators as the input signal voltage matches the reference voltage. That results in one to 10 LEDs illuminating as the signal voltage increases.

WIDE-RANGE RF POWER METER

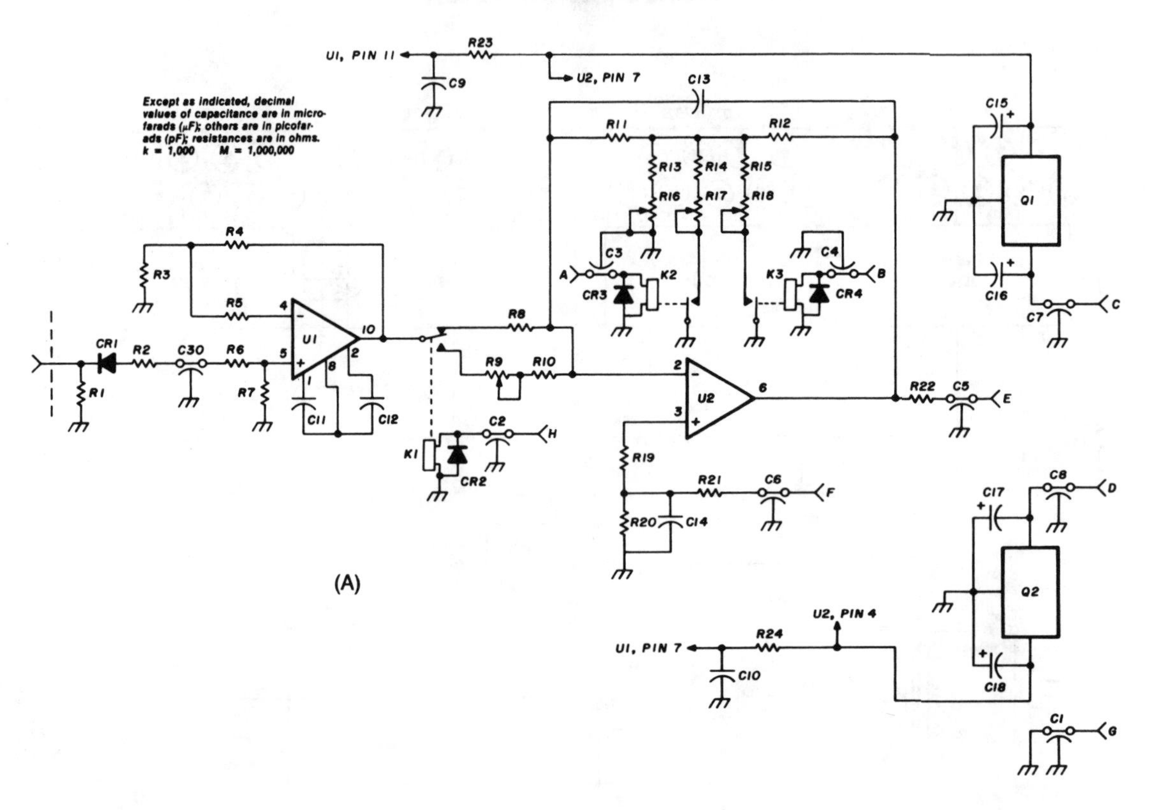

table 1. RF power meter and power supply parts list

C1 thru C8	1000 pF feedthru (Erie, Cambion)
C9,10,15,18	1 μF 10wvdc tantalum
C11-12	0.1 μF metalized film
C13	500 pF disc
C14	0.01 μF disc ceramic
C16,17	2.2 μF 25 wvdc tantalum
C19,21	100 μF 15 wvdc electrolytic
C20	500 μF 15 wvdc electrolytic
C22,23	0.01 μF disc
C30	100pF chip capacitor
CR1	HSCH-3486 Hewlett-Packard
CR2,3,4,9,10	1N914 or equivalent
CR5,6,7,8	1N4003 or equivalent
K1	SPDT reed Magnecraft W172-DIP5 (internal diode — CR2 not used)
K2,3,4,5	SPST reed EAC EAC Z610-ND
M1	1 mA DC meter with dB scale
Q1,4	78LO5 regulator
Q2	79LO5 regulator
Q3	78L12 regulator
R1,2	50 ohm 1/8 watt carbon film

All resistors 1% metal film 1/4 watt	
R3,6,14,22	1k
R5,7	100k
R10	120k
R4	150k
R8,19	4.99k
R11,12	20k
R13	2.74k
R15	165 ohm

All resistors 5% carbon film 1/4 watt	
R20	100 ohm
R21	1 megohm
R23, 24	10 ohm
R27	1.5k
R9	50k Panasonic CEG54 trimpot
R16	500 ohm Panasonic CEG52 trimpot
R17	200 ohm Panasonic CEG22 trimpot
R18	100 ohm Panasonic CEG12 trimpot
R25,26	10k potentiometer

S1	DP6T rotary switch
T1,T2	6.3 VAC transformers
U1	ICL76508CPD Intersil
U2	LM11CLH National
Box	CU-124 BUD
Chassis	9 1/2 x 5 x 2 chassis BUD Ac-403

HAM RADIO

Fig. 13-42

WIDE-RANGE RF POWER METER *(Cont.)*

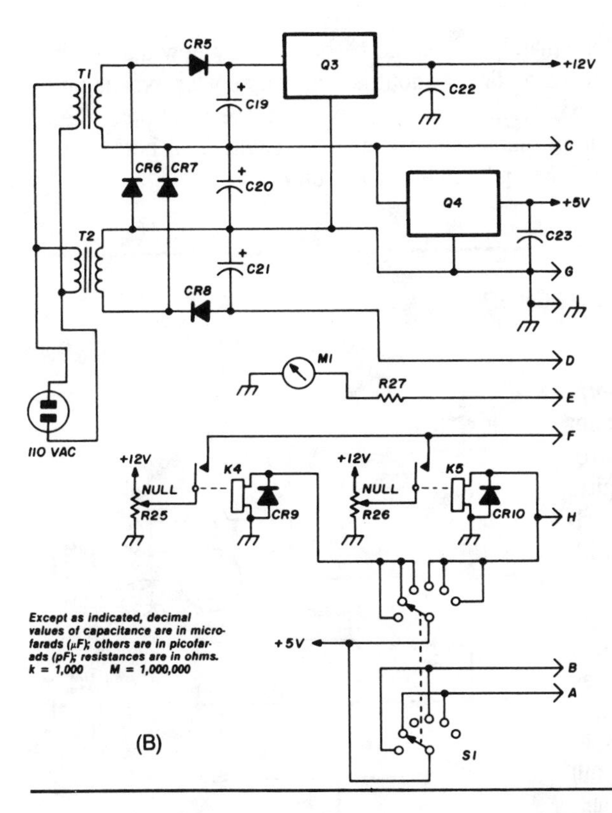

The Hewlett-Packard HSCH-3486 zero-bias Schottky diode is used as the detector. To avoid using a modulation method of detection, a chopper-stabilized op amp is used. The chopper op amp basically converts the input dc voltage to ac, amplifies it, and converts it back to dc. Amplifying the dc output from the detector 150 times with a chopper op amp puts the signal at a level that simpler op amps, such as the LM11, can handle. Offset voltages in the amplifier are nulled with two pots—one for the high range and one for the lower three ranges.

LED PEAK METER

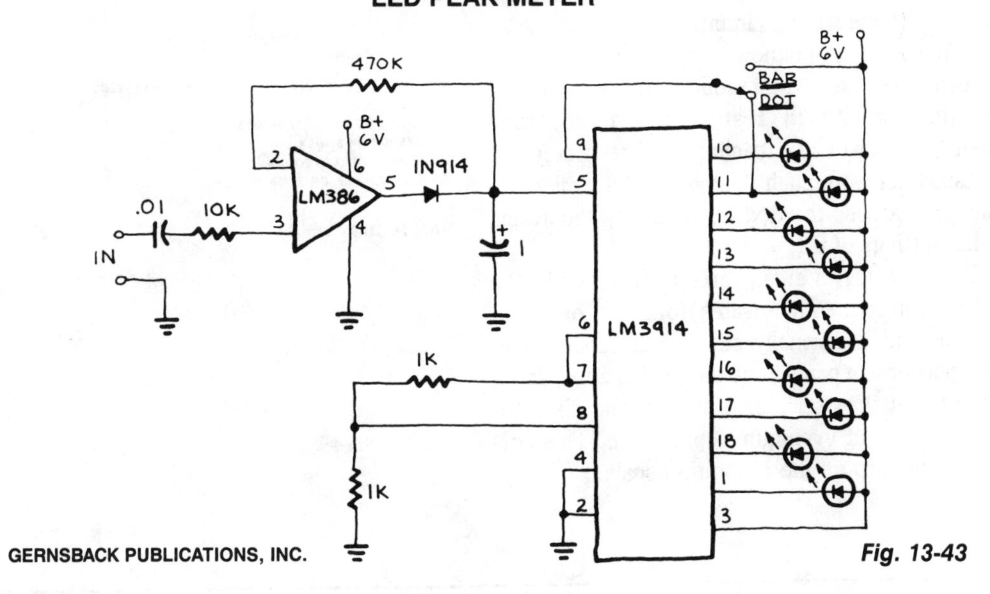

GERNSBACK PUBLICATIONS, INC.

Fig. 13-43

The circuit includes a peak detector that immediately drives the readout to any new higher signal level and slowly lowers it after the signal drops to zero. The readout is a moving dot or expanding bar display. The cicuit can be expanded for a longer bar readout. Tapping five or more LED peak meters into a frequency equalizer or series of audio filters should give a unique result. The bottom LED remains on with no signal at the input, thus providing a pilot light for the unit.

LC CHECKER

The circuit is based on the *grip-dip* or *absorption effect,* which occurs when a parallel resonant circuit is coupled to an oscillator of the same frequency. Q1 operates in a conventional Colpitts oscillator circuit at a fixed frequency of approximately 4 MHz. A meter connected in series with the transistor's base-bias resistor serves as the dip or absorption indicator.

The variable measuring circuit consists of C1, C2, and L2 and is connected to panel terminals as shown. L2 is loosely coupled to L1 in the oscillator circuit. This measuring circuit is tuned to the oscillator frequency with variable capacitor C2 set at full capacitance. When power is applied to the oscillator, the meter shows a dip caused by power absorption from the measuring circuit.

Connecting an unknown capacitor across the test terminals lowers the resonant frequency of the measuring circuit. To restore resonance, tune capacitor C2 lower in capacitance. The meter will dip again when you reach this point. Determine the capacitance across the text terminals by calibrating the dial settings of C2.

Capacitor C4, a small variable trimmer in the oscillator circuit, compensates for drift or other variations and is normally set at half capacitance. The capacitor is a panel control, labeled zero, and it is used to set the oscillator exactly at the dip point when C2 is set at maximum capacitance. This corresponds to zero on the calibration scale.

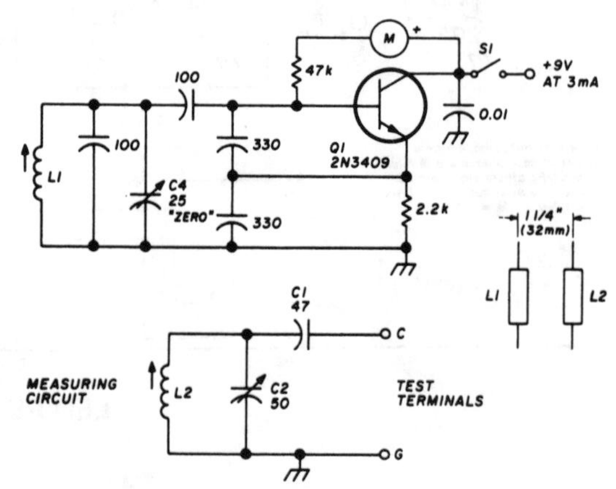

L1- 30T. NO.28E CLOSE-WOUND ON 3/8" (10mm) SLUG-TUNED FORM. APPROX. 7μH.
L2-50T. SAME AS L1. APPROX. 30μH.
Q1- 2N3904 OR SIMILAR.
M- 0 TO 100 OR 0 TO 200μA.
S1- SPST TOGGLE OR SLIDE SWITCH.

HAM RADIO

Fig. 13-44

TACHOMETER AND DIRECTION-OF-ROTATION CIRCUIT

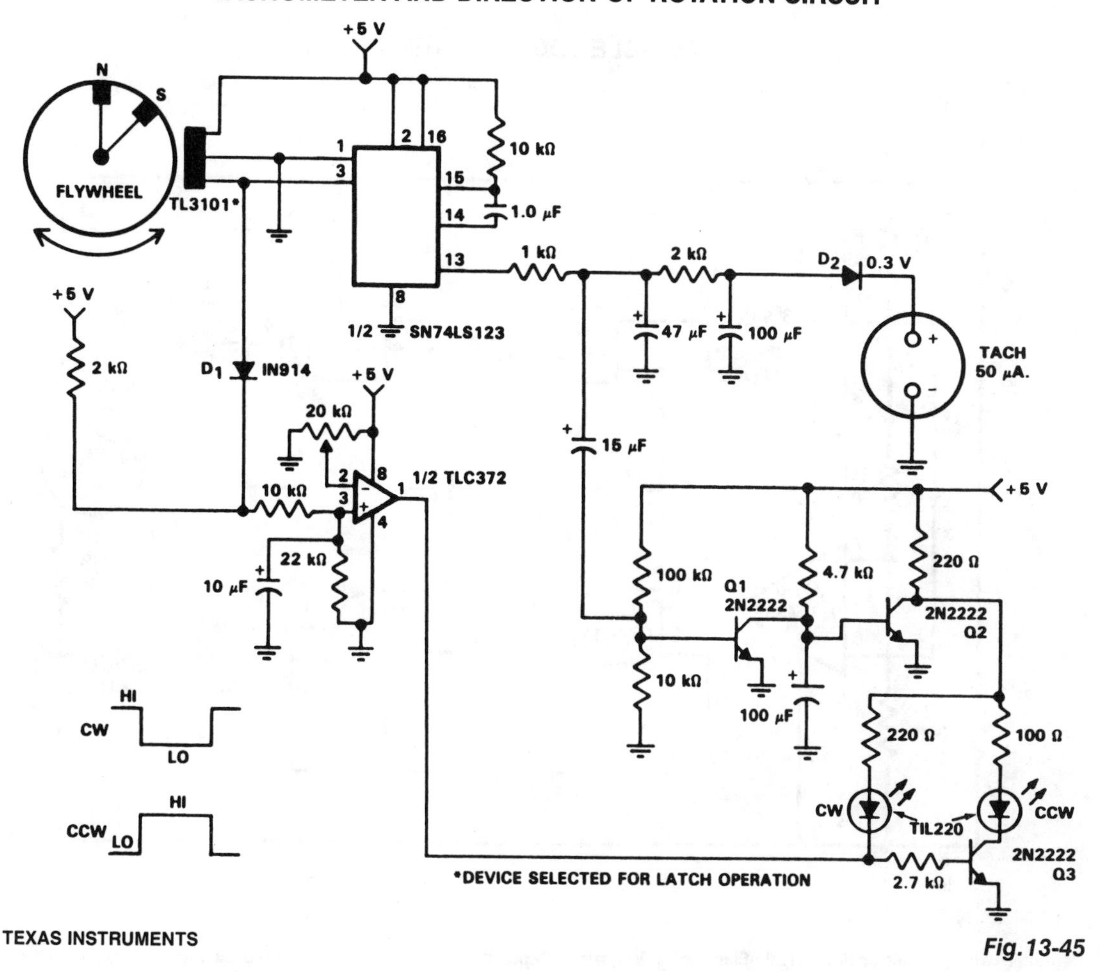

TEXAS INSTRUMENTS

Fig. 13-45

In machine and equipment design, some applications require measurement of both the shaft speed and the direction of rotation. Figure 13-45 shows the circuit of a tachometer, which also indicates the direction of rotation. The flywheel has two magnets embedded in the outer rim about 45° apart. One magnet has the north pole toward the outside and the other magnet has the south pole toward the outside rim of the flywheel. Because of the magnet spacing, a short on pulse is produced by the TL3101 in one direction and a long on pulse in the other direction. A 0- to 50-μA meter is used to monitor the flywheel speed while the LEDs indicate the direction of rotation. The direction-of rotation circuit can be divided into three parts:

- TLC372 device for input conditioning and reference adjustment.
- Two 2N2222 transistors, which apply the V_{CC} to the two LEDs when needed.
- The two TIL220 LEDs, which indicate clockwise (CW) or counterclockwise (CCW) direction of rotation.

AUDIBLE LOGIC TESTER

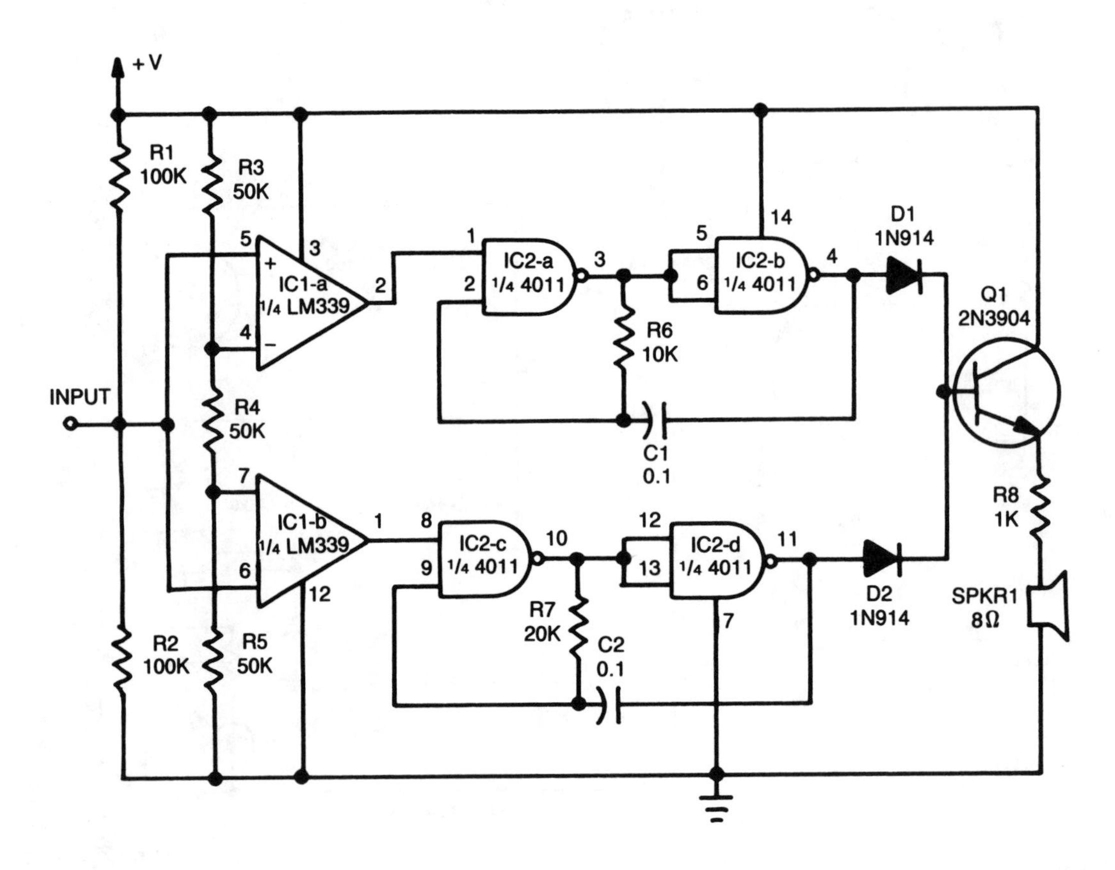

Fig. 13-46

This tester provides an audible indication of the logic level of the signal presented to its input. A logic high is indicated by a high tone, a logic low is indicated by a low tone, and oscillation is indicated by an alternating tone. The input is high impedence, so it will not load down the circuit under test. The tester can be used to troubleshoot TTL or CMOS logic. The input consists of two sections of an LM339 quad comparator. IC1a increases when the input voltage exceeds 67% of the supply voltage. The other comparator increases when the input drops below 33% of the supply.

The tone generators consist of two gated astable multivibrators. The generator built around IC2a and IC2b produces the high tone. The one built around IC2c and IC2d produces the low tone. Two diodes, D1 and D2, isolate the tone-generator outputs. Transistor Q1 is used to drive a low-impedance speaker.

LOW-CURRENT MEASUREMENT SYSTEM

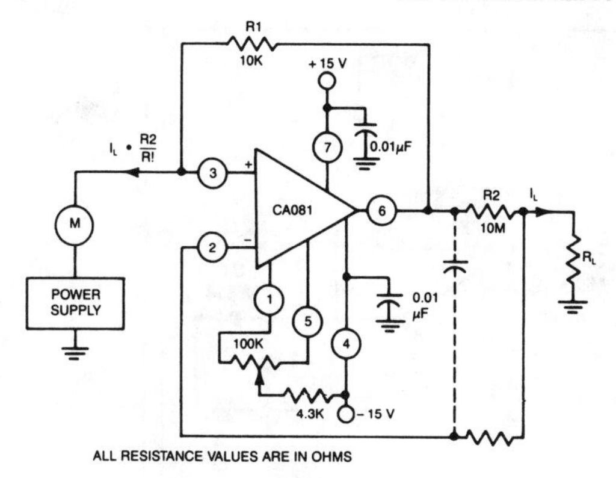

This circuit uses a CA018 BiMOS op amp. Low current, supplied at input potential as power supply to load resistor R_L, is increased by R2/R1, when load current I_L is monitored by power supply meter M. Thus, if I_L is 100 nA, with values shown, I_L presented to supply will be 100 μA.

GE/RCA ***Fig. 13-47***

SIMPLE CONTINUITY TESTER

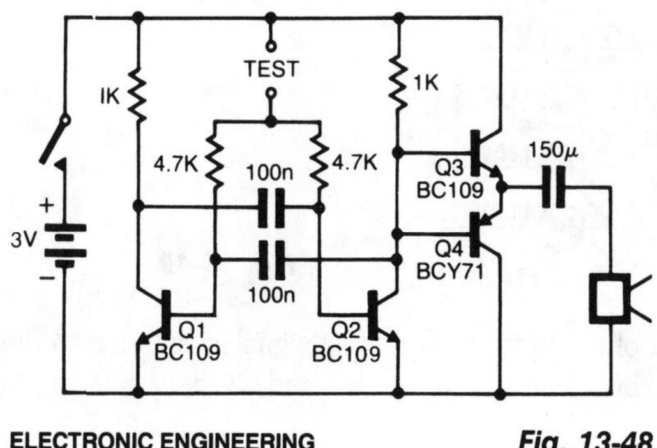

ELECTRONIC ENGINEERING ***Fig. 13-48***

The pitch of the tone is dependent upon the resistance under test. The tester will respond to resistances of hundreds of kilohms, yet it is possible to distinguish differences of just tens of ohms in low-resistance circuits. Q1 and Q2 form a multivibrator, the frequency of which is influenced by the resistance between the test points. The output stage, Q3 and Q4, will drive a small loudspeaker or a telephone earpiece. The unit is powered by a 3-V battery, and draws very little current when not in use.

SOUND-LEVEL METER

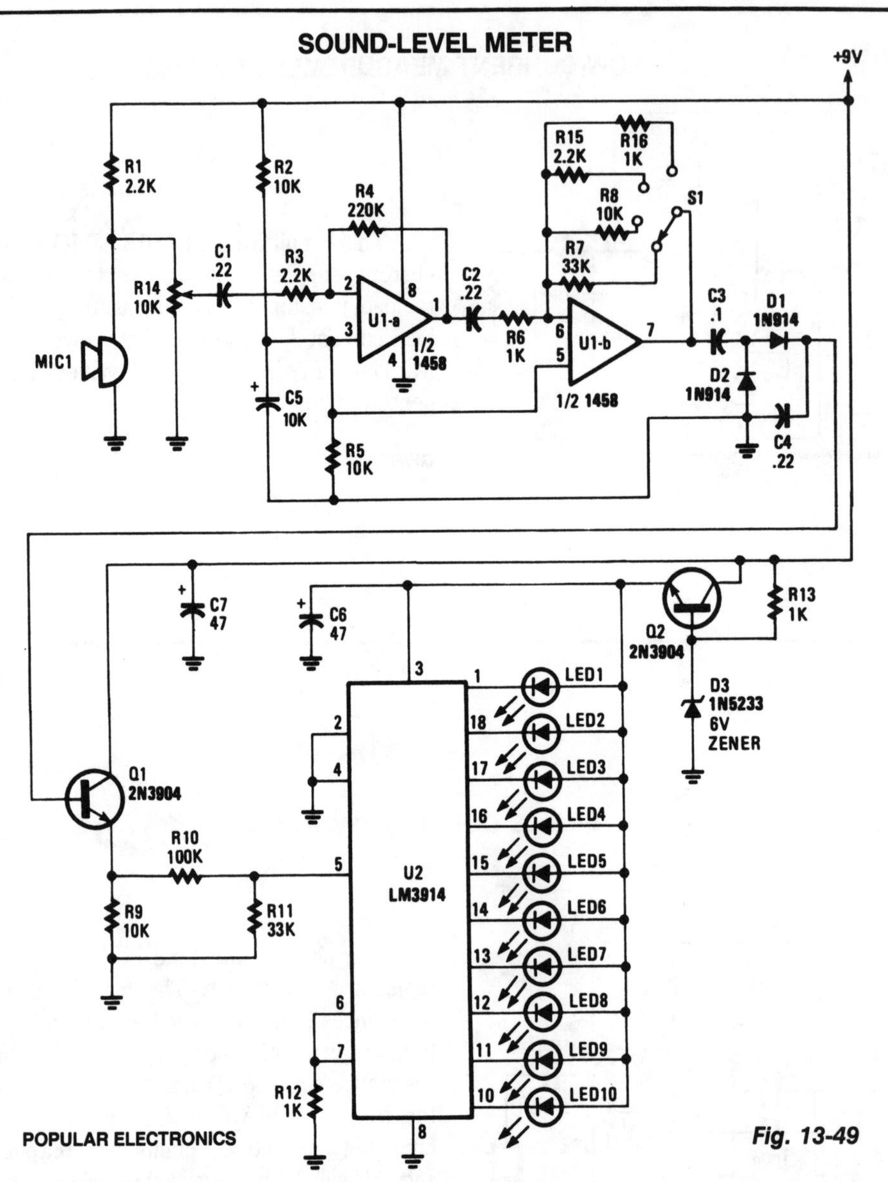

POPULAR ELECTRONICS

Fig. 13-49

Sounds are picked up by MIC1 and fed to the input of the first op amp. The signal is then fed to the input of a second op amp U1b, where it is boosted again by a factor of between 1 and 33, depending upon the setting of range switch S1.

With the range switch set in the A position, R6 is 1kΩ and R7 is 33kΩ, so that stage has a gain of 33. In the B position, the gain is 10 Ω; in the C position, the gain is 22Ω; and in the D position the gain is 1 Ω.

As the signal voltage fed to the input of U2 at pin 5 varies, one of ten LEDs will light to correspond with the input-voltage level. At the input's lowest operating level, U2 produces an output at pin 1, causing LED1 to light. The highest input level presented to the input of U2, about 1.2 V, causes LED10 to turn on.

LED PANEL METER

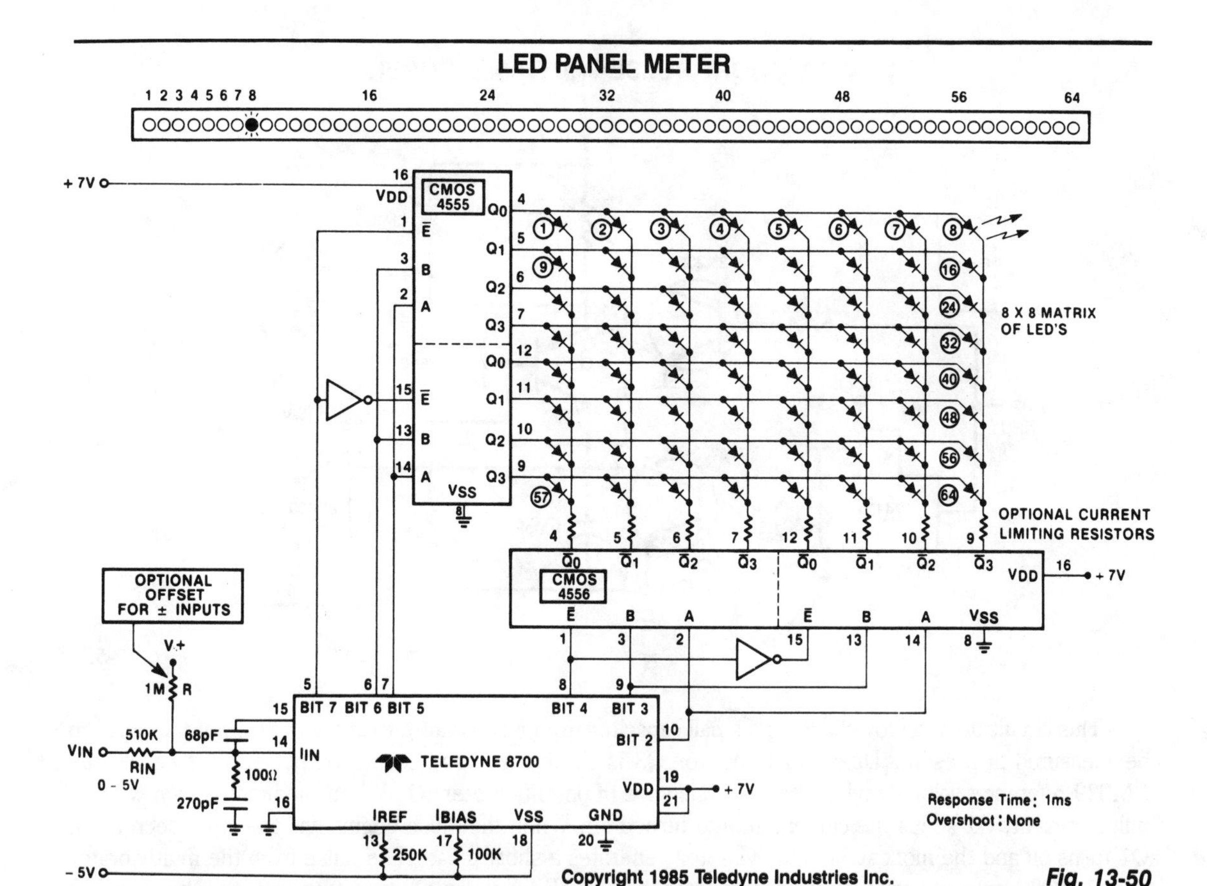

Copyright 1985 Teledyne Industries Inc.

Fig. 13-50

OPTICAL PICK-UP TACHOMETER

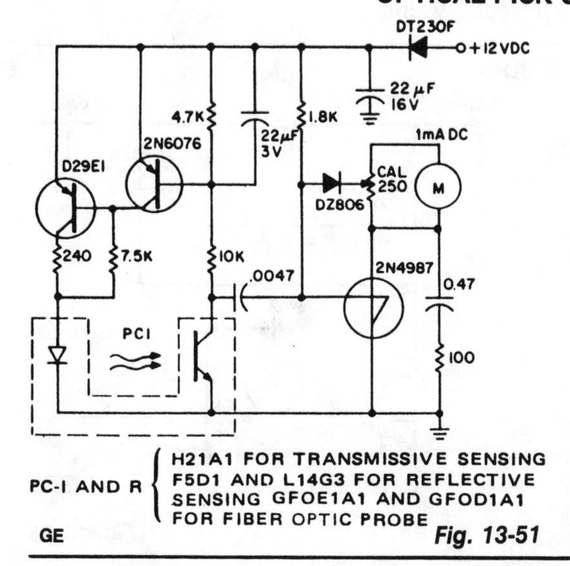

PC-I AND R { H21A1 FOR TRANSMISSIVE SENSING F5D1 AND L14G3 FOR REFLECTIVE SENSING GFOE1A1 AND GFOD1A1 FOR FIBER OPTIC PROBE

GE

Fig. 13-51

Remote, noncontact, measurement of the speed of rotating objects is the purpose of this simple circuit. Linearity and accuracy are extremely high and normally limited by the milliammeter used and the initial calibration. This circuit is configured to count the leading edge of light pulses and to ignore normal ambient light levels. It is designed for portable operation because the tachometer is not sensitive to supply voltage within the supply voltage tolerance. Full scale at the maximum sensitivity of the calibration resistance is read at about 300 light pulses per second. A digital voltmeter can be used on the 100-mV full-scale range, in place of the milliammeter. Shunt its input with a 100-Ω resistor in parallel with a 100-μF capacitor. This RC network replaces the filtering supplied by the analog meter.

VERY SHORT PULSE-WIDTH MEASURER

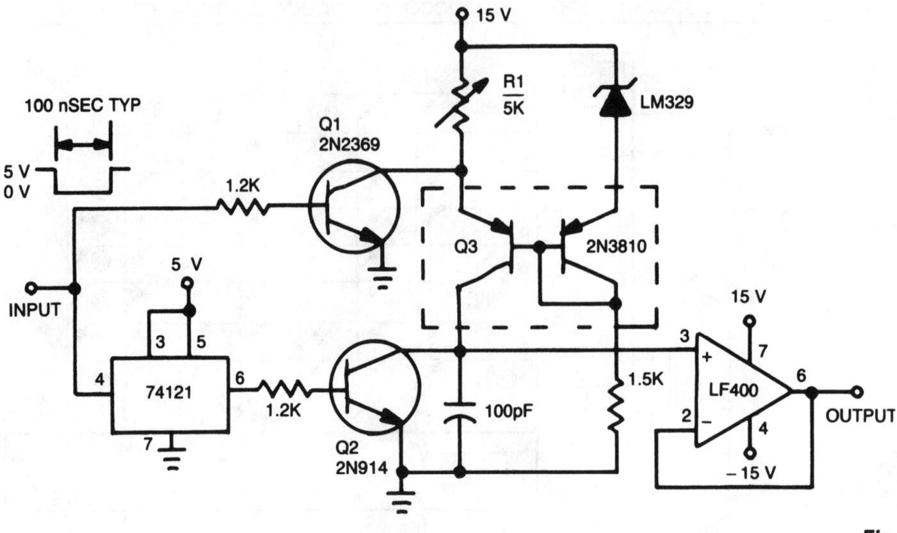

EDN *Fig. 13-52*

This circuit operates by changing a small capacitor from a constant-current source when the pulse to be measured is present. Dual pnp transistor Q3 is the current source; its output current equals the LM329 reference voltage divided by the resistance of potentiometer R1. When the input is high with no pulse present, Q1 keeps the current source turned off. When the pulse begins and the input decreases, Q1 turns off and the monostable multivibrator generates a short pulse. The pulse from the multivibrator turns on Q2, removing any residual charge from the 100-pF capacitor. Q2 then turns off, and the capacitor begins to charge linearly from the current source. When the input pulse ends, the current source turns off, and the voltage on the capacitor is proportional to the pulse width.

QRP SWR BRIDGE

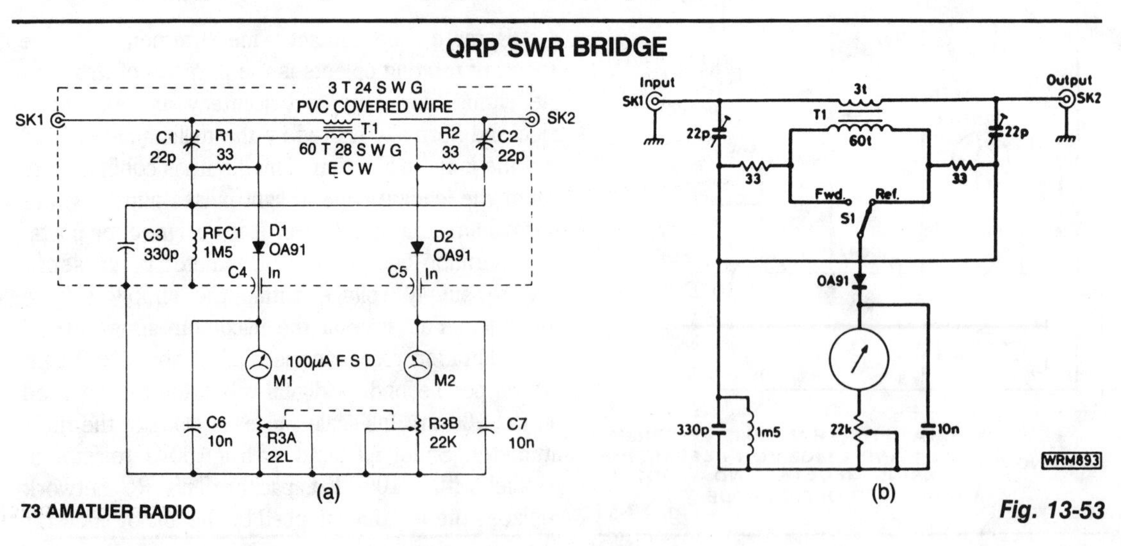

73 AMATUER RADIO *Fig. 13-53*

The design shown is a simple unit for QRP operation on all authorized frequencies up to 30 MHz, based on a toroidal transformer T1. The secondary winding of T1 samples a small amount of RF power, both forward and reflected, which is divided by the bridge circuit and rectified by diodes D1 and D2. Forward and reflected readings are obtained simultaneously on the two meters (M1 and M2), and the bridge is matched and balanced at the required load impedance by C1 and C2. See Fig. 13-53 for an alternative, less expensive, single-meter version. The bridge also measures forward power.

PEAK-dB METER

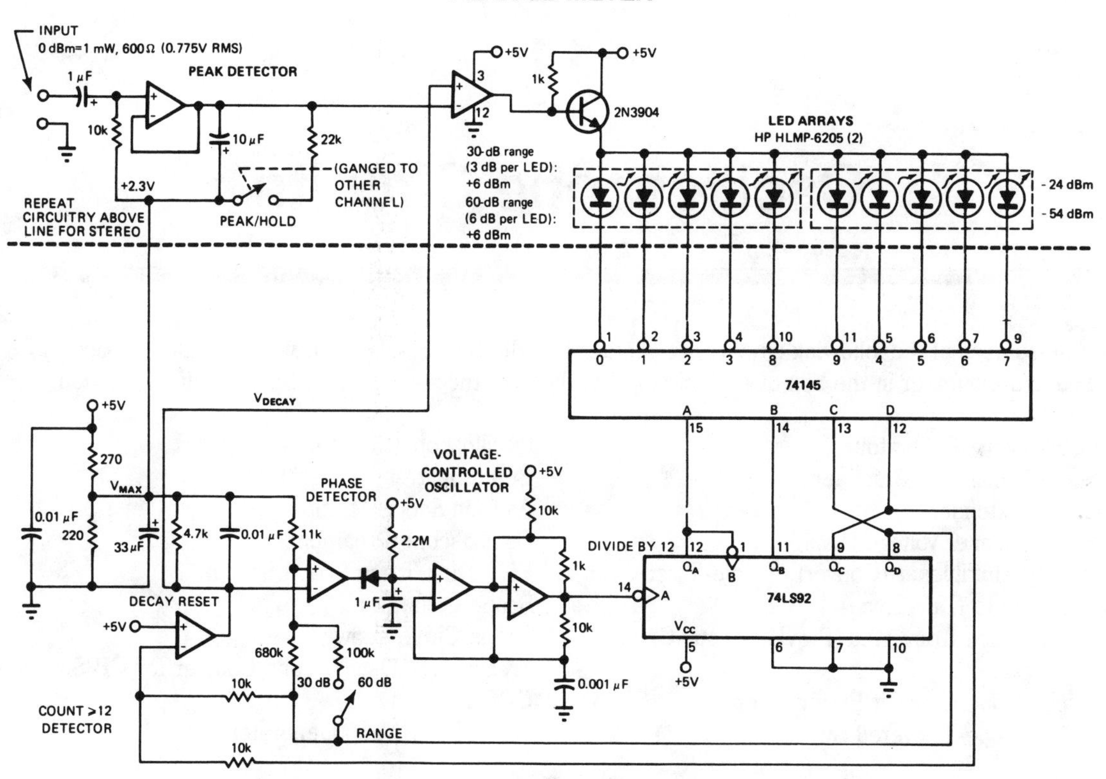

EDN *Fig. 13-54*

This circuit compares a rectified input, V_{IN}, with a voltage that decays exponentially across a 4.7-kΩ resistor and a 0.01-μF capacitor. Comparing the exponentially decaying voltage with the rectified input provides a peak-level indication that requires no adjustment. A phase-locked loop controls the scan rate so that each LED represents 6 dB in the 30-dB range.

14

Oscilloscope Circuits

The sources of the following circuits are contained in the Sources section, which begins on page 217. The figure number in the box of each circuit correlates to the source entry in the Sources section.

Oscilloscope Calibrator
Scope Sensitivity Amplifier
Scope Extender
Eight-Channel Voltage Display
Analog Multiplexer (Converts Single-Trace Scope to Four-Trace)
Oscilloscope Converter (Provides Four-Channel Displays)
Oscilloscope/Counter Preamplifier
Oscilloscope-Triggered Sweep
Oscilloscope Monitor
Beam Splitter for Oscilloscope
Add-On Scope Multiplexer
Oscilloscope Preamplifier
FET Dual-Trace Scope Switch
Scope Calibrator
Scope Circle Drawer
Transmitter-Oscilloscope Coupler for CB Signals
CRO Doubler
10.7-MHz Sweep Generator

OSCILLOSCOPE CALIBRATOR

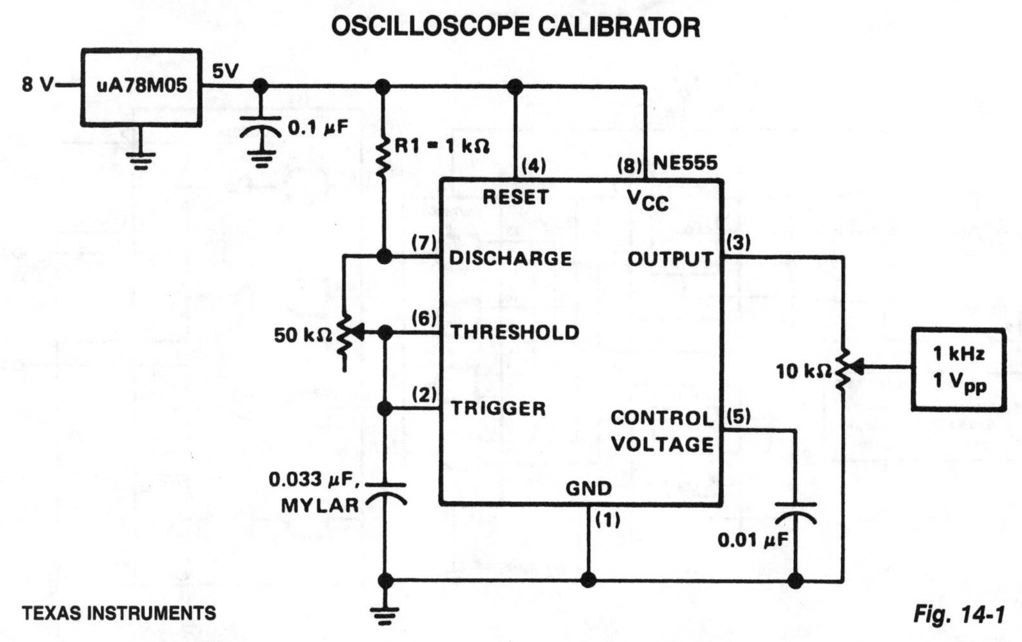

TEXAS INSTRUMENTS

Fig. 14-1

The calibrator can be used to check the accuracy of a time-base generator, as well as to calibrate the input level of amplifiers. The calibrator consists of an NE555 connected in the astable mode. The oscillator is set to exactly 1 kHz by adjusting potentiometer P1 while the output at pin 3 is being monitored against a known frequency standard or frequency counter. The output level, likewise, is monitored from potentiometer P2's center arm to ground with a standard instrument. P2 is adjusted for 1 V pk-pk at the calibrator output terminal. During operation, the calibrator output terminal will produce a 1-kHz, square-wave signal at 1 V pk-pk with about 50% duty cycle. For long-term oscillator frequency stability, C1 should be a low-leakage mylar capacitor.

SCOPE SENSITIVITY AMPLIFIER

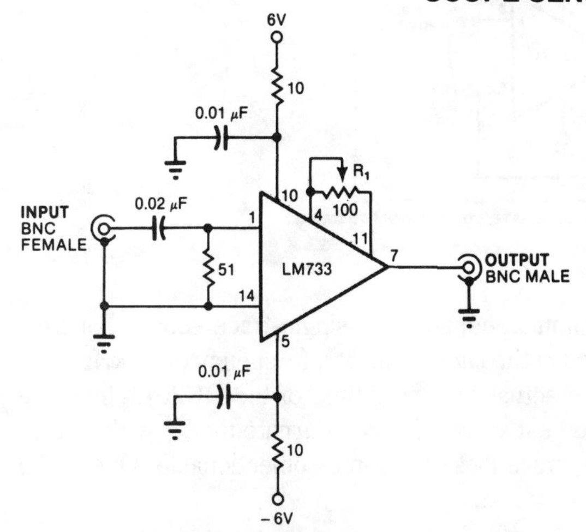

This circuit provides 20 ±0.1 dB voltage gain from 0.5 to 25 MHz and ±3 dB from 70 kHz to 55 MHz. An LM733 video amplifier furnishes a low input-noise spec, 10-μV typical, measured over a 15.7-MHz bandwidth. The scale factor of the instrument can be preserved by using a trimmer R1 or a selected precision resistor, to set the circuit's voltage gain to exactly 100.

EDN

Fig. 14-2

SCOPE EXTENDER

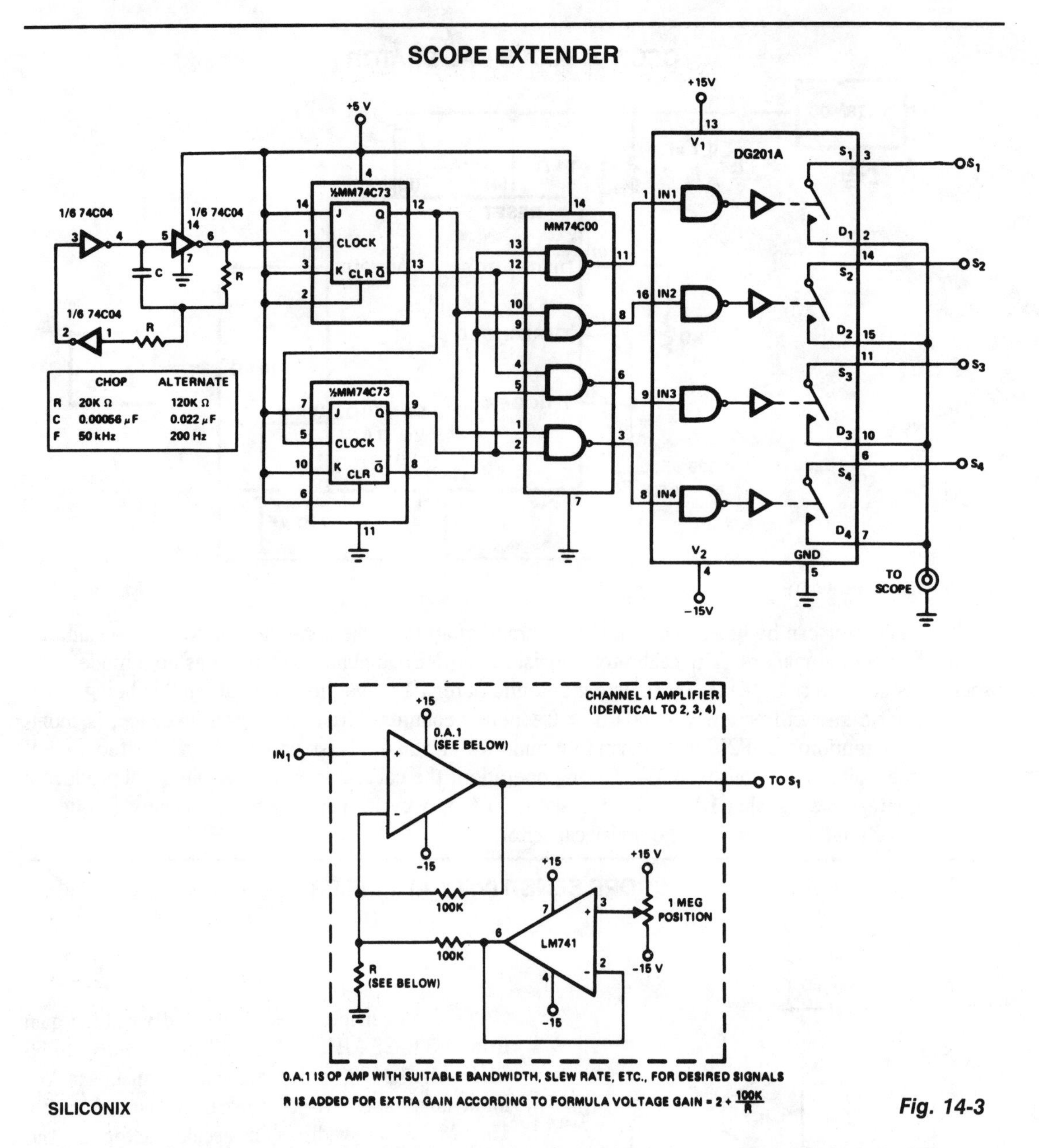

SILICONIX

Fig. 14-3

The adapter allows four inputs to be displayed simultaneously on a single-trace scope. For low-frequency signals, less than 500 Hz, the adapter is used in the *chop* mode at a frequency of 50 kHz. The clock can be run faster, but switching glitches and the actual switching time of the DG201A limit the maximum frequency to 200 kHz. High frequencies are best viewed in the alternate mode, with a clock frequency of 200 Hz. When the clock is below 100 Hz, trace flicker becomes objectionable. One of the four inputs is used to trigger the horizontal trace of the scope.

EIGHT-CHANNEL VOLTAGE DISPLAY

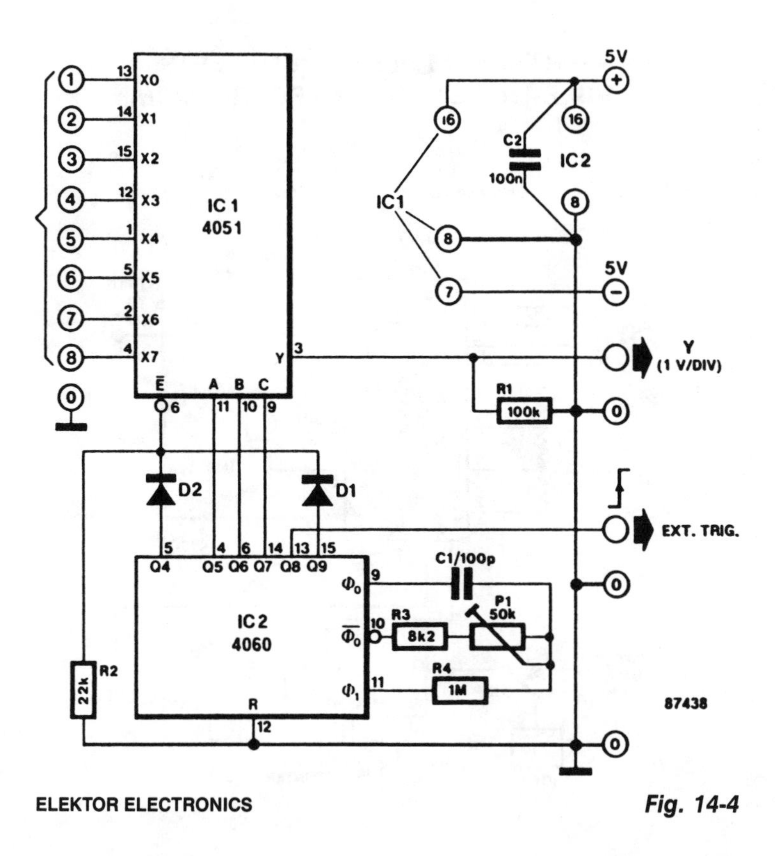

ELEKTOR ELECTRONICS

Fig. 14-4

This circuit turns a common oscilloscope into a versatile eight-channel display for direct voltages. The trend of each of the eight input levels is readily observed, albeit that the attainable resolution is not very high.

The circuit diagram shows the use of an eight-channel analog multiplexer IC1, which is the electronic version of an eight-way rotary switch with contacts X0 through X7 and pole Y. The relevant channel is selected by applying a binary code to the A-B-C inputs. For example, binary code 011 (A-B-C) enables channel 7 (X6 Y). The A-B-C inputs of IC1 are driven from three successive outputs of binary counter IC2, which is set to oscillate at about 50 kHz with the aid of P1. Because the counter is not reset, the binary state of outputs Q5, Q6, and Q7 steps from 0 to 7 in a cyclic manner. Each of the direct voltages at input terminals 1 to 8 is therefore briefly connected to the Y input of the oscilloscope. All eight input levels can be seen simultaneously by setting the timebase of the scope, in accordance with the time it takes the counter to output states 0 through 7, on outputs Q5, Q6, and Q7.

The timebase on the scope should be set to 0.5 ms/div, and triggering should occur on the positive edge of the external signal. Set the vertical sensitivity to 1 V/div. The input range of this circuit is from −4 V to +4 V; connected channels are terminated in about 100 kΩ.

ANALOG MULTIPLEXER (CONVERTS SINGLE-TRACE SCOPE TO FOUR-TRACE)

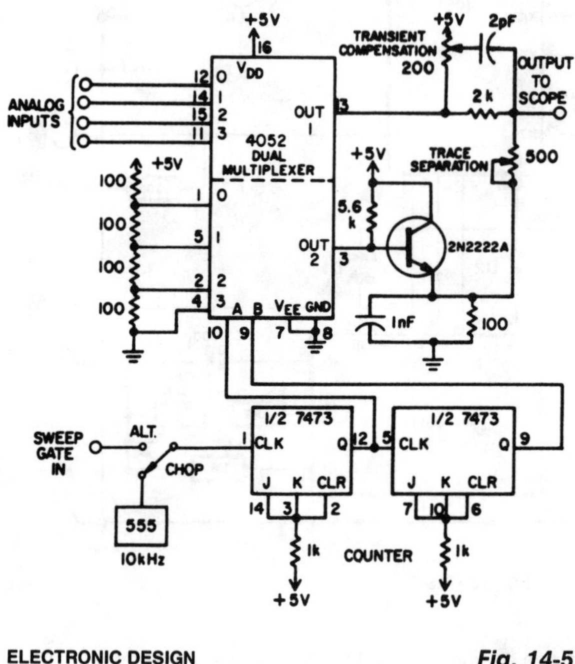

ELECTRONIC DESIGN

Fig. 14-5

This adapter circuit, based on a dual four-channel analog multiplexer, handles digital signals to at least 1 MHz, and analog signals at least through the audio range. The dual multiplexer's upper half selects one input for display. The lower half generates a staircase to offset the baselines of each channel, keeping them separate on the screen. The emitter-follower buffers the staircase, which is then summed with the selected signal. A two-bit binary counter addresses the CMOS 4052 multiplexer.

OSCILLOSCOPE CONVERTER (PROVIDES FOUR-CHANNEL DISPLAYS)

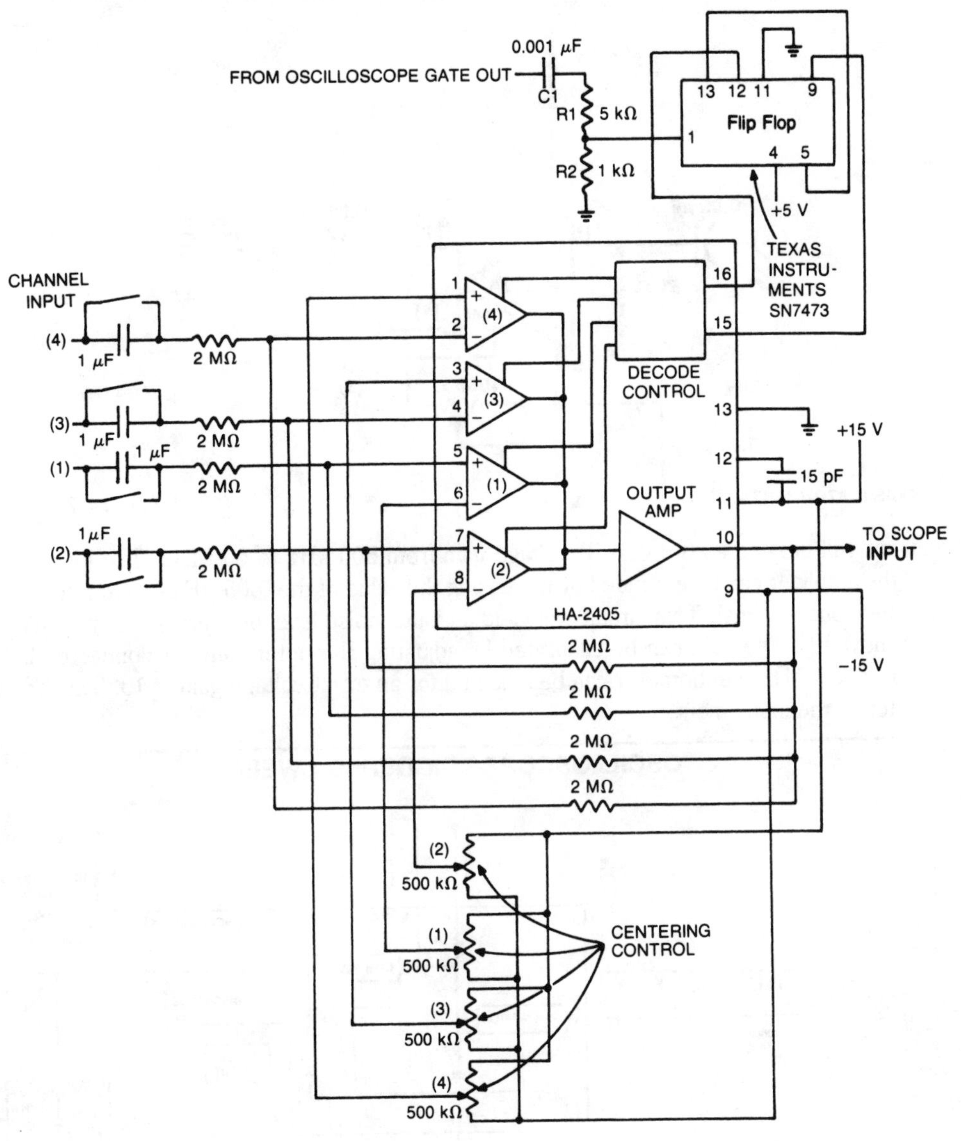

ELECTRONICS

Fig. 14-6

The monolithic quad operational amplifier provides an inexpensive way to increase display capability of a standard oscilloscope. Binary inputs drive the IC op amp; a dual flip-flop divides the scope's gate output to obtain channel-selection signals. All channels have centering controls for nulling offset voltage. A negative-going scope gate signal selects the next channel after each trace. The circuit operates out to 5 MHz.

OSCILLOSCOPE/COUNTER PREAMPLIFIER

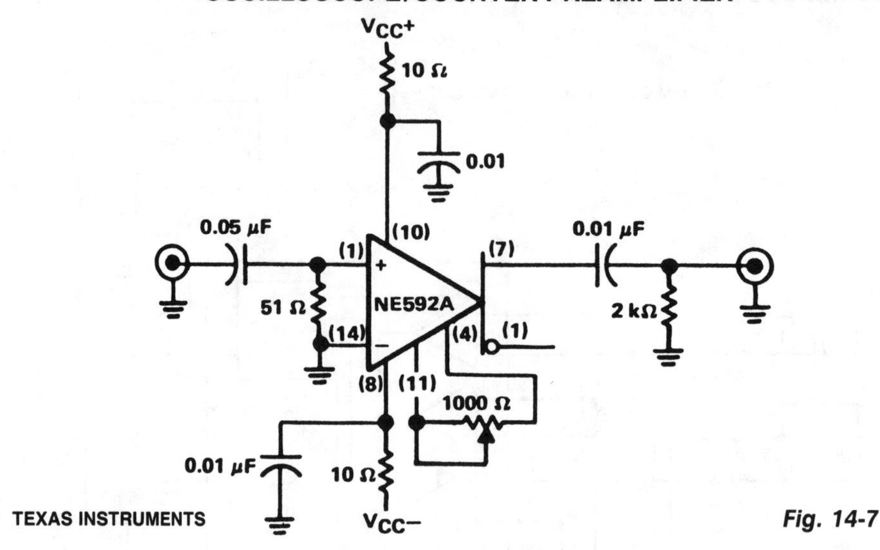

TEXAS INSTRUMENTS

Fig. 14-7

The circuit will provide a 20 ±0.1-dB voltage gain from 500 kHz to 50 MHz. The low-frequency response of the amplifier can be extended by increasing the value of the 0.05-μF capacitor connected in series with the input terminal. This circuit will yield an input-noise level of approximately 10 μV over a 15.7-MHz bandwidth. The gain can be calibrated by adjusting the potentiometer connected between pins 4 and 11. The 1-kΩ potentiometer can be adjusted for an exact voltage gain of 10. This preserves the scale factor of the instrument.

OSCILLOSCOPE-TRIGGERED SWEEP

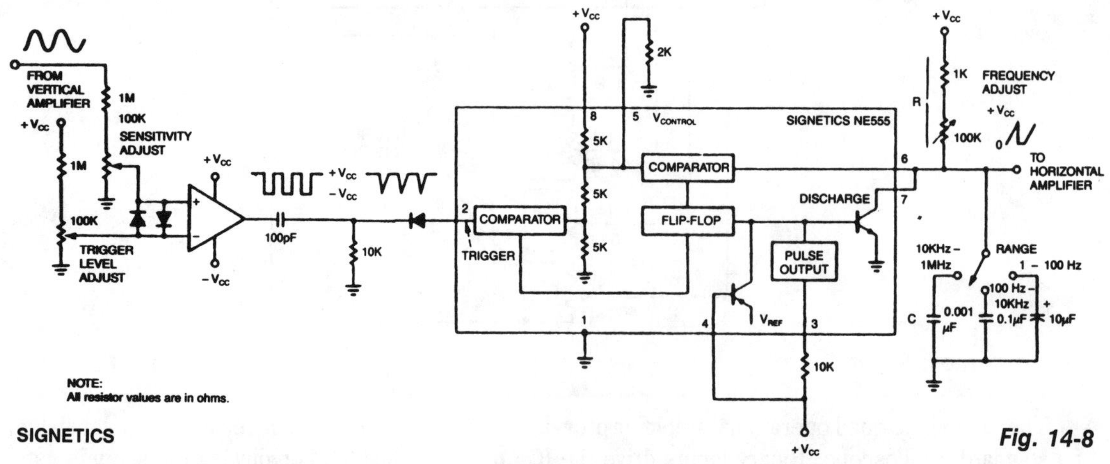

SIGNETICS

Fig. 14-8

The circuit's input op amp triggers the timer, sets its flip-flop and cuts off its discharge transistor so that capacitor C can charge. When the capacitor's voltage reaches the timer's control voltage of 0.33 V_{CC}, the flip-flop resets and the transistor conducts, discharging the capacitor. Greater linearity can be achieved by substituting a constant-current source for frequency-adjust resistor R.

OSCILLOSCOPE MONITOR

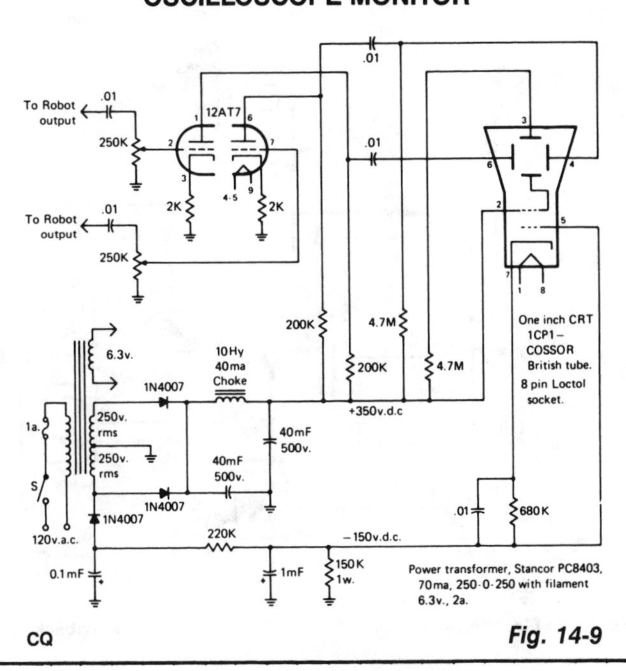

CQ *Fig. 14-9*

BEAM SPLITTER FOR OSCILLOSCOPE

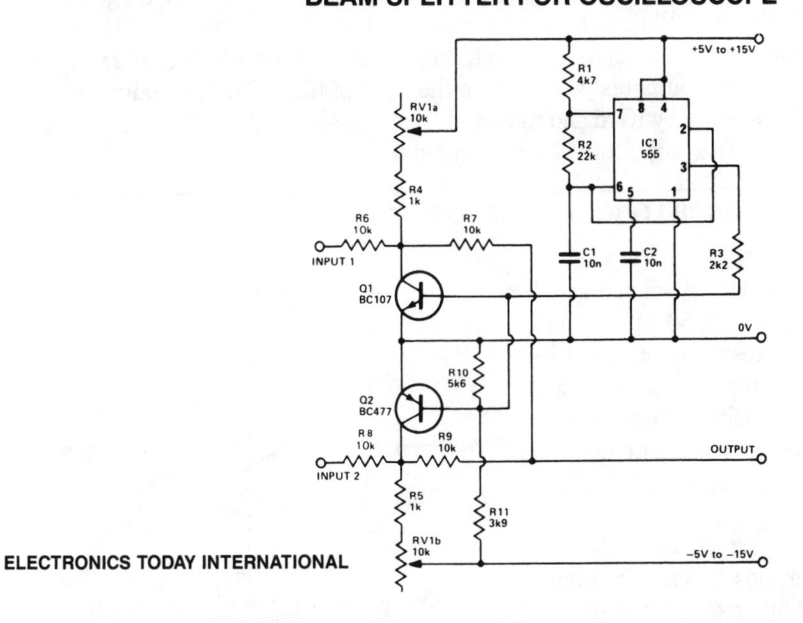

ELECTRONICS TODAY INTERNATIONAL *Fig. 14-10*

The basis of the beam splitter is a 555 timer connected as an astable multivibrator. Signals at the two inputs are alternately displayed on the oscilloscope with a clear separation between them. The output is controlled by the tandem potentiometer RV1a/b, which also varies the amplitude of the traces.

ADD-ON SCOPE MULTIPLEXER

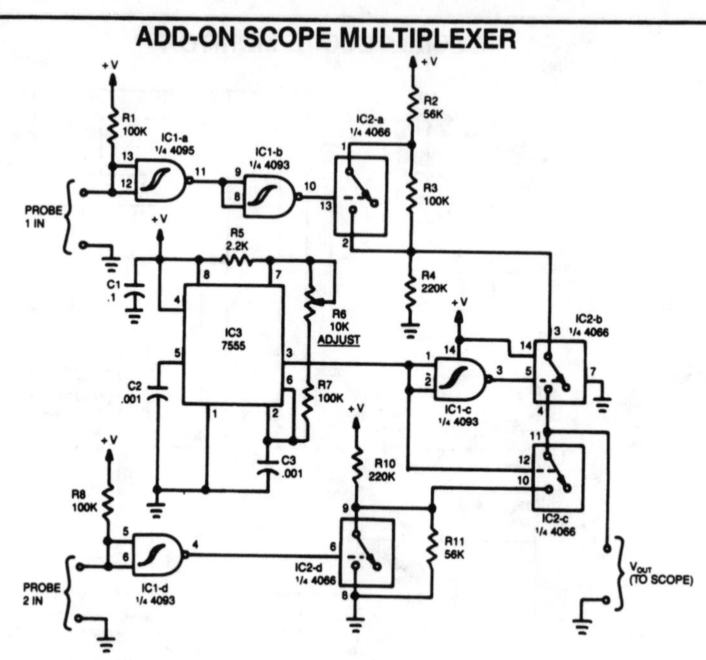

Fig. 14-11

The operation of the unit revolves around three ICs: a 4093 quad NAND Schmitt trigger, a 4066 quad analog switch, and a 7555 timer. When a high is fed to probe 1 in, it is inverted to IC1a and once again by IC1b so that the input to IC2a is high. That high causes the *switch contacts* in IC2a to close. With the *contacts* closed, a high-level output is presented to the input of IC2b. The high output is fed to probe 2 in. That signal is then inverted by IC1d and routed to IC2d, causing its *contacts* to open, and the unit to output a logic-level high. The output of IC2d is then fed to IC2c.

OSCILLOSCOPE PREAMPLIFIER

This circuit provides about 20-dB voltage gain with a frequency range from 0.5 to 50 MHz. You can extend the low-frequency response of this circuit by increasing the value of the 0.05-μF capacitor—or try removing the capacitor. This circuit delivers a particularly small level of input noise, measured at approximately 20 μV over a bandwidth range of 15 MHz.

Calibrate the gain by adjusting the gain potentiometer connected between pins 3 and 10, then adjust the 1-kΩ trimmer potentiometer for an exact voltage gain of 10; this helps preserve the scale factor of the oscilloscope.

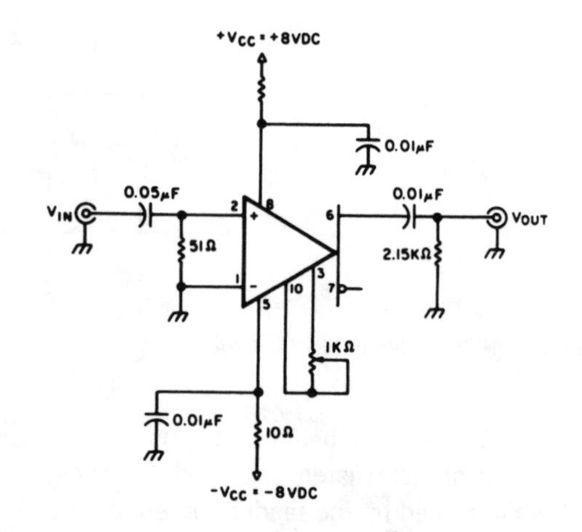

73 AMATEUR RADIO

Fig. 14-12

FET DUAL-TRACE SCOPE SWITCH

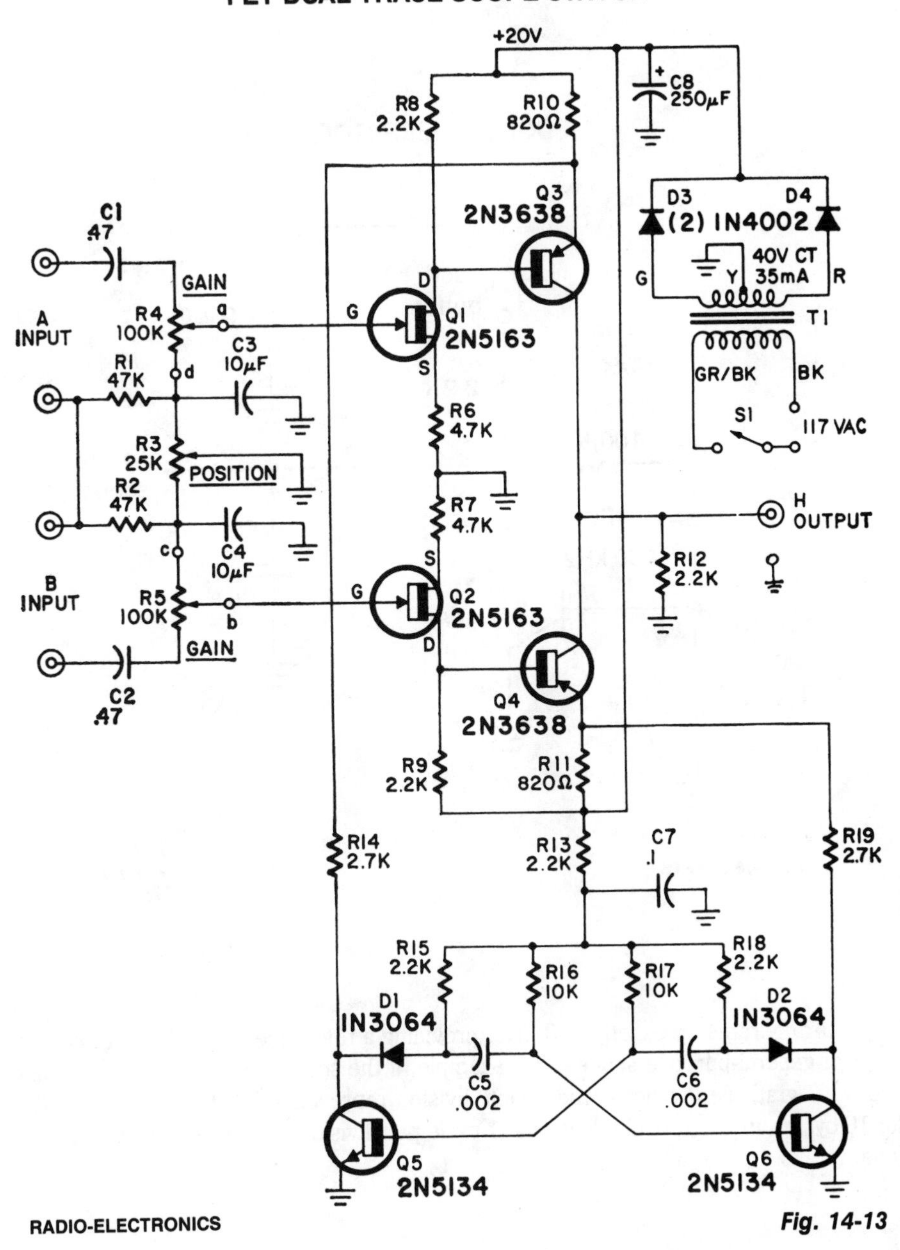

Fig. 14-13

The switcher output goes to the single vertical input of the scope, and a sync line from one of the inputs is taken to the scope's external-sync input. Frequency response of the input amplifiers is 300 kHz over the range of the gain controls. With the gain controls wide open so that no attenuation of the signal occurs, the frequency response is up to 1 MHz.

SCOPE CALIBRATOR

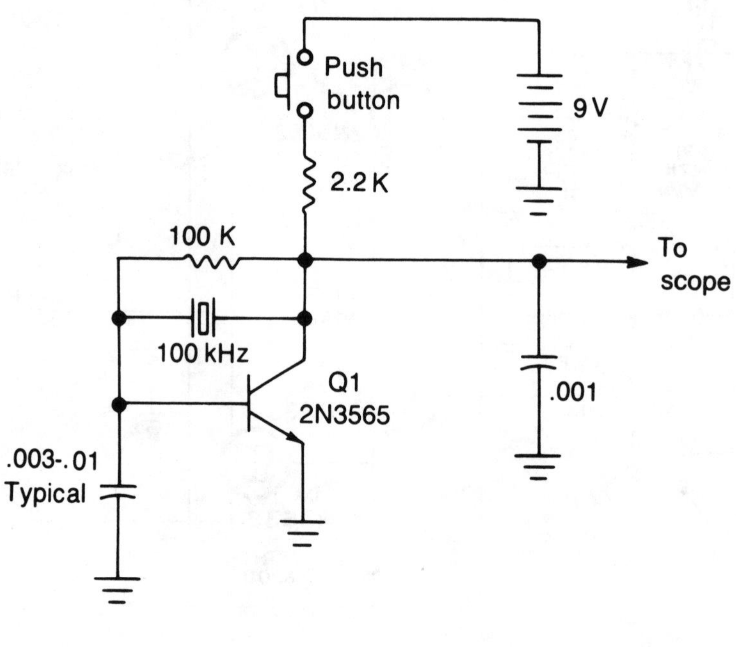

WILLIAM SHEETS

Fig. 14-14

The calibrator operates on exactly 100 kHz, providing a reference for calibrating the variable time base oscillator of general-purpose scopes. For example, if the scope is set so that one cycle of the signal fills exactly 10 graticule divisions, then each division represents 1 MHz, or 1 μs. If the scope is adjusted for 10 cycles on 10 graticule divisions (1 cycle per division), then each division represents 100 kHz or 10 μs.

SCOPE CIRCLE DRAWER

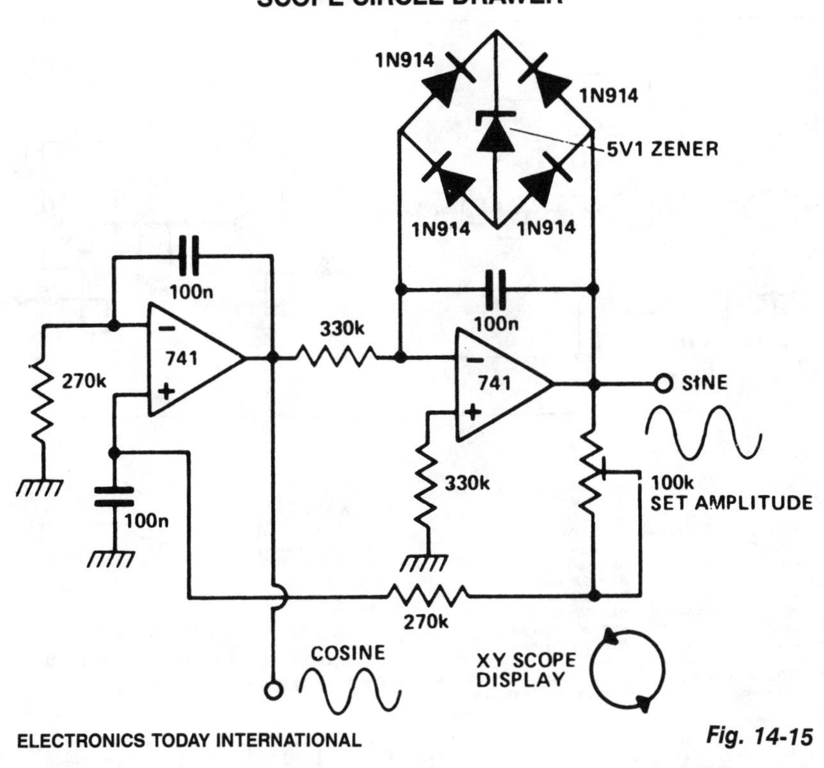

ELECTRONICS TODAY INTERNATIONAL

Fig. 14-15

The circuit is that of a quadrature sine and cosine oscillator. To generate circular displays, connect the two outputs to the X and Y inputs of the oscilloscope.

TRANSMITTER-OSCILLOSCOPE COUPLER FOR CB SIGNALS

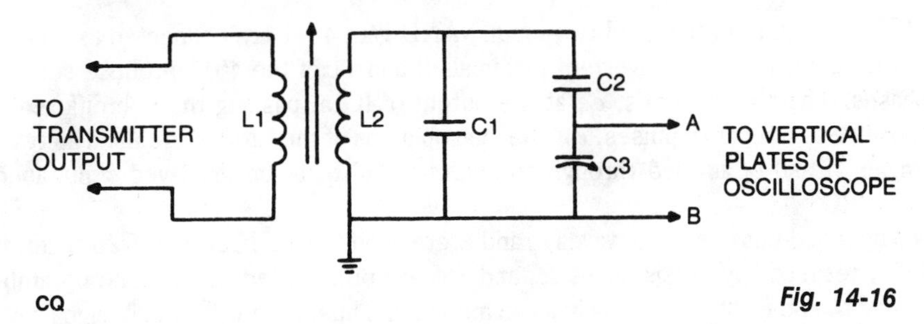

CQ

Fig. 14-16

To display an RF signal, connect L1 to the transmitter and points A and B to the vertical plates of the oscilloscope. Adjust L1 for minimum SWR and C3 for the desired trace height on the CRT. L2 = 4 turns #18 on 3/4" slug tuned RF coil form, L1 = 3 turns #22 adjacent to grounded end of L2. C1, and C2 = 5pF, C3 = 75 pF trimmer.

CRO DOUBLER

Parts List (fig. 1)

IC1	4011	R3	20k
IC2	LM358	R4	20k
		R5	200k
C1	0.001 μF	R6	200k
	0.01 μF	R7	50k
	0.1 μF	R8	50k
	1.0 μF	R9	50k pot
	10 μF	R10	50k pot
C2	25 μF	R11	100 ohm
C3	25 μF	R12	300 ohm pot
C4	0.001 μF	R13	50k pot
R1	500-ohm + 50-k pot	Vcc	6 volts
R2	500 ohm	S	multipole switch

$R_{12} < R_{11}$ $R_{12} > R_{11}$

Oscillograms of the displayed input signals e_1 (sine wave) and e_2 (square wave).

Probe adjusting circuit for CRO.

$C_4R_{13} < R_1C_1$

$C_4R_{13} > C_1R_1$

$C_4R_{13} = C_1R_1$

Displayed square waves on CRO screen.

HAM RADIO

Fig. 14-17

IC1a, IC1b, and IC1c of the quad two-input NAND gate 4011 are connected as an astable multivibrator; IC1d is connected as an inverter. Terminals 3 and 11 of the 4011 produce square waves with opposite phases. The square waves, e_p, at the output of IC1a, passing through differentiator C4R13, then form positive and negative pulses, e_t. The dual op amps of the LM358 are used as two gated amplifiers for singles e_1 and e_2 and fed through terminals 2 and 6, to be displayed simultaneously on the screen of the CRO.

The two opposite-phase square waves $\bar{e}_p$ and e_p are used to gate IC2a and IC2b at terminals 3 and 5 of the LM358, respectively. Resistances R_9 and R_{10} are preadjusted so that one op amp is driven to saturation while the other works normally as an amplifier. Thus, they will amplify signals e_1 and e_2 alternately, and two separate traces will be displayed on the screen. Resistance R_{12} can be varied to adjust the vertical separation of the two traces.

Select a suitable value for C_1 with switch S, and adjust the pot of R1. The frequency of square waves can be varied from 1 to 10^6 Hz. This process is necessary to stablize the waveforms displayed on the screen. A common supply of 6 V is used in the circuit.

10.7-MHz SWEEP GENERATOR

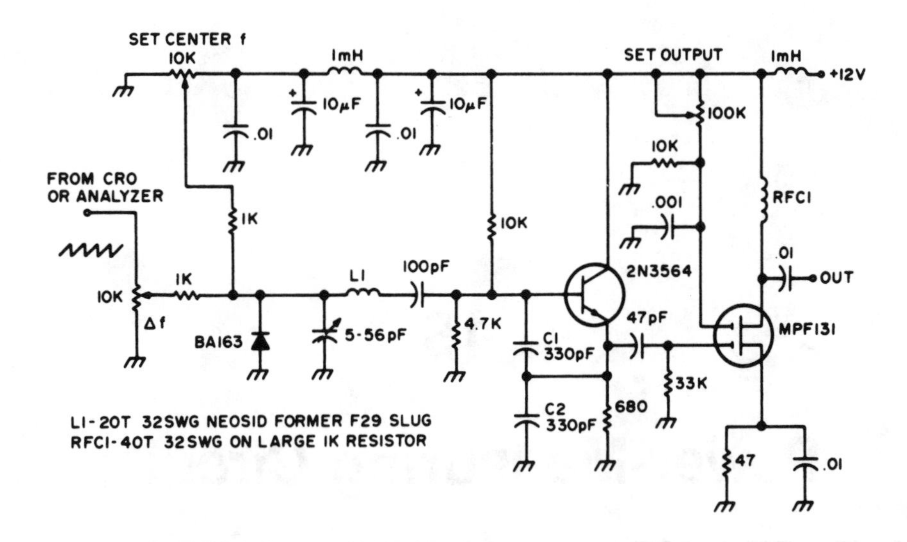

73 AMATEUR RADIO

Fig. 14-18

This circuit is used to observe the response of an IF amp or a filter. It can be used with an oscilloscope or, for more dynamic range, with a spectrum analyzer.

15

Power-Measuring Circuits

The sources of the following circuits are contained in the Sources section, which begins on page 217. The figure number in the box of each circuit correlates to the source entry in the Sources section.

EXTENDED-RANGE VU METER (DOT MODE)

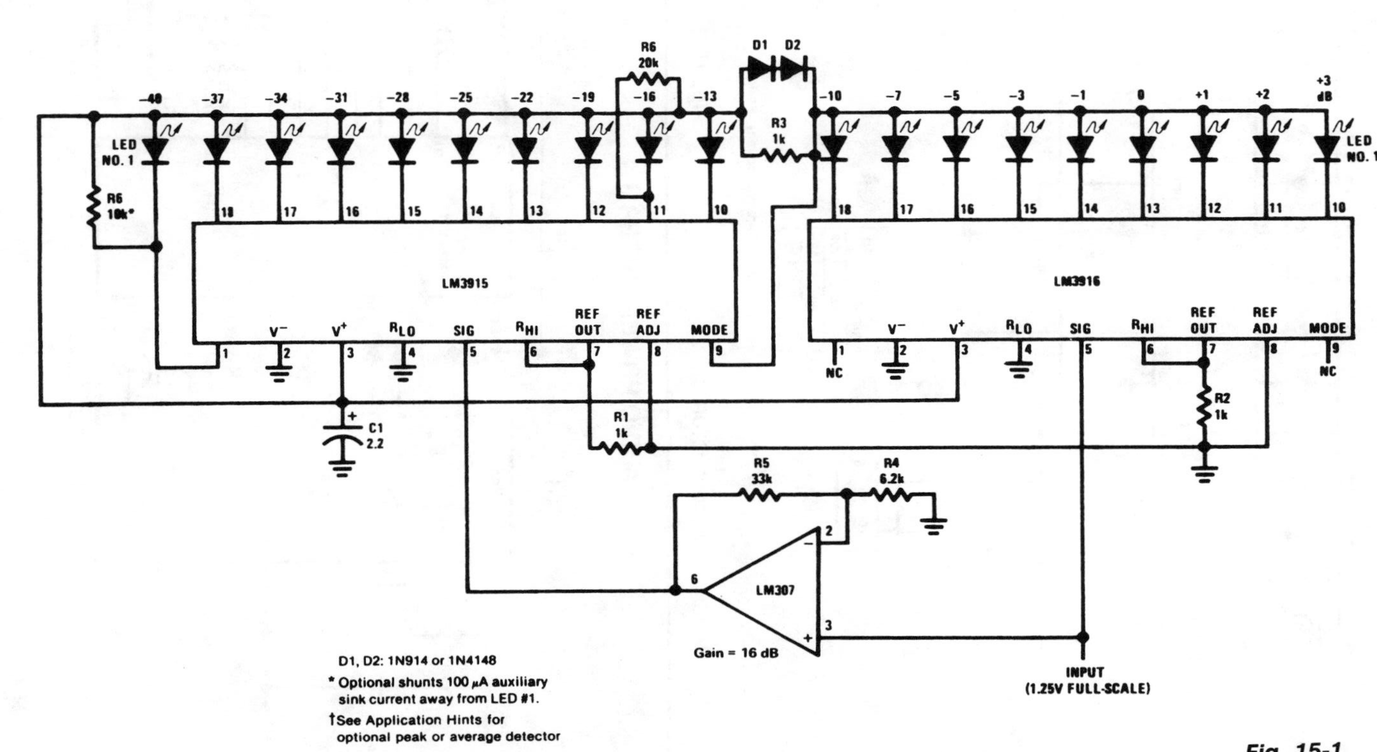

D1, D2: 1N914 or 1N4148

* Optional shunts 100 μA auxiliary sink current away from LED #1.

†See Application Hints for optional peak or average detector

Fig. 15-1

NATIONAL SEMICONDUCTOR

AUDIO POWER METER I

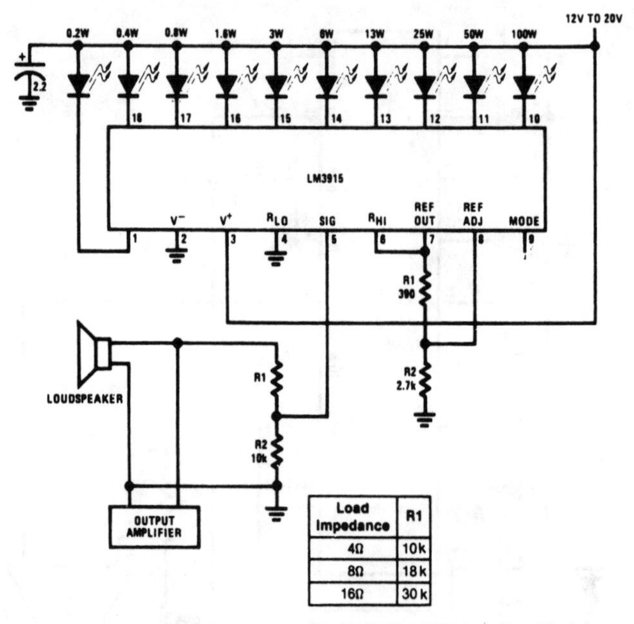

Load Impedance	R1
4Ω	10k
8Ω	18k
16Ω	30k

See Application Hints for optional Peak or Average Detector

NATIONAL SEMICONDUCTOR

Fig. 15-2

AUDIO POWER METER II

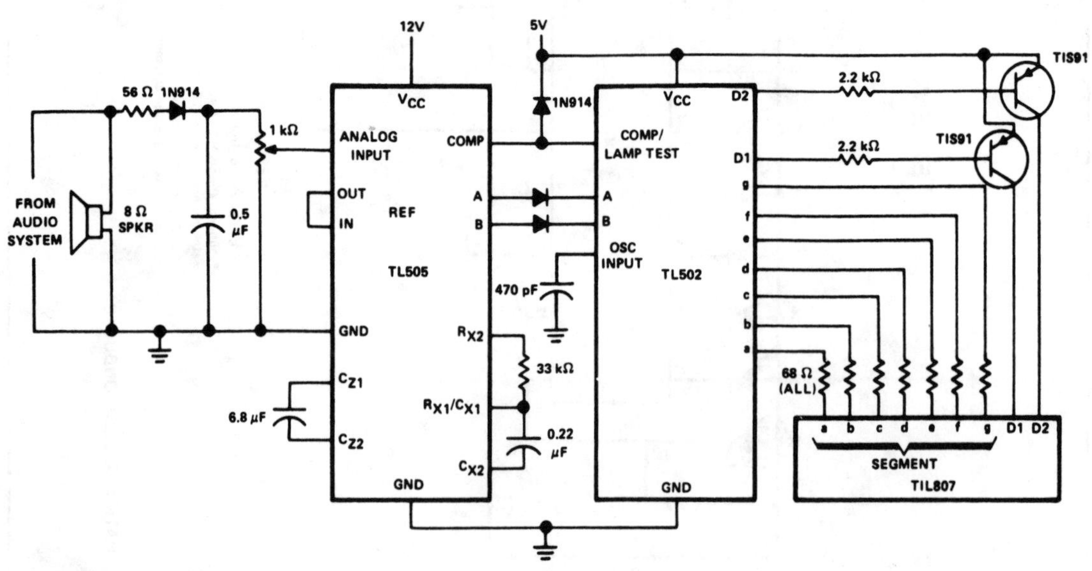

MOTOROLA

Fig. 15-3

POWER METER (1 kW FULL SCALE)

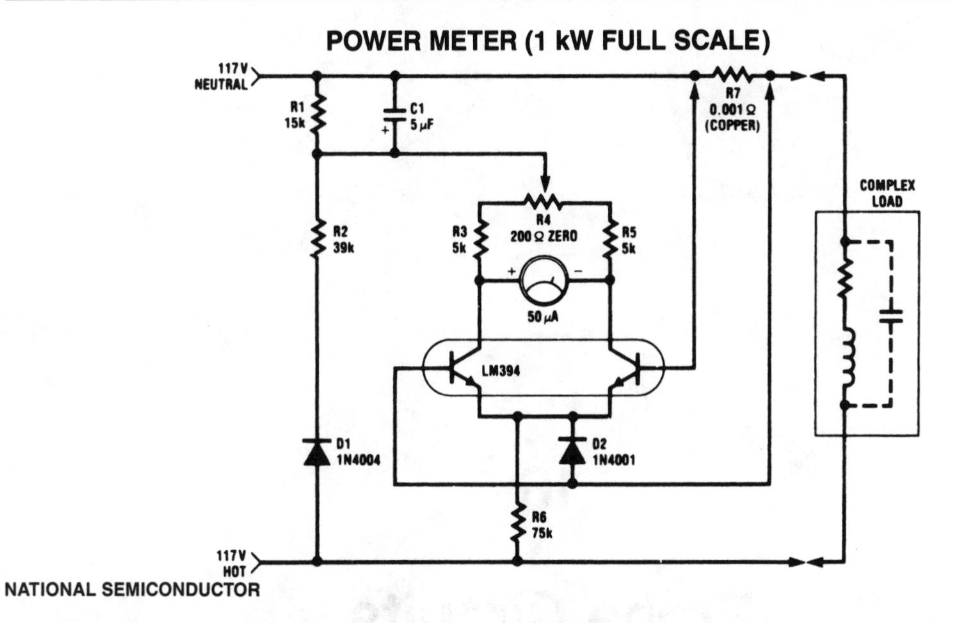

NATIONAL SEMICONDUCTOR

Fig. 15-4

The circuit is intended for 117 Vac ± 50 Vac operation, but it can be easily modified for higher or lower voltages. It measures true (nonreactive) power being delivered to the load and requires no external power supply. Idling power drain is only 0.5 W. Load current sensing voltage is only 10 mV, keeping load voltage loss to 0.01%. Rejection of reactive load currents is better than 100:1 for linear loads. Nonlinearity is about 1% full scale when using a 50-μA meter movement.

60-MHz POWER-GAIN TEST CIRCUIT

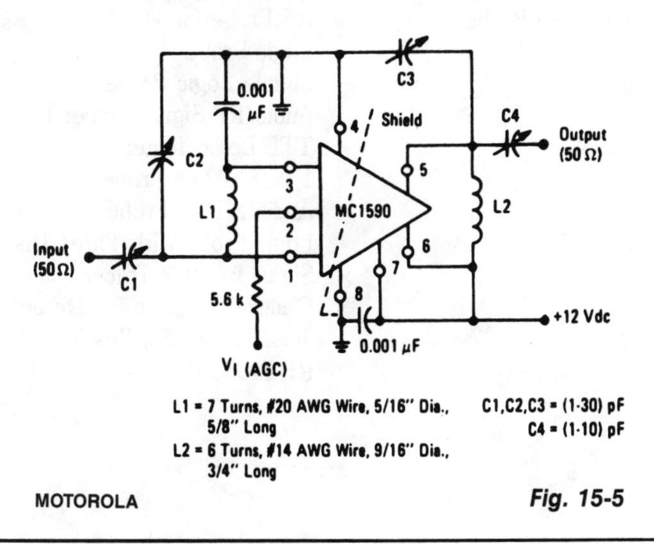

MOTOROLA

Fig. 15-5

16

Probe Circuits

The sources of the following circuits are contained in the Sources section, which begins on page 217. The figure number in the box of each circuit correlates to the source entry in the Sources section.

Digital Logic Probe
650-MHz Amplifying Prescaler Probe
Battery-Powered Ground-Noise Probe
Injector/Tracer
Low-Input Capacitance Buffer
RF Probe I
CMOS Universal Logic Probe
4- to 220-V Test Probe
FET Probe
pH Probe and Detector
Signal Injector/Tracer
Logic Probe
Logic Test Probe with Memory
Clamp-On Current-Probe Compensator
Microvolt Probe
CMOS Logic Probe
RF Probe for VOM
Logic Probe
Simple Logic Probe
Audio/RF Signal-Tracer Probe
TTL Logic Tester
1,000-MΩ dc Probe
Audible TTL Probe
Logic Probe with Three Discrete States
Signal Injector-Tracer
General Purpose RF Detector
Tone Probe (For Testing Digital ICs)
RF Probe II

DIGITAL LOGIC PROBE

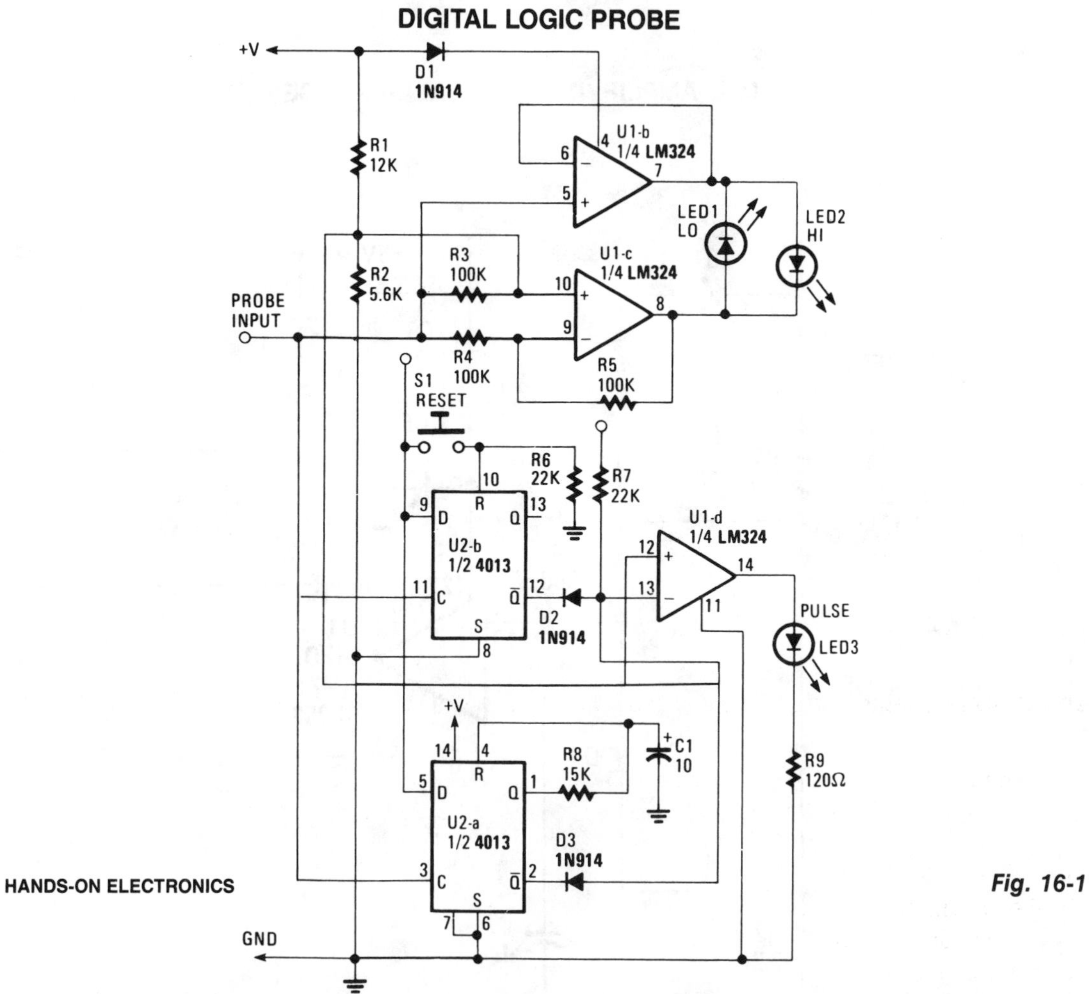

HANDS-ON ELECTRONICS

Fig. 16-1

The probe relies on the power supply of the CUT (circuit under test). The input to the probe, at probe tip, is fed along two paths. One path flows to the clock inputs of U2a and U2b. The other path feeds both the inverting input of U1c, which is set up as an inverting-mode integrator, and the noninverting input of U1b, which is configured as a noninverting unity-gain amplifier, in a logic-low state.

That low, below the reference set at pin 10, causes U1b's output at pin 7 to become high. With Ub1 outputting low and U1c outputting high, LED1 is forward-biased, and lights. LED2, reverse-biased, remains dark. Suppose that the logic level on the same pin becomes high. That high is applied to pin 5 of U1b, causing its output to be high. LED2 is now forward-biased and lights, while LED1 is reverse-biased and becomes dark.

Assume that a clock frequency is sensed at the probe input; LED1 and LED2 alternately light, and depending on the frequency of the signal, can appear constantly lit. That frequency, which is also applied to the clock input of both flip-flops, causes the *Q* outputs of U2a and U2b to simultaneously alternate between high and low. Each time that the *Q* outputs of the two flip-flops decrease, the output of U1d increases, lighting LED3, indicating that a pulse stream has been detected.

650-MHz AMPLIFYING PRESCALER PROBE

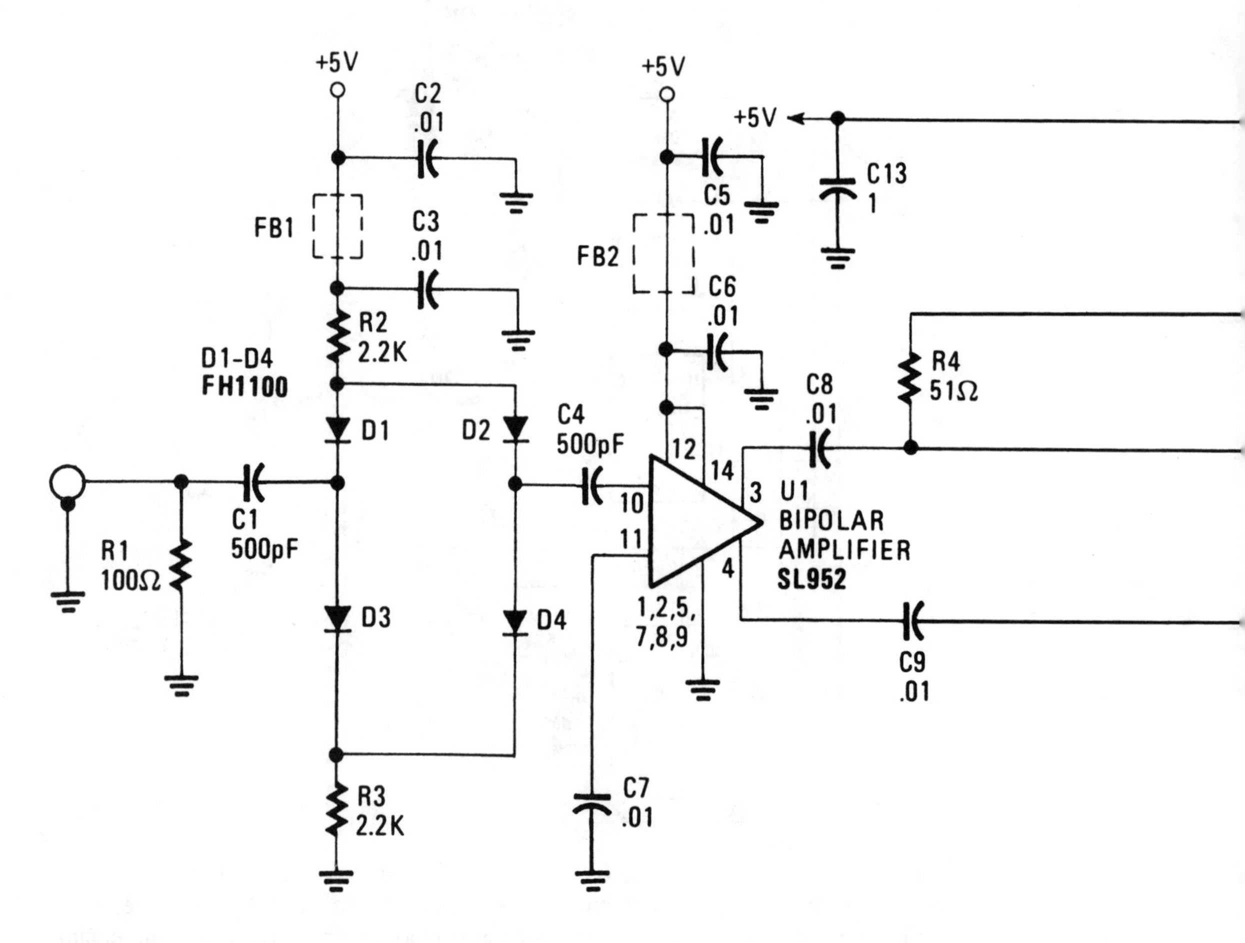

HANDS-ON ELECTRONICS *Fig. 16-2*

The 650-MHz prescaler probe's input is terminated by resistor R1 and is fed through C1 to the diode limiter (composed of D1 through D4). Those diodes are forward-biased by the +5-V supply for small-input signals and, in turn, they feed the signal to U1. However, for larger input signals, diodes D1 through D4 will start to turn off, passing less of the signal, and thus, attenuating it. But even in a full off state, the FH1100-type diodes will always pass a small part of the input to U1 because of capacitive leakage within the diodes. Integrated circuit U1, a Plessey SL952 bipolar amplifier, capable of 1-GHz operation, provides 20 to 30 dB of gain. The input signal is supplied to pin 10, U1 with the other input (pin 11) is bypassed to ground. The output signal is taken at pin 3 and pin 4, with pin 3 loaded by R4 and pin 4 by R5.

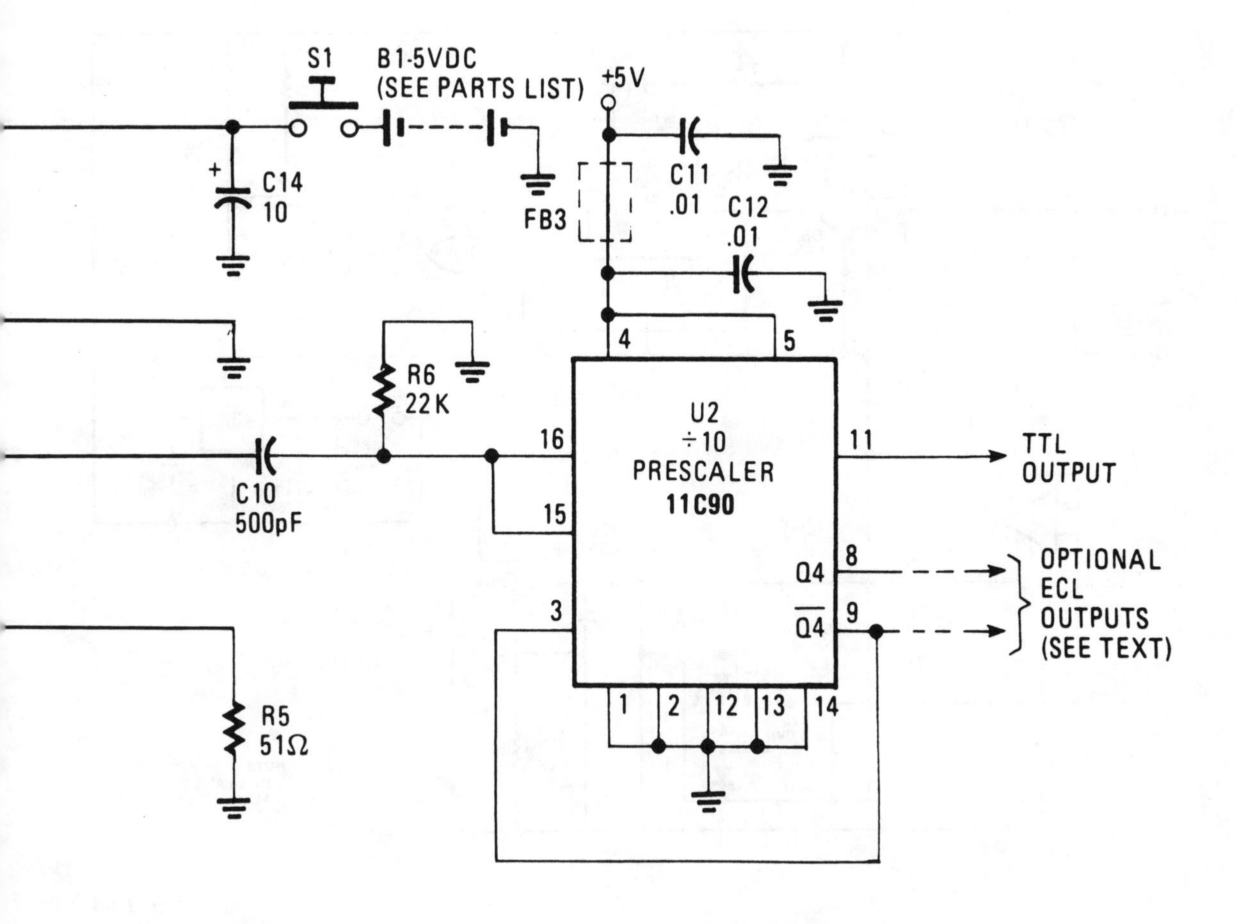

Integrated circuit 11C90, U2, is a high-speed prescaler that is capable of 650-MHz operation, configured for a divide-by-10 format. A reference voltage (internally generated) appears at pin 15 and is tied to pin 16, the clock input. This centers the capacitive-coupled input voltage from U1 around the switching threshold-voltage level. An ECL-to-TTL converter in U1 provides level conversion to drive TTL input counters by typing pin 13 low. Therefore, no external ECL-to-TTL converter is required at the pin 11 output. On the other hand, ECL outputs are available at U2, pin 8 (Q4) and at pin 9 (Q4), if desired. In that circuit configuration, pin 13 is left open, and U2 will use less power.

BATTERY-POWERED GROUND-NOISE PROBE

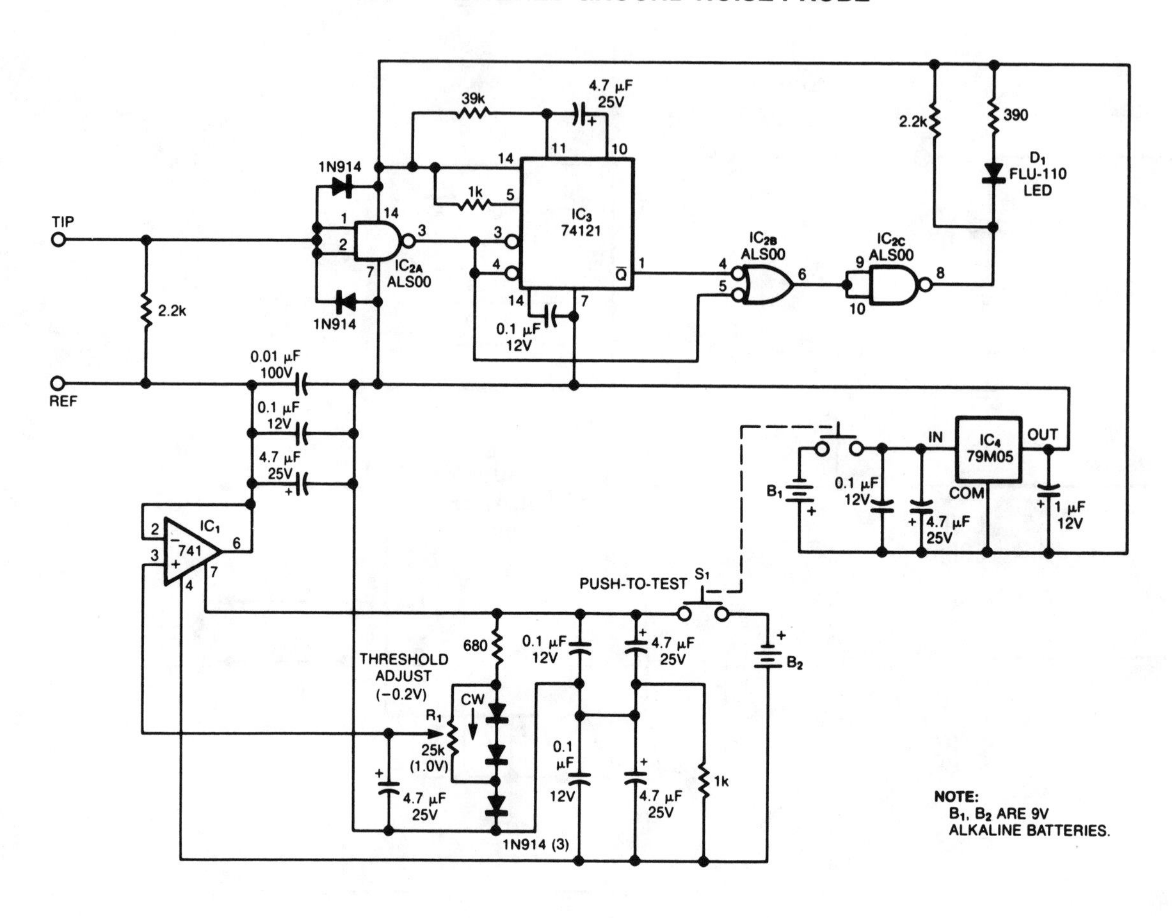

EDN *Fig. 16-3*

Oscilloscope measurements of ground noise can be unreliable because noise can enter your circuit via the scope's three-pronged power plug. You can avoid this problem by using this ground-noise tester. Powered by two 9-V batteries, the circuit dissipates power only while push-to-test switch S1 is depressed. Noise pulses that reach IC2A's switching threshold of about 1.5 to 1.8 V create a logic transition that triggers the monostable multivibrator (IC3), which stretches the pulse to produce a visible blink from LED D1. You set the noise reference level by adjusting threshold-adjust potentiometer R1, which lets the circuit respond to minimum pulse amplitudes ranging from about 0 to 1 V. For convenience, you can use a one-turn potentiometer for R1 and calibrate the dial by applying an adjustable dc voltage, monitored by an accurate voltmeter.

INJECTOR/TRACER

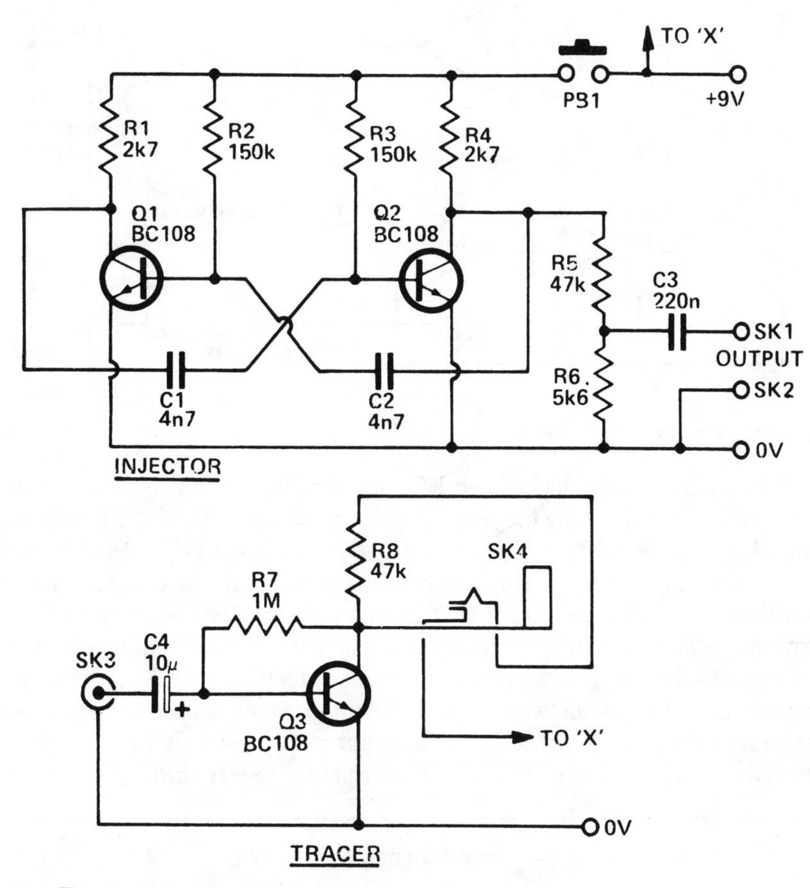

The circuit diagrams for both parts of the injector/tracer. Note that SK4 is used to apply power to the amplifier section.

ELECTRONICS TODAY INTERNATIONAL **Fig. 16-4**

The unit has a separate amplifier and oscillator section, which allows them to be used separately, if need be. The injector is a multivibrator running at 1 kHz, with R5 and R6 dividing down the output to a suitable level (≈ 1 V). The tracer is a single-stage amplifier that drives the high-impedance earpiece. C4 decouples the input.

LOW-INPUT CAPACITANCE BUFFER

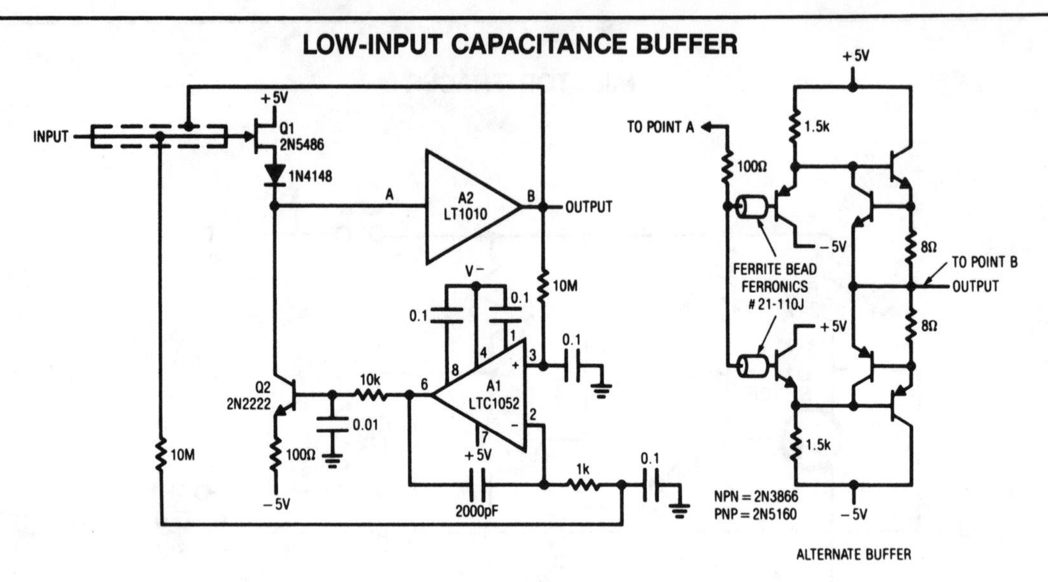

LINEAR TECHNOLOGY CORP. *Fig. 16-5*

Q1 and Q2 constitute a simple, high-speed FET input buffer. Q1 functions as a source follower, with the Q2 current-source load setting the drain-source channel current. The LT1010 buffer provides output-drive capability for cables or for whatever load is required. The LTC1052 stabilizes the circuit by comparing the filtered circuit output to a similarly filtered version of the input signal. The amplified difference between these signals is used to set Q2's bias, and hence Q1's channel current. This forces Q1's V_{GS} to whatever voltage is required to match the circuit's input and output potentials. The diode in Q1's source line ensures that the gate never forward biases and the 2000-pF capacitor at A1 provides stable loop compensation. The RC network in A1's output prevents it from seeing high-speed edges coupled through Q2's collector-base junction. A2's output is also fed back to the shield around Q1's gate lead, bootstrapping the circuit's effective input capacitance to less than 1 pF.

RF PROBE I

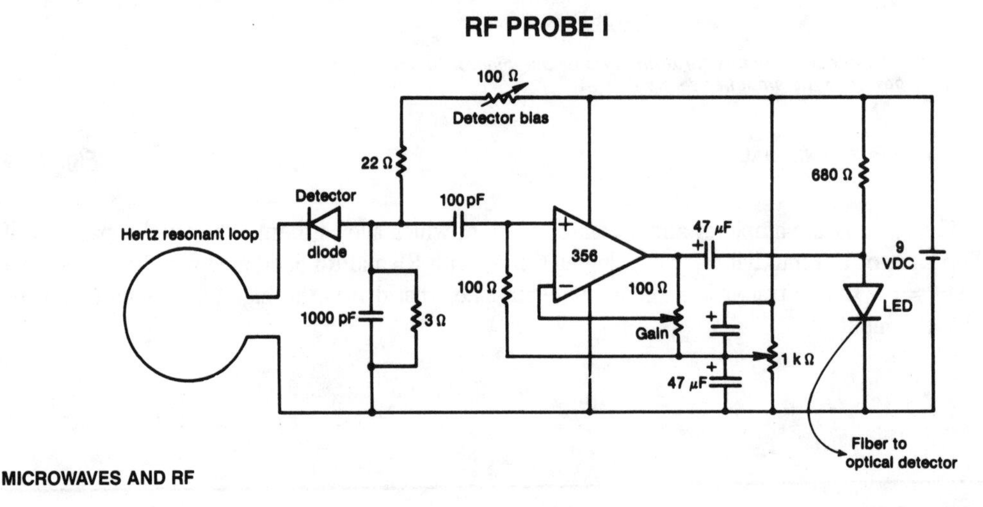

MICROWAVES AND RF *Fig. 16-6*

RF PROBE I *(Cont.)*

This RF probe is coupled with a fiber-optic cable to the test equipment. It utilizes inexpensive components to improve the probe performance at UHF frequencies. The receiving antenna in this probe feeds an envelope-detector diode. After amplification by the LF356 op amp, the low-frequency output modulates the LED, which in turn feeds the optical fiber. The design facilitates the use of a single battery for the op amp, with voltage splitting by means of the 1-kΩ potentiometer and the miniature 47-μF tantalum capacitors to provide decoupling. The gain control is easily adjusted to give the best dynamic range for a specific LED.

CMOS UNIVERSAL LOGIC PROBE

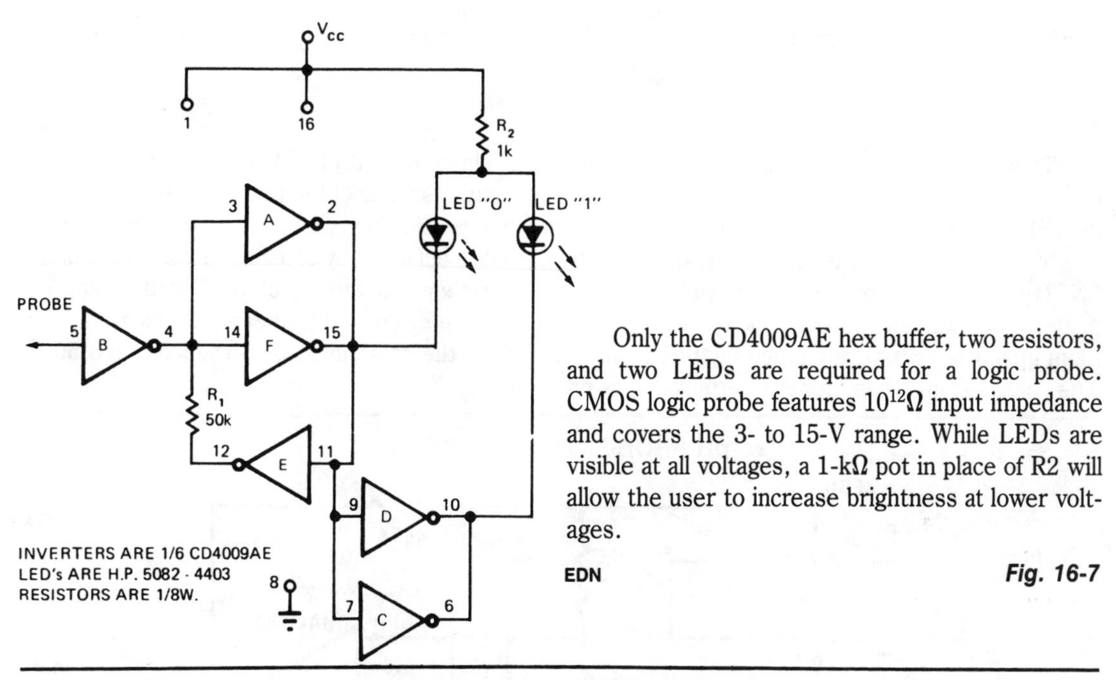

Only the CD4009AE hex buffer, two resistors, and two LEDs are required for a logic probe. CMOS logic probe features $10^{12}\Omega$ input impedance and covers the 3- to 15-V range. While LEDs are visible at all voltages, a 1-kΩ pot in place of R2 will allow the user to increase brightness at lower voltages.

EDN *Fig. 16-7*

4- TO 220-V TEST PROBE

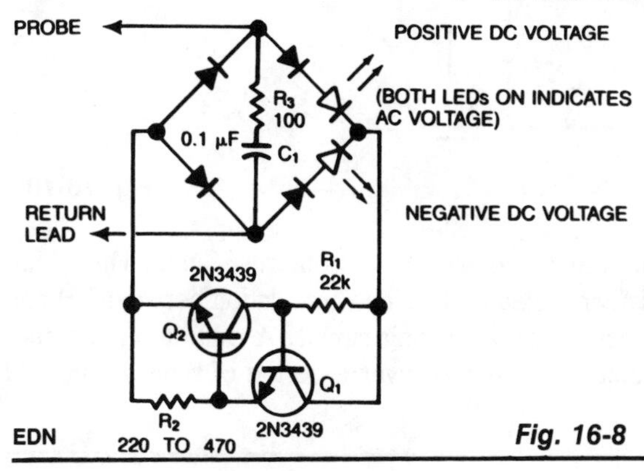

Fig. 16-8

Using inexpensive components, you can fit a simple probe circuit into a pencil-sized enclosure. When both LEDs are on, the probe indicates the presence of an ac voltage; either LED alone indicates the presence and polarity of a dc voltage. The diode-bridge arrangement allows one-way current source R1, R2, Q1, and Q2 to light either LED (or both) when the probe is activated by a test voltage. Diodes provide the necessary peak-inverse voltage rating; R3 and C1 provide a spike-suppression network to protect the current-source transistors.

FET PROBE

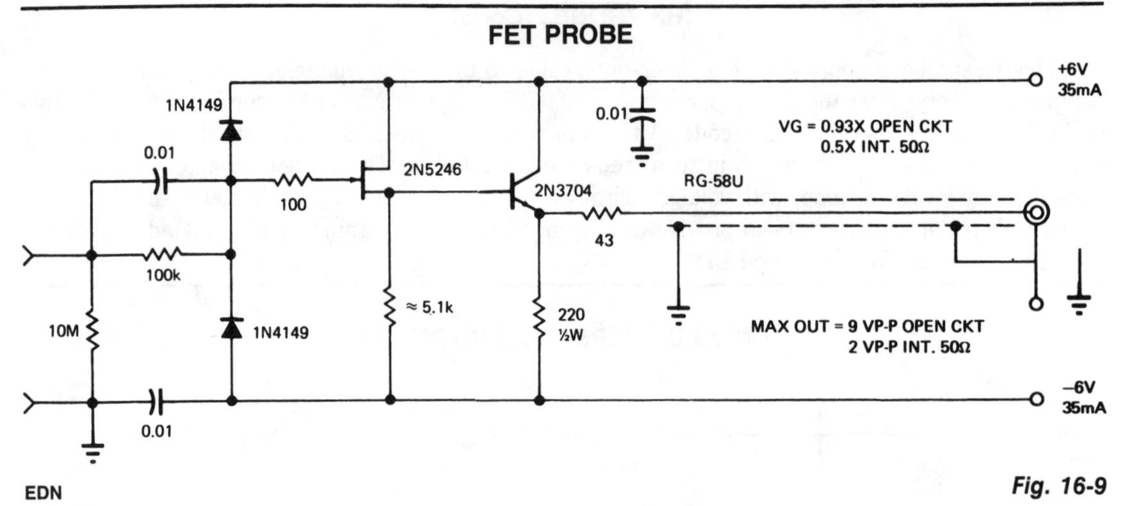

EDN

Fig. 16-9

This FET probe has an input impedance of 10 MΩ shunted by 8 pF. Eliminating the protective diodes reduces the capacitance to about 4 pF. The frequency response of the probe extends from dc to 20 MHz (−1 dB), although higher frequency operation is possible through optimized construction and use of a UHF-type transistor. Zero dc offset at the output is achieved by selecting a combination of a 2N5246 and a source resistor, which yields a gate-source bias equal to the V_{BE} of the 2N3704 at approximately 0 V. At medium frequencies, the probe can be used unterminated for near-unity gain; for optimum impedance converter probe high-frequency response, the cable must be terminated into 50 Ω. The voltage gain, when properly terminated, is precisely 0.5 *X*.

pH PROBE AND DETECTOR

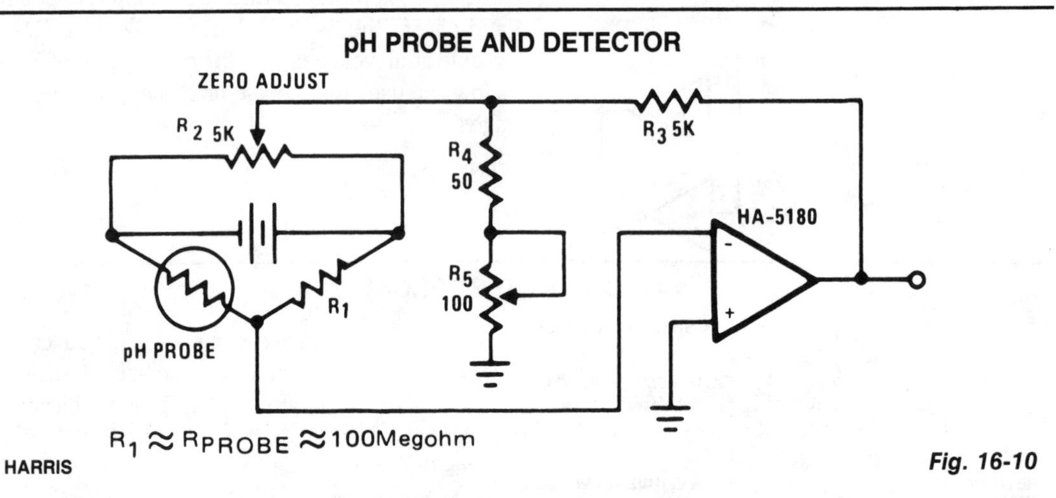

HARRIS

Fig. 16-10

The greatest sensitivity is achieved if R_1 is approximately equal to the probe resistance. The circuit can be *zeroed* with R2, while the full-scale voltage is controlled by R5. The correlation between pH and output voltage might not be linear, which would necessitate a shaping circuit. A calibration scheme, using solutions of known pH, might be adequate and more reliable over a period of time because of probe variance.

SIGNAL INJECTOR/TRACER

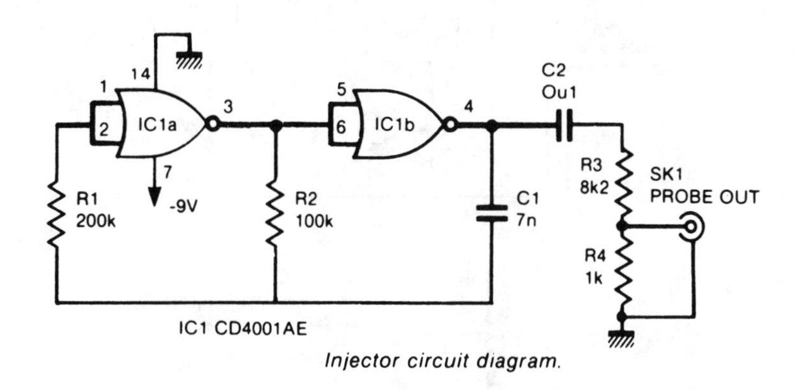

Injector circuit diagram.

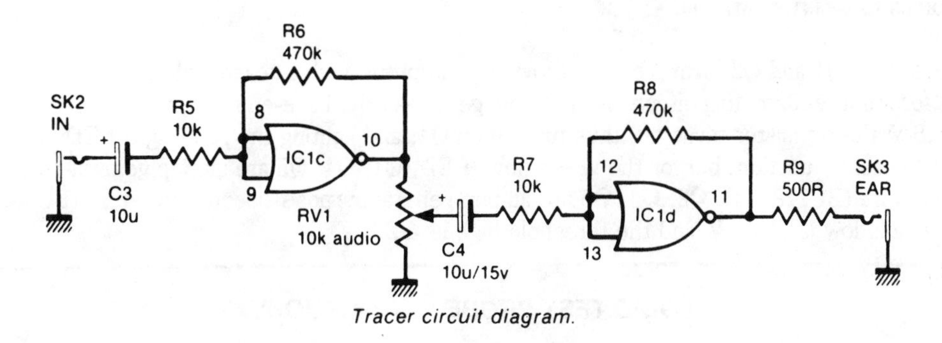

Tracer circuit diagram.

CANADIAN PROJECTS ***Fig. 16-11***

The injector is a CMOS oscillator with period approximately equal to $1.4 \times C_1 \times R_2$ seconds. The values are given for 1-kHz operation. Resistors R3 and R4 divide the output to 1 V. Whereas the oscillator employs the gates in their digital mode, the tracer used them in a linear fashion by applying negative feedback from output to input. They are used in much the same way as op amps. The circuit uses positive ground. It offers an advantage at the earphone output because one side of the earphone must be connected to ground via the case. Use of a positive ground allows the phone to be driven by the two n-channel transistors inside the CD4001, which are arranged in parallel and are thus able to handle more current for better volume.

LOGIC PROBE

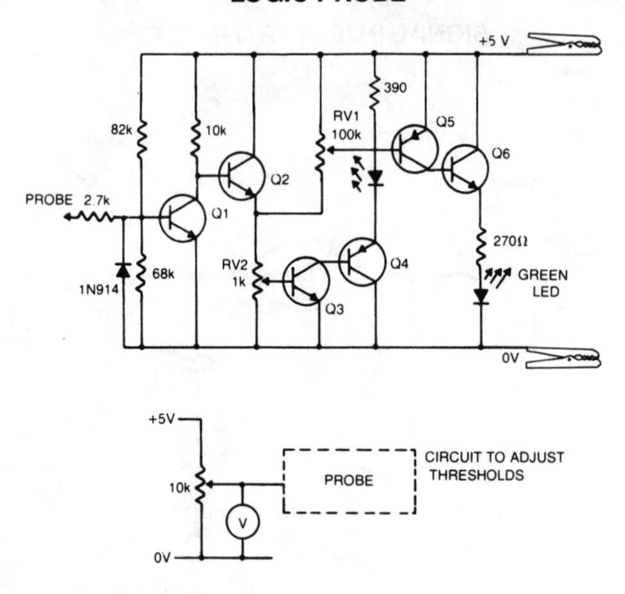

ELECTRONICS TODAY INTERNATIONAL *Fig. 16-12*

Transistors Q1 and Q2 form a buffer, providing the probe with a reasonable input impedance. Q3 and Q4 form a level detecting circuit. As the voltage across the base-emitter junction of the Q3 rises above 0.6 V, the transistor turns on, thus turning on Q4, and lighting the red (high) LED. Q5 and Q6 perform the same function, but for the green (low) LED. Q1, Q4, Q5 are all pnp general-purpose silicon transistors (BC178 etc). Q2, Q3, Q6 are all pnp general purpose silicon transistors (BC 108 etc.) The threshold low is ≤0.8 V, and the threshold high is ≥2.4 V.

LOGIC TEST PROBE WITH MEMORY

There are two switches: a memory disable switch and a pulse polarity switch. Memory disable is a pushbutton that resets the memory to the low state when it is depressed. Pulse polarity is a toggle switch that selects whether the probe responds to a high-level or pulse (+5 V) or a low-level or pulse (ground). Use IC logic of the same type as is being tested.

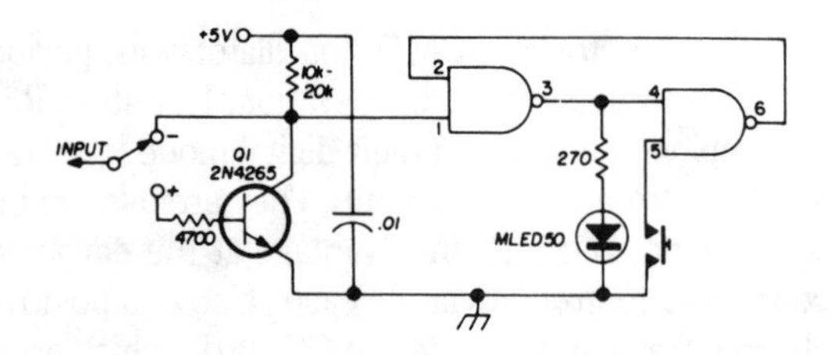

HAM RADIO *Fig. 16-13*

CLAMP-ON CURRENT-PROBE COMPENSATOR

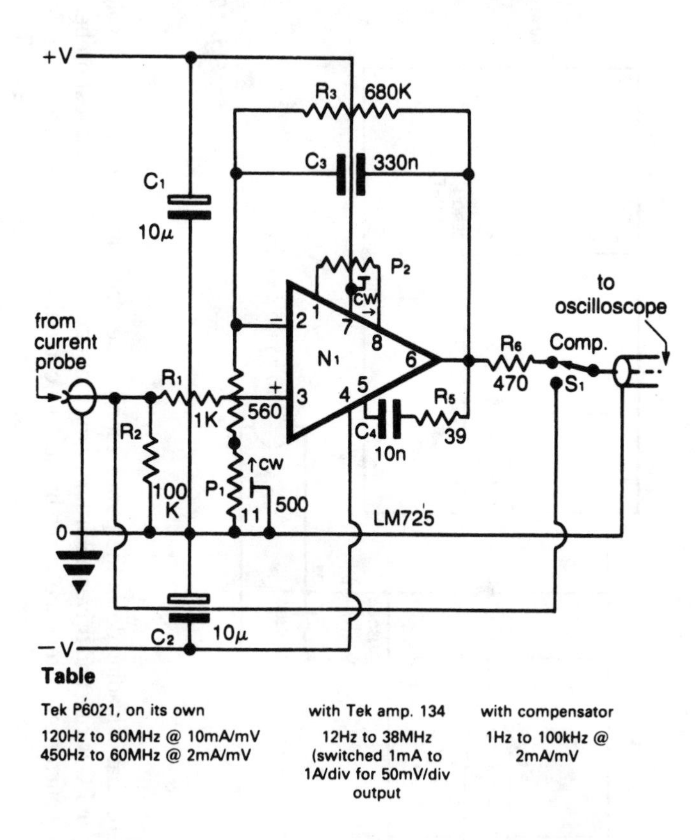

Table

Tek P6021, on its own	with Tek amp. 134	with compensator
120Hz to 60MHz @ 10mA/mV 450Hz to 60MHz @ 2mA/mV	12Hz to 38MHz (switched 1mA to 1A/div for 50mV/div output	1Hz to 100kHz @ 2mA/mV

ELECTRONIC ENGINEERING *Fig. 16-14*

A clamp-on "current probe," such as the Tektronix P6021, is a useful means of displaying current waveforms on an oscilloscope. Unfortunately, the low-frequency response is somewhat limited, as shown in the table.

The more sensitive range on the P6021 is 2 mA/mV, but it has a roll-off of 6 dB per octave below 450 Hz. The compensator counteracts the low-frequency attenuation, and this is achieved by means of C_3 and R_4+P_1 in the feedback around op amp N1. The latter is a low-noise type, such as the LM725 shown; even so, it is necessary at some point to limit the increasing gain with decreasing frequency. Otherwise, amplifier noise and drift will overcome the signal. The values shown for C_3R_3 give a lower limit below 1 Hz. A test square wave of ± 10 mA is fed to the current probe so that P1 can be adjusted for minimum droop or overshoot in the output waveform. At high frequencies, the response begins to fall off at 100 kHz.

MICROVOLT PROBE

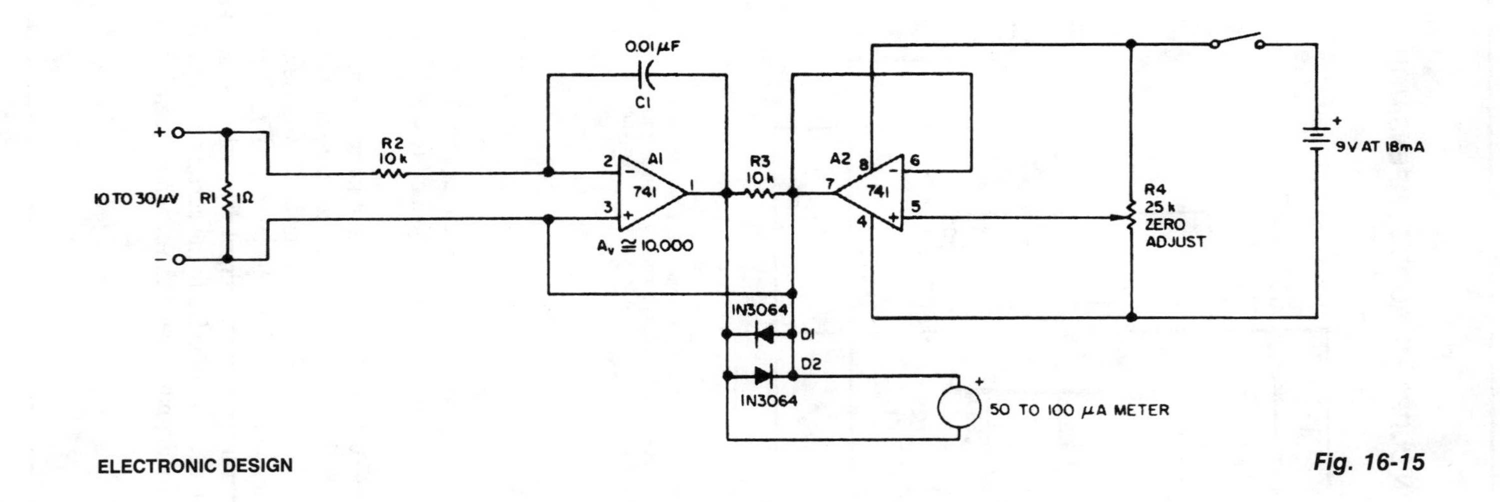

ELECTRONIC DESIGN

Fig. 16-15

The current tracer helps locate a defective IC that is loading down the power supply. The tracer amplifies the small voltage drop caused by current flow along a fraction of an inch of PC wiring and drives an ordinary microammeter. Needle-point test probes are used to contact the edge of a PC trace and to follow the current to determine which branch the current takes. One-half of a dual 741 op amp forms a dc amplifier with ac feedback to prevent oscillations and hum-pickup problems. It drives a 50- to 100-μA meter. The other op amp provides a center tap for the 9-V battery supply and zero adjustment with R4. Two diodes protect the meter. Resistor R1 eliminates the necessity for shorting the probes when the meter is zeroed. The value of 1 Ω is large when compared with the resistance of the meter leads, plus the bridged portion of PC wiring.

CMOS LOGIC PROBE

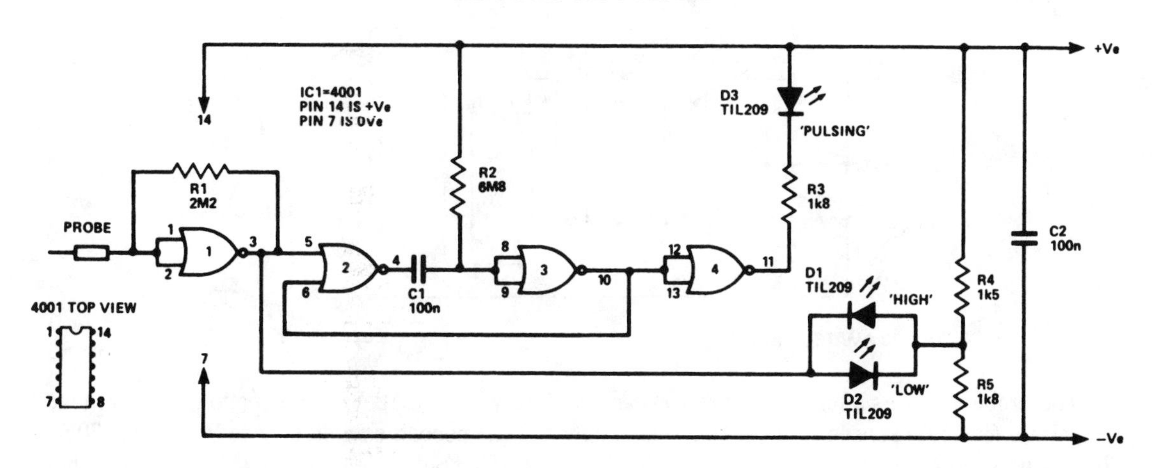

ELECTRONICS TODAY INTERNATIONAL

Fig. 16-16

The logic probe can indicate four input states, as follows: floating input—all LEDs off; logic 0 input—D2 switched on (D3 will briefly flash on); logic 1 input—D1 switched on; pulsing input—D3 switched on, or pulsing in the case of a low-frequency input signal (one or both of the other indicators will switch on and show if one input state predominates).

RF PROBE FOR VOM

PARTS LIST FOR RF PROBE FOR VOM

C1—500-pF, 400-VDC capacitor
C2—0.001-uF, disc capacitor
D1—1N4149 diode
R1—15,000-ohm, ½-watt resistor

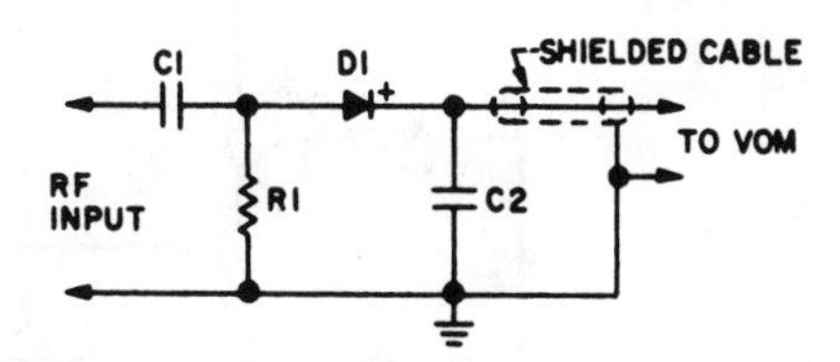

101 ELECTRONIC PROJECTS

Fig. 16-17

This probe makes possible relative measurements of RF voltages to 200 MHz on a 20,000 ohms-per-volt multimeter. RF voltage must not exceed the breakdown rating of the 1N4149 (approximately 100 V).

LOGIC PROBE

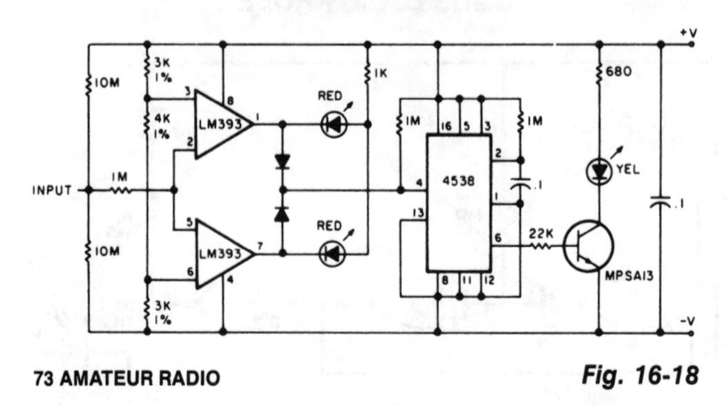

73 AMATEUR RADIO

Fig. 16-18

The probe indicates a high or low at 70% and 30% of V+ (5 to 12 V). One section of the voltage comparator (LM393) senses V in over 70% of supply and the second section senses V in under 30%. These two sections direct-drive the appropriate LEDs. The pulse detector is a CMOS one shot (MC14538), triggered on the rising edge of the LM393 outputs through 1N4148 diodes. With the RC values shown, it triggered reliably at greater than 30 kHz on both sine and square waves.

SIMPLE LOGIC PROBE

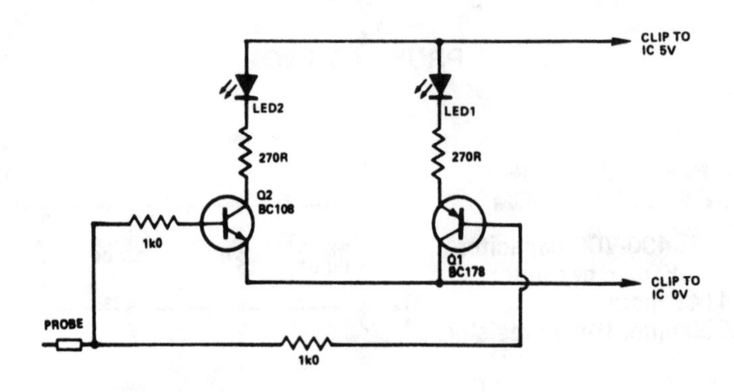

ELECTRONICS TODAY INTERNATIONAL

Fig. 16-19

If the probe is connected to logic 0, Q1 will be turned on and light D1. At logic 1, Q2 will be turned on and light D2. For Q1 and Q2, any npn or pnp transistors will do. Also, D1 and D2 can be any LED.

AUDIO/RF SIGNAL-TRACER PROBE

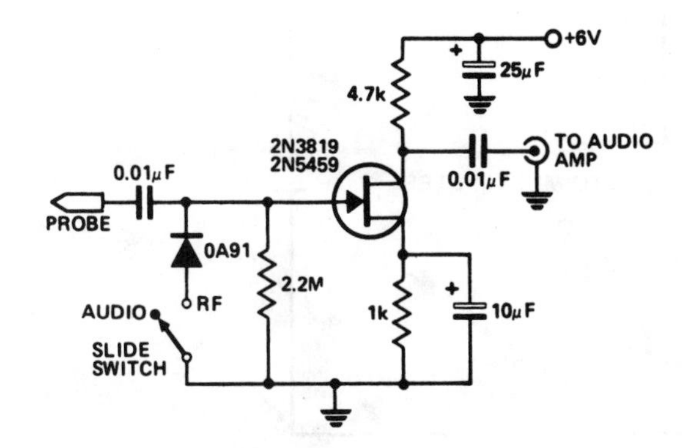

ELECTRONICS TODAY INTERNATIONAL *Fig. 16-20*

This economical signal tracer is useful for servicing and aligning receivers and low-power transmitters. When switched to RF, the modulation on any signal is detected by the diode and amplified by the FET. Use twin-core shielded lead to connect the probe to an amplifier and to feed 6 V to it.

TTL LOGIC TESTER

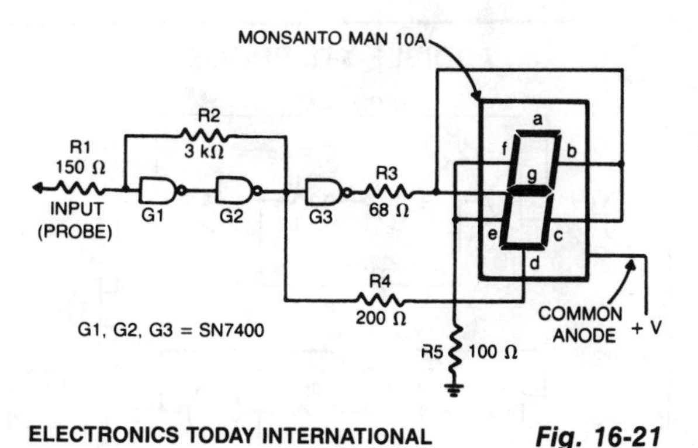

ELECTRONICS TODAY INTERNATIONAL *Fig. 16-21*

Gates G1 and G2 together with resistors R1 and R2 form a simple voltage monitor that has a trip point of 1.4 V. Gate G3 is simply an inverter. The display section of the tester consists of a common anode alphanumeric LED and current-limiting resistors. It indicates whether the input voltage is above or below 1.4 V, and displays an H or an L (for high or low logic-level) respectively.

1,000-MΩ dc PROBE

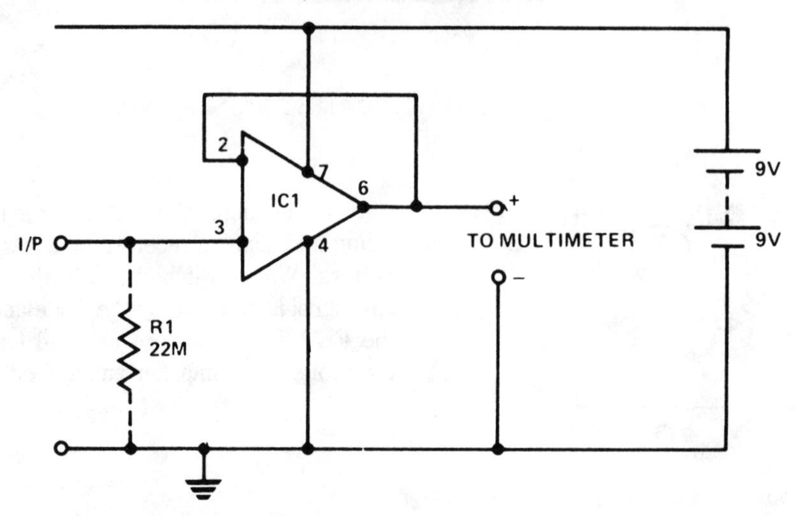

ELECTRONICS TODAY INTERNATIONAL *Fig. 16-22*

A 741 op amp is used with 100% ac and dc feedback to provide a typical input impedance of 10^{11} Ω and unity gain. To avoid hum and RF pickup, the input leads should be kept as short as possible and the circuit should be mounted in a small grounded case. Output leads can be long because the output impedance of the circuit is a fraction of an ohm. With no input the output level is indeterminate. Including R1 in the circuit though lowers the input impedance to 22 MΩ.

AUDIBLE TTL PROBE

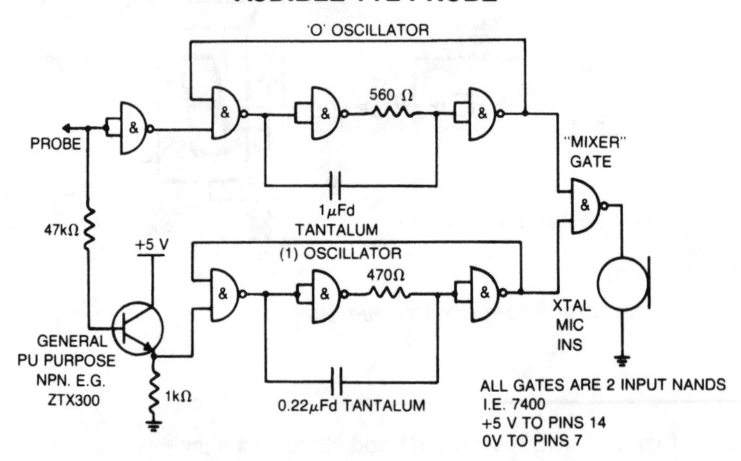

ELECTRONICS TODAY INTERNATIONAL *Fig. 16-23*

When the probe is in contact with a TTL low (0) the probe emits a low note. With a TTL high (1), a high note is emitted. Power is supplied by the circuit under test.

LOGIC PROBE WITH THREE DISCRETE STATES

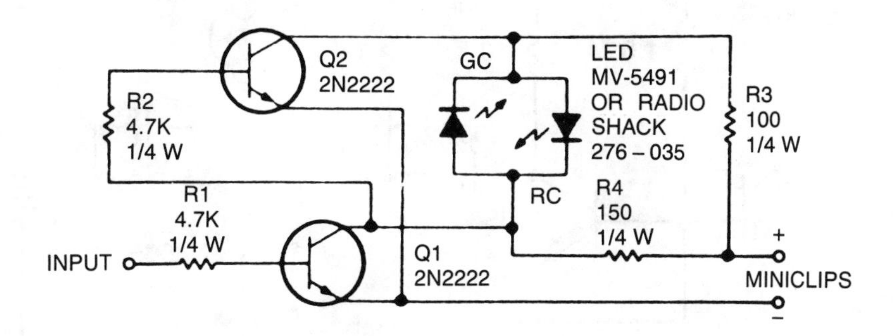

HAM RADIO *Fig. 16-24*

The circuit uses a dual LED. When power is applied to the probe through the power leads, and the input is touched to a low level or ground, Q1 is cut off. This will cause Q2 to conduct because the base is positive, with respect to the emitter. With Q1 cut off and Q2 conducting, the green diode of the dual LED will be forward biased, yielding a green output. Touching the probe tip to a high level will cause Q1 and Q2 to complement, and the red diode will be forward biased, yielding a red output from the LED. An alternating signal will cause alternating conduction of the red and green diodes and will yield an indication approximately amber. Thus, both static and dynamic signals can be traced with the logic probe.

SINGLE INJECTOR-TRACER

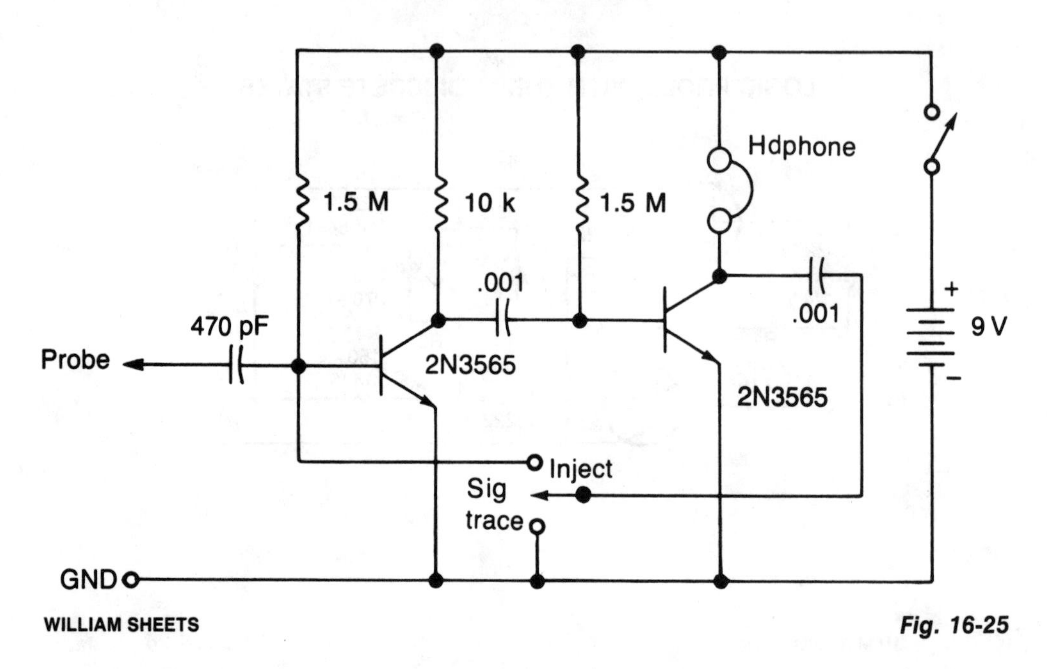

WILLIAM SHEETS *Fig. 16-25*

This circuit will provide a nominal square-wave output in the audio range in the "Inject" mode; the harmonics of which should be heard at several MHz. In the "Trace" mode, the nonlinear operation of the amplifier will detect modulated RF signals, which will be filtered by the 0.001-μF capacitor and heard in the headphones.

GENERAL-PURPOSE RF DETECTOR

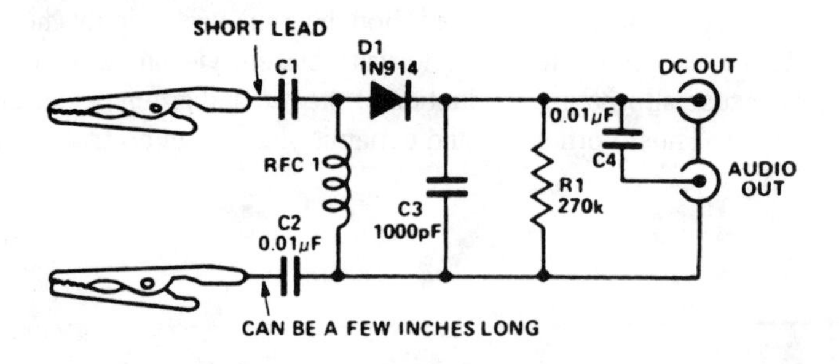

ELECTRONICS TODAY INTERNATIONAL *Fig. 16-26*

This circuit provides a dc output to a meter and an audio output (if necessary) for checking transmitters or modulated signals. It can be used also as a field-strength meter or transmitter monitor.

TONE PROBE (FOR TESTING DIGITAL ICs)

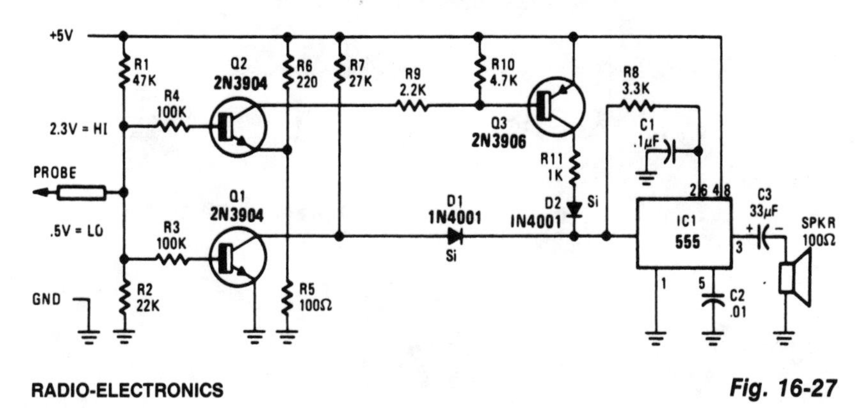

RADIO-ELECTRONICS *Fig. 16-27*

The tone probe uses sound to tell the status of the signal being probed. The probe's input circuit senses the condition of the signal and produces either a low-pitched tone for low-level signals (less than 0.8 V) or a high-pitched tone for high-level signals (greater than 2 V).

RF PROBE II

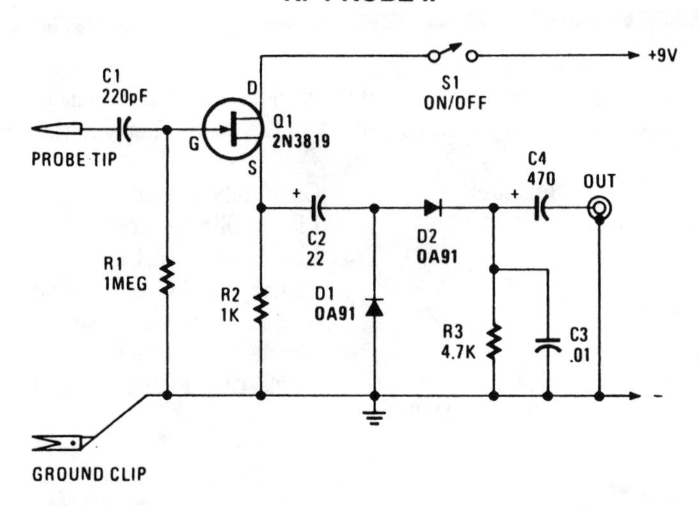

HANDS-ON ELECTRONICS *Fig. 16-28*

Transistor Q1, configured as a source-follower buffer stage, offering a bit under unity voltage gain, gives the unit a high-impedance input of about 1 MΩ shunted by about 10 pF, which puts only minimal loading on the equipment being tested. C1 serves as input dc blocking capacitor. The Q1 output is coupled by C2 to a simple AM detector circuit made up of D1, D2, R3 and C3. Capacitor C4 provides output dc blocking. Total current consumption should be somewhere around 1 mA. The circuit responds to frequencies from 100 kHz to well over 50 MHz.

17

Resistance and Continuity-Measuring Circuits

The sources of the following circuits are contained in the Sources section, which begins on page 217. The figure number in the box of each circuit correlates to the source entry in the Sources section.

Simple Ratiometric Resistance Measurer
Audio Continuity Tester
Linear-Scale Ohmmeter I
Bridge Circuit
Linear-Scale Ohmmeter II
Ohmmeter
Cable Tester
Adjustable, Audible Continuity Tester for Delicate
 Circuits

Continuity Tester
Linear Ohmmeter
Low-Resistance Continuity Tester
"Buzz-Box" Continuity and Coil Checker
Simple Continuity Tester for PCBs
Simple Continuity Tester
Single-Chip Resistance Checker

SIMPLE RATIOMETRIC RESISTANCE MEASURER

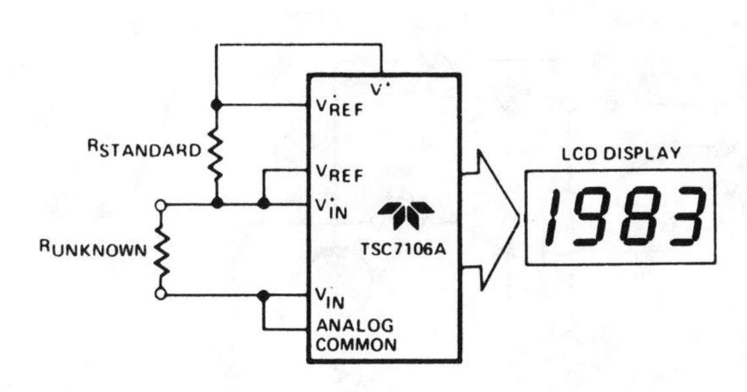

TELEDYNE

Fig. 17-1

The unknown resistance is put in series with a known standard and a current passed through the pair. The voltage developed across the unknown is applied to the input and the voltage across the known resistor applied to the reference input. If the unknown equals the standard, the display will read 1000. The displayed reading can be determined from the following expression:

$$\text{Displayed Reading} = \frac{R_{\text{unknown}}}{R_{\text{standard}}} \times 1000$$

The display will overrange for R_{unknown}, $\geq 2 \times R_{\text{standard}}$.

AUDIO CONTINUITY TESTER

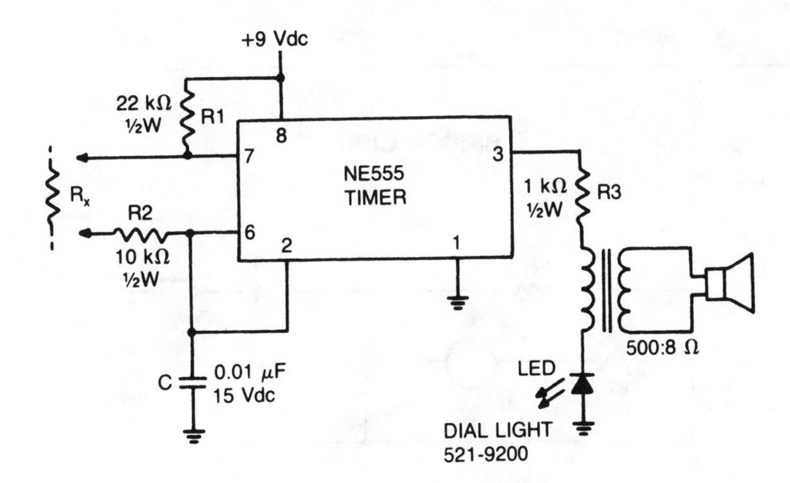

ELECTRONICS

Fig. 17-2

This low-current audio continuity tester indicates the unknown resistance value by the frequency of the audio tone. A high tone indicates a low resistance, and a tone of a few pulses per second indicates a resistance as high as 30 MΩ.

LINEAR-SCALE OHMMETER I

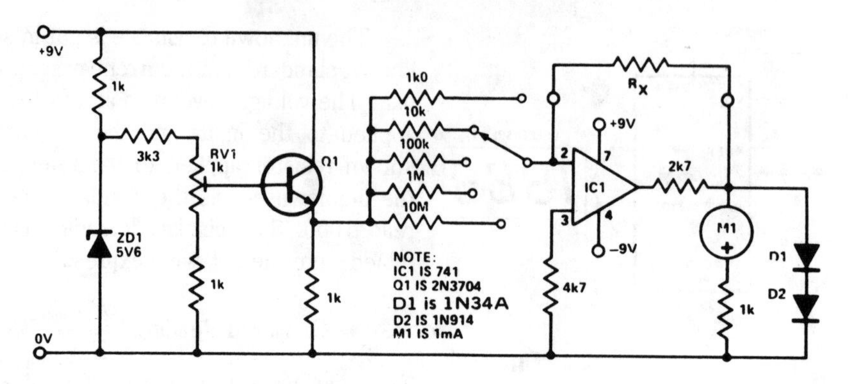

ELECTRONICS TODAY INTERNATIONAL ***Fig. 17-3***

One preset resistor is used for all the ranges to simplify the setup. Diode clamping is included to prevent damage to the meter if the unknown resistor is higher than the range selected. When the meter has been assembled, a 10-kΩ precision resistor is placed in the test position, R_x; the meter is set to the 10-kΩ range and RV1 is adjusted for full-scale deflection.

BRIDGE CIRCUIT

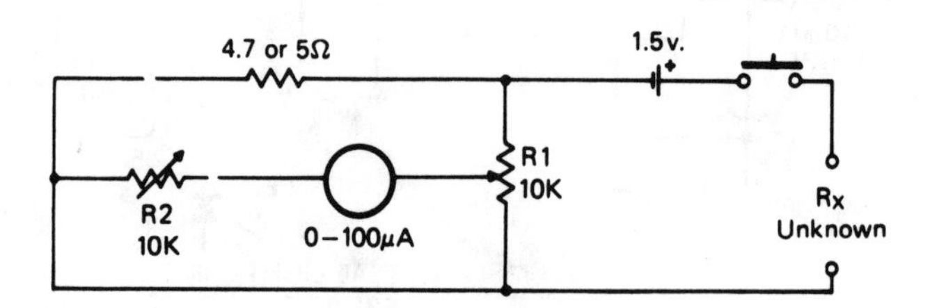

CQ ***Fig. 17-4***

This bridge circuit measures resistances from about 5 Ω down to about 0.1 Ω.

LINEAR-SCALE OHMMETER II

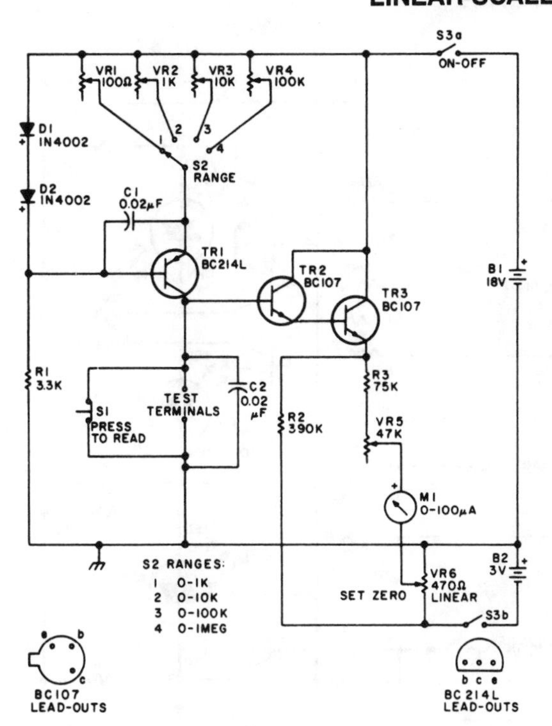

This circuit is designed to provide accurate measurement and a linear-resistance scale at the high end. The circuit has four ranges. Another meter with a current range of 10 µA to 10 mA and sensitivity of 10,000 Ω/V is needed for setting up.

73 AMATEUR RADIO *Fig. 17-5*

OHMMETER

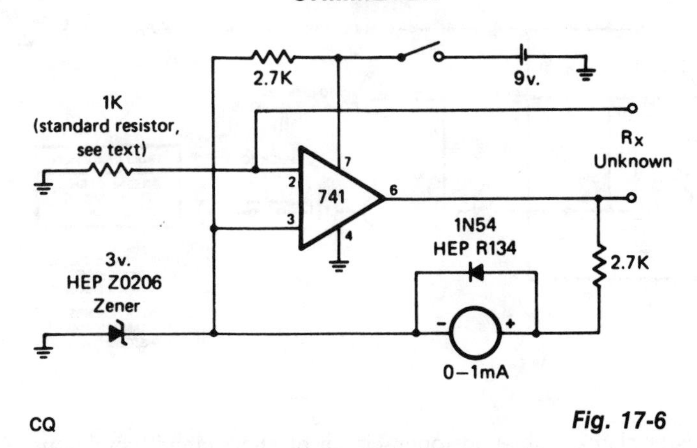

CQ *Fig. 17-6*

This circuit has a linear-reading scale, requires no calibration, and requires no zero adjustment. It can be made multirange by switching in different standard resistors.

CABLE TESTER

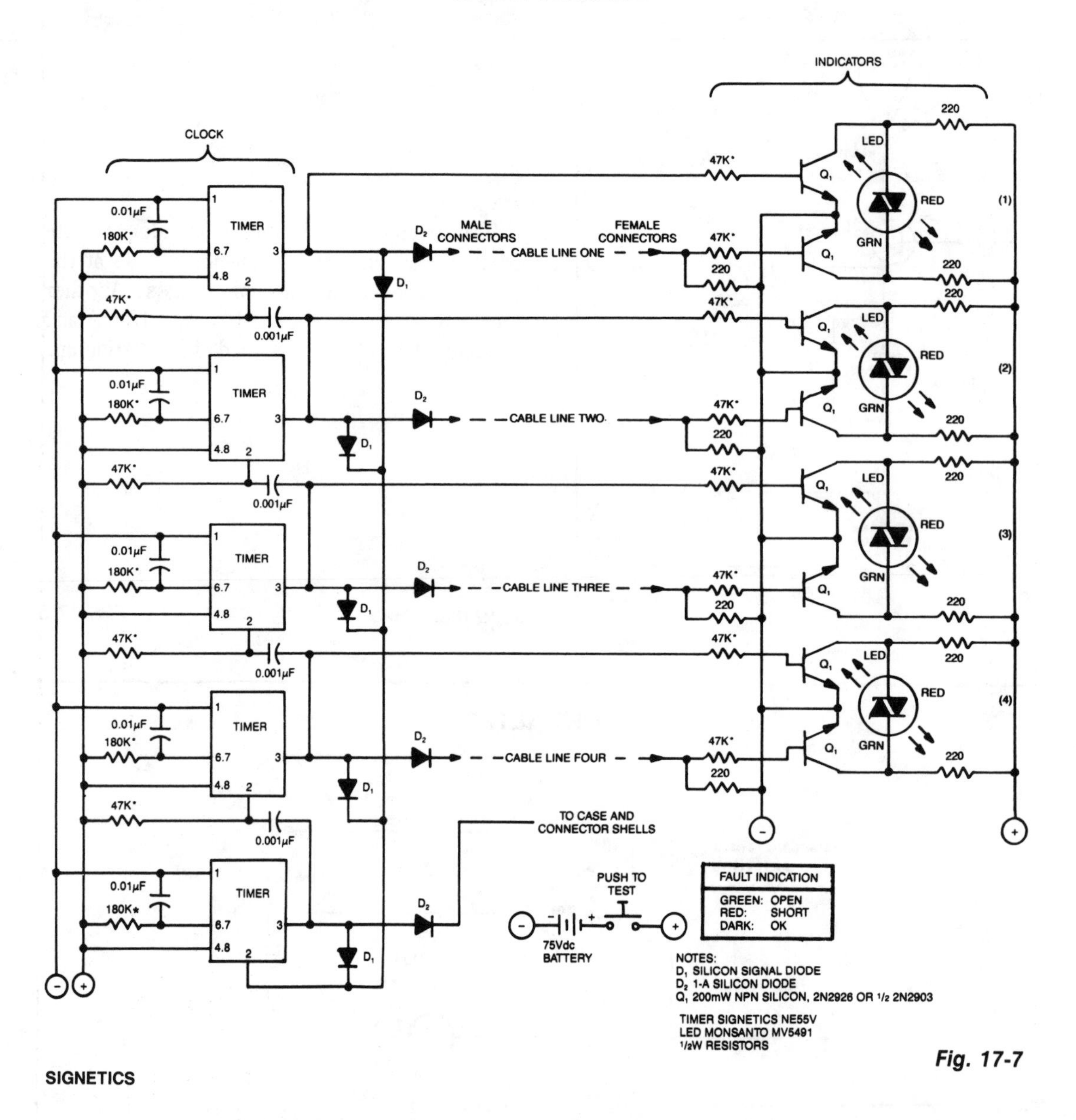

SIGNETICS

Fig. 17-7

This compact tester checks cables for open-circuit or short-circuit conditions. A differential transistor pair at one end of each cable line remains balanced as long as the same clock pulse generated by timer IC appears at both ends of the line. A clock pulse, just at the clock end of the line, lights a green LED, and a clock pulse, only at the other end, lights a red LED.

ADJUSTABLE, AUDIBLE CONTINUITY TESTER FOR DELICATE CIRCUITS

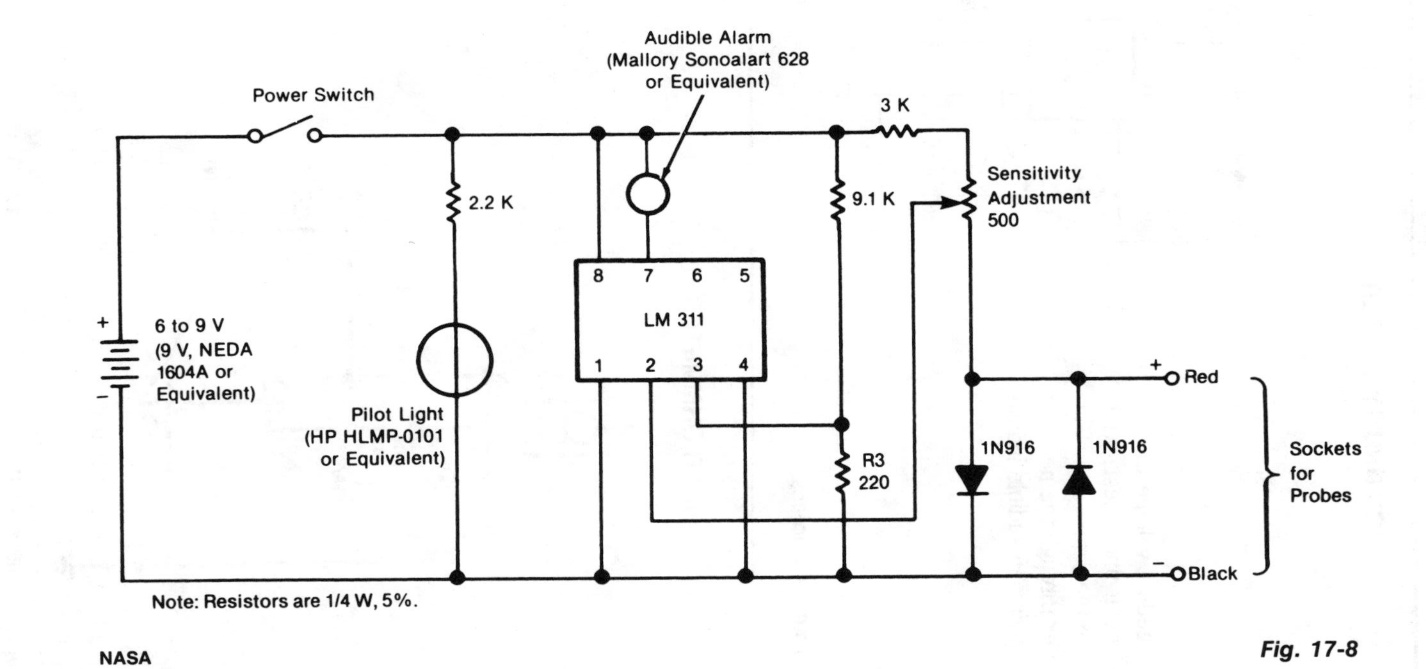

NASA

Fig. 17-8

The tester gives an audible indication, making it unnecessary for the user to look directly at the instrument to observe a meter reading. In addition, the current and voltage of the tester are strictly limited. It can apply no more than 0.6 Vdc and no more than 3 mA through the probes. It can therefore be used safely on circuit boards in which semiconductor components have been installed, and on complementary metal oxide/semiconductor integrated circuits, which are highly susceptible to damage during testing. The tester can be adjusted to indicate continuity below any resistance value up to 35 Ω. For example, if the user sets the tester to 30 Ω, the unit will emit an audible tone whenever the resistance between the probes is 30 Ω or less; if, for example, the resistance is 30.2 Ω, the unit will remain silent.

CONTINUITY TESTER

The continuity tester feeds a voltage through the positive probe to the circuit under test, while the negative probe serves as the return line. Voltage that returns to the tester through the negative probe triggers the circuit, giving an audible indication of continuity.

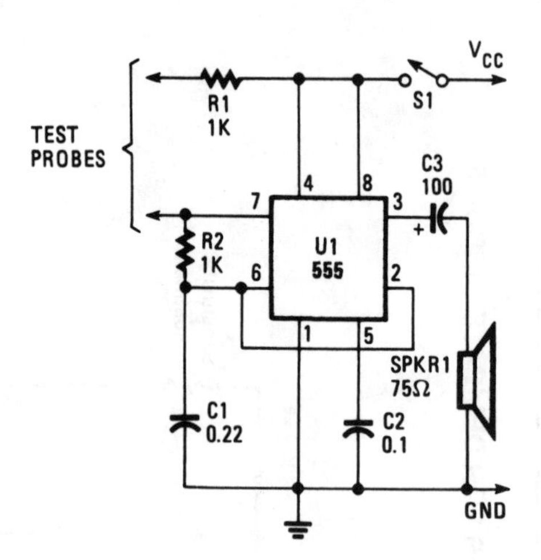

HANDS-ON ELECTRONICS/POPULAR ELECTRONICS

Fig. 17-9

LINEAR OHMMETER

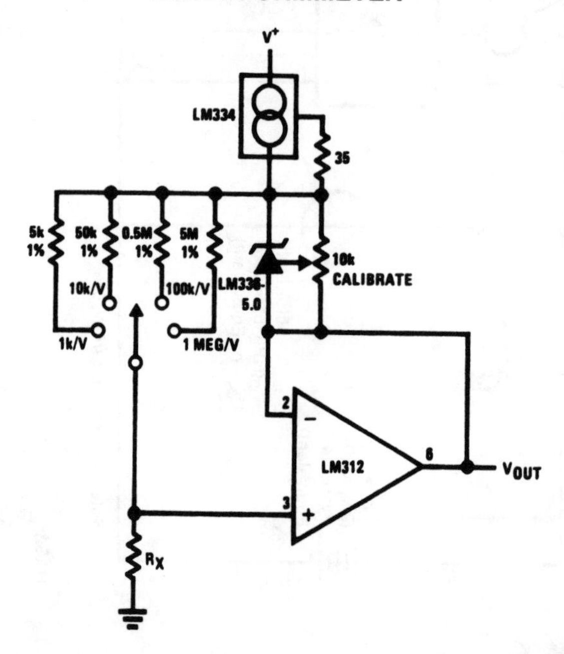

NATIONAL SEMICONDUCTOR CORP.

Fig. 17-10

LOW-RESISTANCE CONTINUITY TESTER

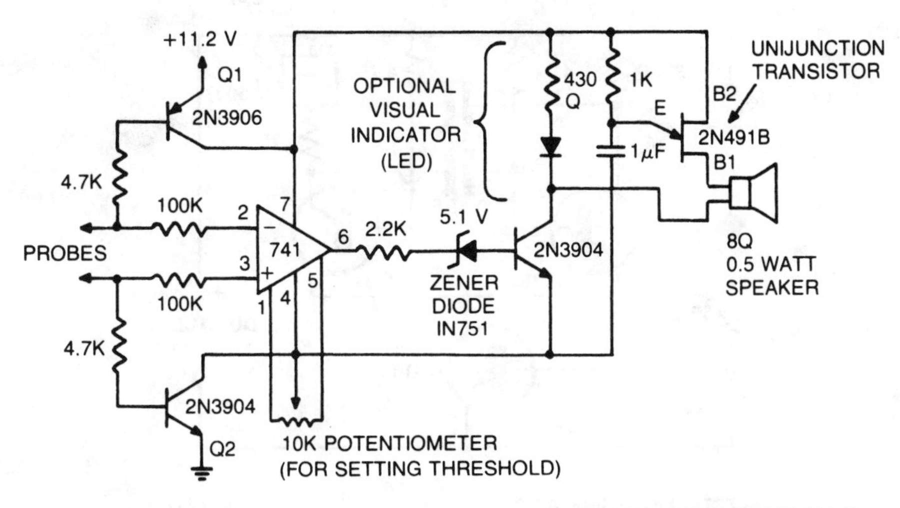

INSULATION/CIRCUITS *Fig. 17-11*

This tester can be used to check IC PC boards. Two 4.7-kΩ resistors and the transistors connected to them prevent current flow through the operational amplifier until the probe circuit is completed. The zener diode in series with the operational-amplifier output prevents audio-oscillator operation until the positive output of the operational amplifier has sufficient amplitude.

"BUZZ-BOX" CONTINUITY AND COIL CHECKER

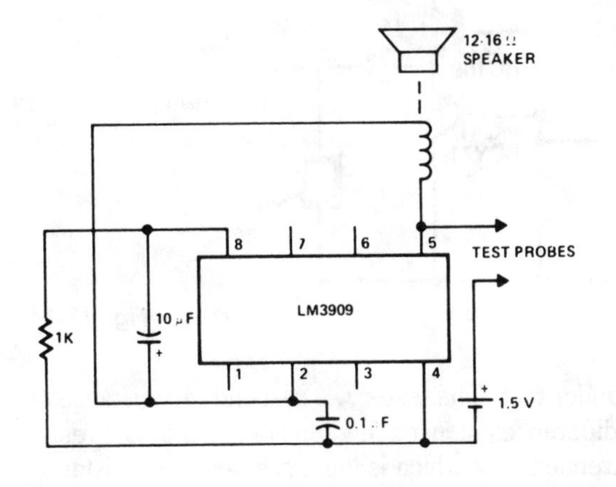

Differences between shorts, coils, and a few ohms of resistance can be heard.

SILICONIX *Fig. 17-12*

SIMPLE CONTINUITY TESTER FOR PCBs

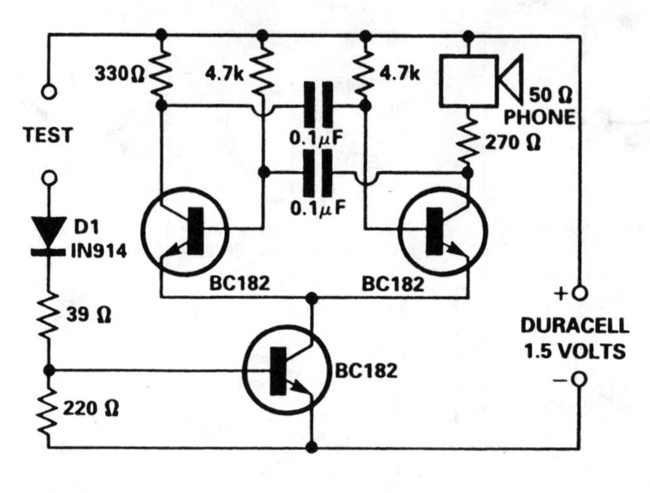

ELECTRONIC ENGINEERING *Fig. 17-13*

This tester is for tracing wiring on PC boards. Resistors below 50 Ω act as a short circuit; above 100 Ω as an open circuit. The circuit is a simple multivibrator switched on by transistor T3. The components in the base of T3 are D1, R1, R2, and the test resistance. A 1.5-V supply, has insufficient voltage to turn on a semiconductor connected to the test terminals.

SIMPLE CONTINUITY TESTER

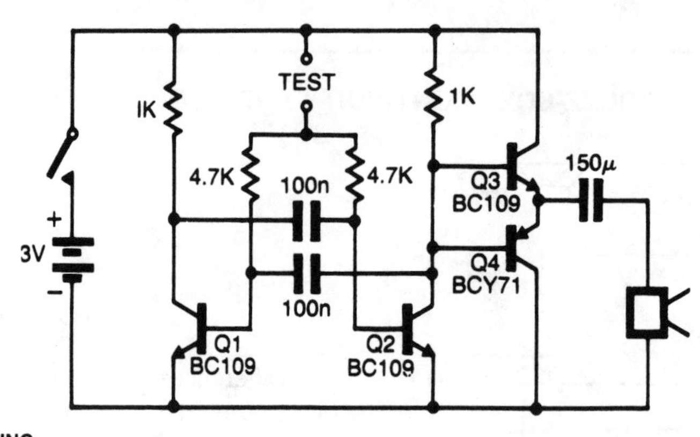

ELECTRONIC ENGINEERING *Fig. 17-14*

The pitch of the tone depends on the resistance under test. The tester will respond to resistance of hundreds of kilohms, yet it is possible to distinguish differences of just a few tens of ohms in low-resistance circuits. Q1 and Q2 form a multivibrator, the frequency of which is influenced by the resistance between the test points. The output stage (Q3 and Q4) will drive a small loudspeaker or an earphone.

SINGLE-CHIP RESISTANCE CHECKER

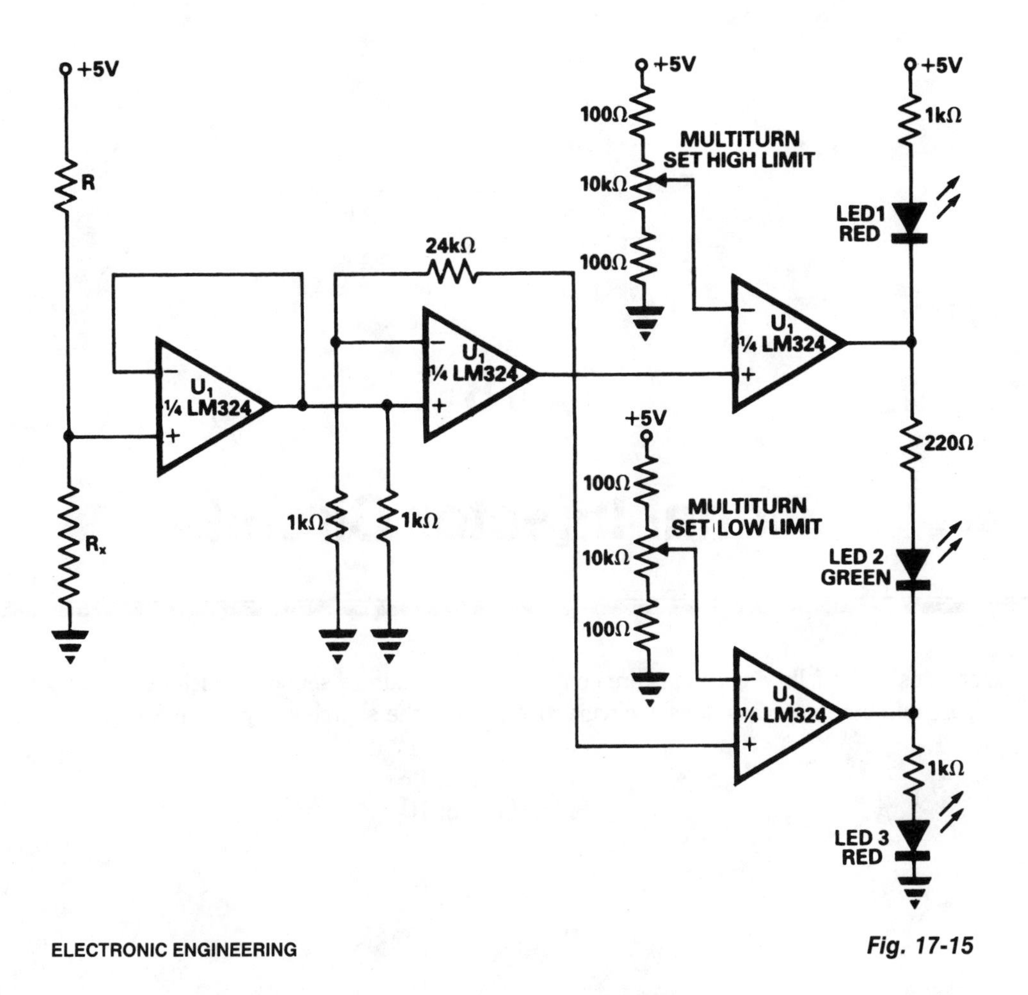

ELECTRONIC ENGINEERING

Fig. 17-15

A simple tester can be used for routine checks for resistance on production lines of relays, coils, or similar components, where frequent changes in resistance to be tested are not required. The tester is built around a single quad op amp chip, the LM324. R, which is chosen to be around 80 times the resistance to be checked, and the 5-V supply form the current source. The first op amp buffers the voltage generated across the resistance under test, R_x. The second op amp amplifies this voltage. The third and fourth op amps compare the amplified voltage with high and low limits. The high and low limits are set on multiturn presets with high and low limit resistors connected in place of R_x. LED 1 (red) lights when the resistance is high. LED 2 (green) shows that the resistance is within limits. LED 3 (red) indicates that the resistance is low.

18

Signal-Injector Circuits

The sources of the following circuits are contained in the Sources section, which begins on page 217. The figure number in the box of each circuit correlates to the source entry in the Sources section.

SIGNAL INJECTOR I

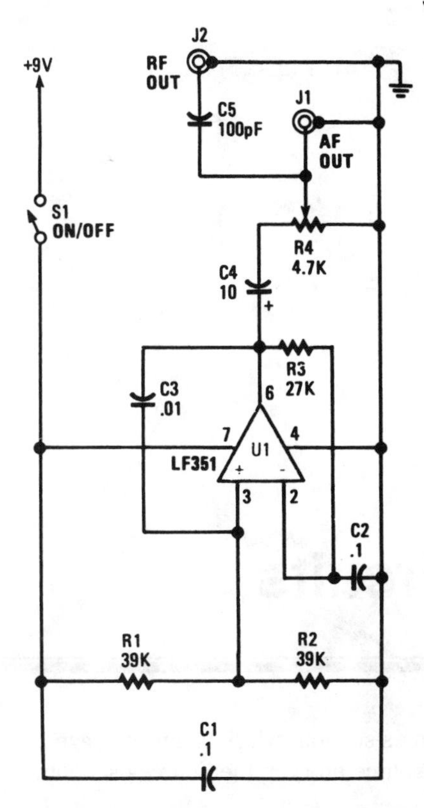

This unit is a single oscillator built around an LF351 JFET-input op amp. Resistors R1 and R2 bias the noninverting input, while R3 biases the inverting input from the output. This layout provides 100% negative feedback, but the decoupling caused by C2 gives reduced feedback and high-voltage gain when dealing with audio frequencies. The fundamental operating frequency is about 800 Hz. Potentiometer R4 is the output-level control. To use it start at the speaker. If no tone is heard, move back to the amplifier input, and listen for the tone. Still, if no tone is heard, continue backtracking from the output to the input, covering all stages in between. The stage where the signal is lost is the one that is not operating.

POPULAR ELECTRONICS/HANDS-ON ELECTRONICS *Fig. 18-1*

SIGNAL INJECTOR II

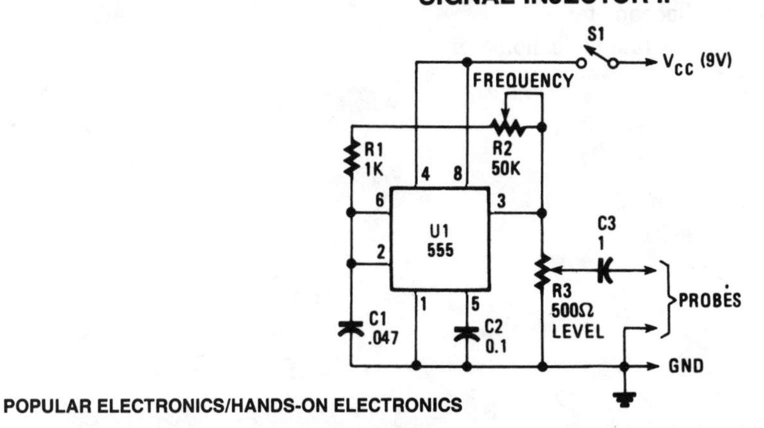

POPULAR ELECTRONICS/HANDS-ON ELECTRONICS *Fig. 18-2*

The unit provides a square-wave output that is rich in harmonic content. The circuit's output frequency can be varied from 50 Hz to 15 kHz. The heart of the circuit is a 555 astable connected in its equal mark/space mode. The frequency is controlled by potentiometer R2 and capacitor C1. Resistor R3 controls the output level with the output ac-coupled through C3.

19

Tachometer Circuits

The sources of the following circuits are contained in the Sources section, which begins on page 217. The figure number in the box of each circuit correlates to the source entry in the Sources section.

LOW-FREQUENCY TACHOMETER

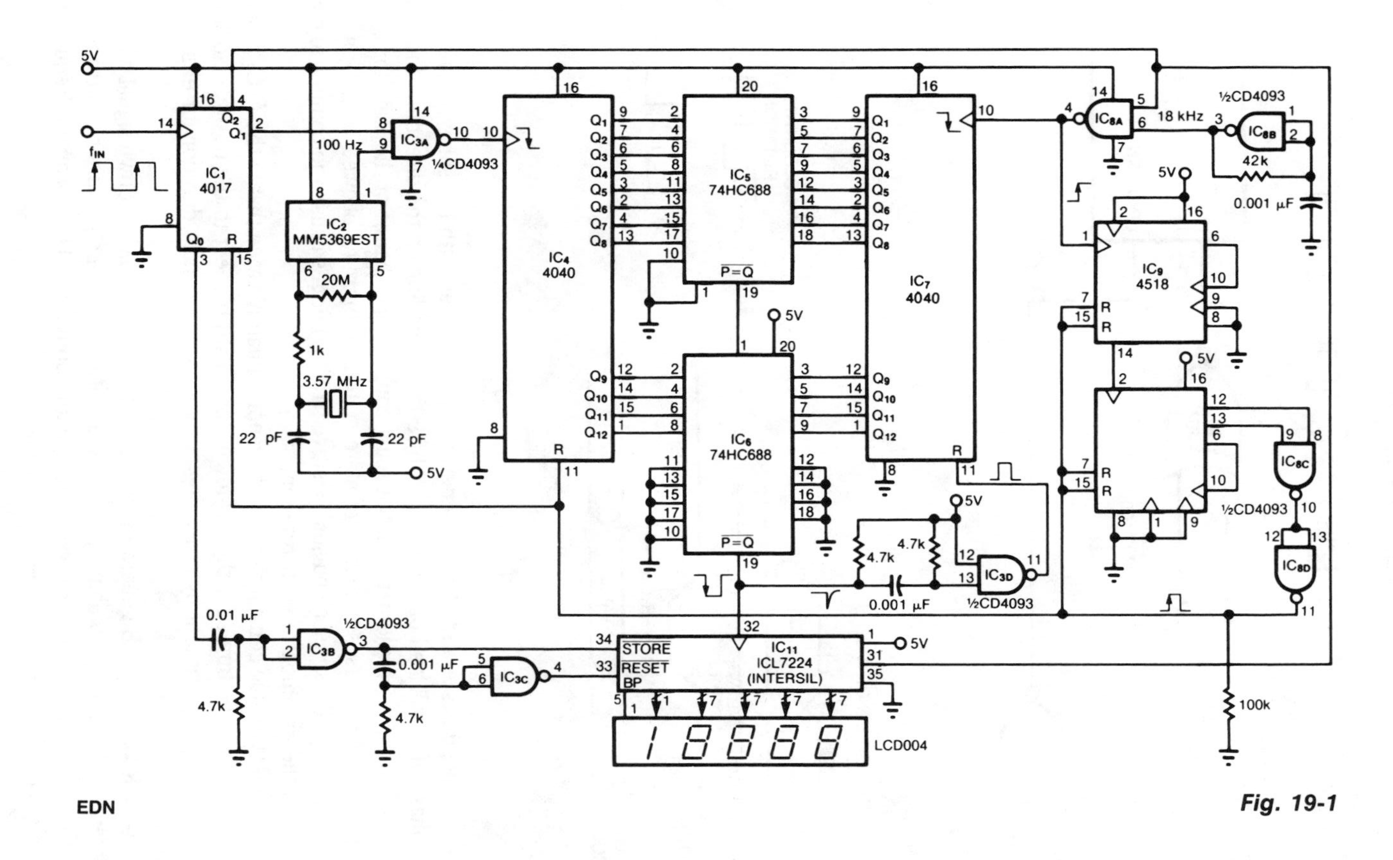

EDN

Fig. 19-1

This tachometer lets you measure heartbeats, respiratory rates, and other low-frequency events that recur at intervals of 0.33 to 40.96 seconds. The circuit senses the period of f_{IN}, computes the equivalent pulses per minute, and updates the LCD accordingly. Although the decimal readout equals 60 f_{IN}, the circuit doesn't actually produce a frequency of 60 f_{IN}. The computation involves counting and comparison techniques and takes 0.33 seconds.

TACHOMETER

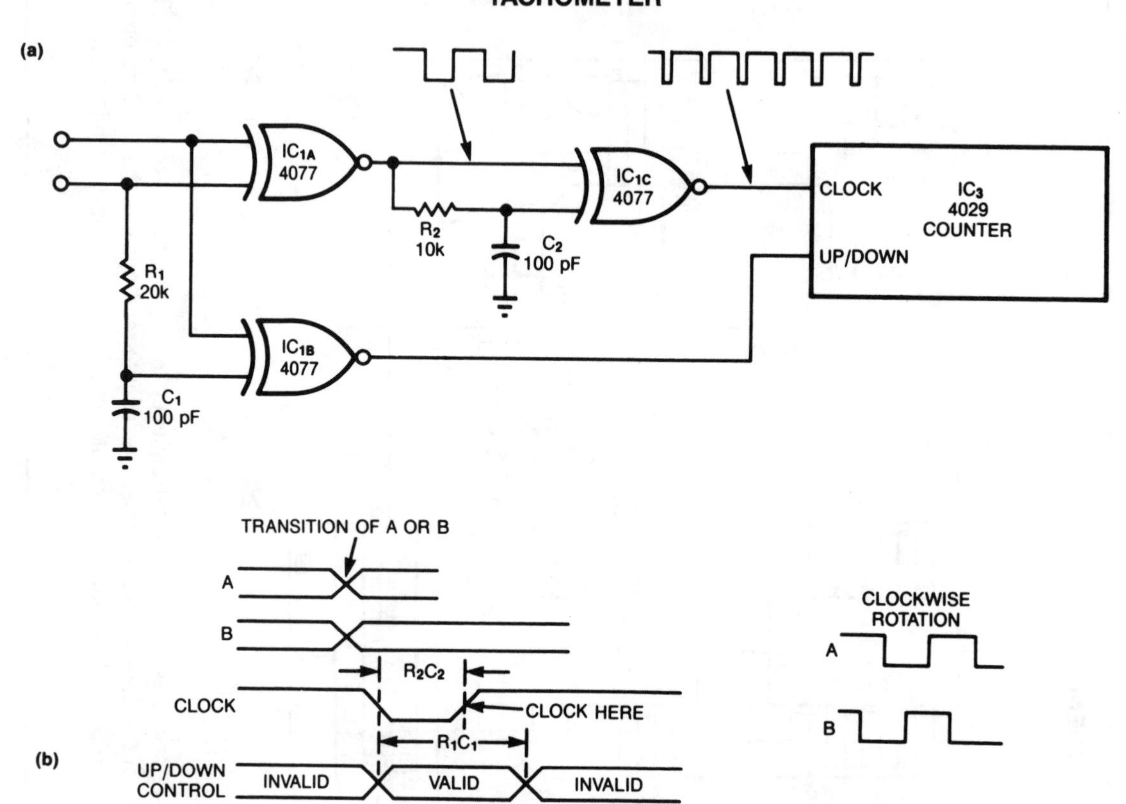

EDN

Fig. 19-2

A standard shaft encoder's *A* and *B* ports generate square waves with the same frequency as the shaft turns. The phase of *A* will lead or lag that of *B* by 90°, depending on the direction of rotation. To obtain maximum resolution, the tachometer circuit must count every change of the state for the *A* and *B* signals. Each such change causes a change of state at IC1A's output, followed by a 1-μs negative pulse at the output of IC1C. These clock pulses' positive (trailing) edges cause the counter to count up or down, according to the direction of shaft rotation.

You should set the R1C1 time constant so that it is approximately twice that of the R_2C_2 product, to ensure adequate setup and hold times for the up/down signal with respect to the positive clock edges. IC1C supports this timing requirement by producing clock pulses of similar duration for either positive or negative transitions or IC1A.

The exclusive-NOR logic of IC1B generates the correct polarity of the up/down signal when necessary, at the positive clock edges, by combining the *A* value with the *B* value just prior to a transition of *A* or *B*. C1 provides memory by sorting the *B* value voltage for about 2 μs. The maximum frequency for *A* or *B* is approximately $(4R_1C_1)^{-1}$.

CALIBRATED TACHOMETER

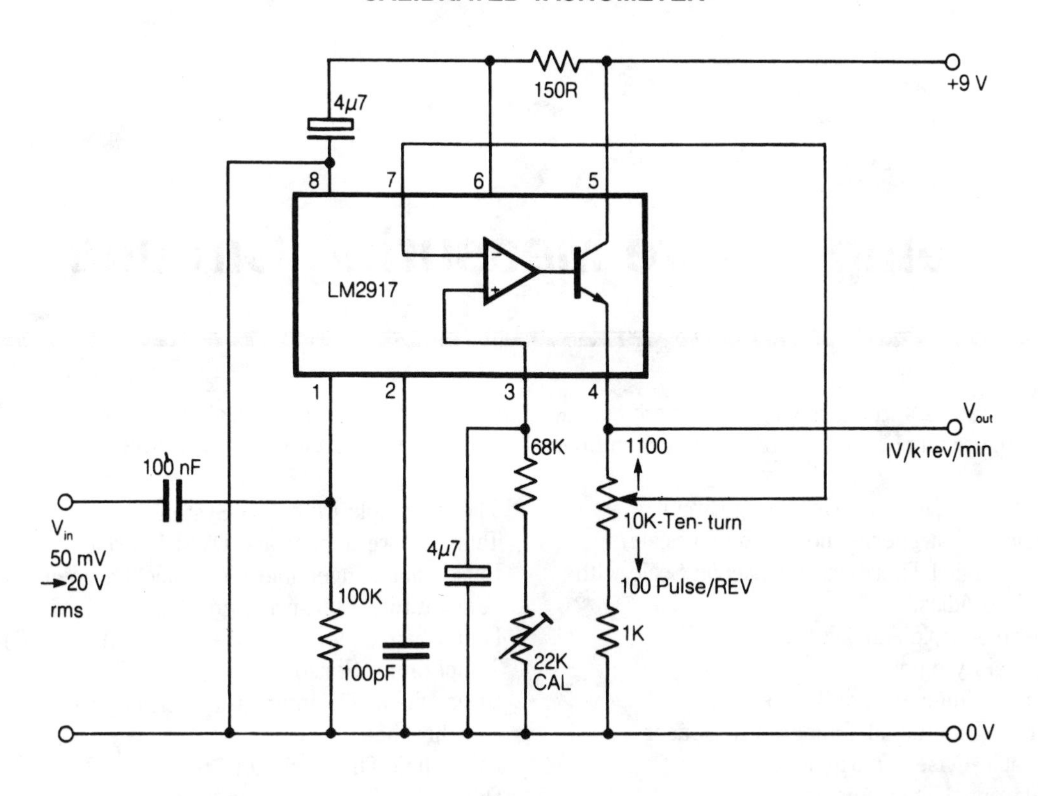

ELECTRONIC ENGINEERING

Fig. 19-3

Here is a simple tachometer circuit for use with a hand-held DVM or portable chart recorder. A novel feature is that the source frequency pulse/rev rate can be directly set on a 10-turn potentiometer to provide a convenient calibration of 1 V per 1000 rev/min. This is particularly useful when measuring a shaft or engine speed by sensing the gear teeth.

The circuit uses an LM2917 IC, which is specifically designed for tachometer applications. The 10-turn potentiometer, which provides the pulse/rev setting, is suitably configured in the output amplifier feedback path. The pulse/rev range is 100 to 1100, so the potentiometer dial mechanism should be set to start at 100 to provide direct calibration.

The IC's internal 7.5-V zener provides stable operation from a 9-V battery. The tachometer accepts an input signal between 50 mv and 20 V rms and has an upper speed limit of 6000 rev/min with the component values shown.

20

Temperature Measuring Circuits

The sources of the following circuits are contained in the Sources section, which begins on page 217. The figure number in the box of each circuit correlates to the source entry in the Sources section.

DIGITAL TEMPERATURE-MEASURING CIRCUIT

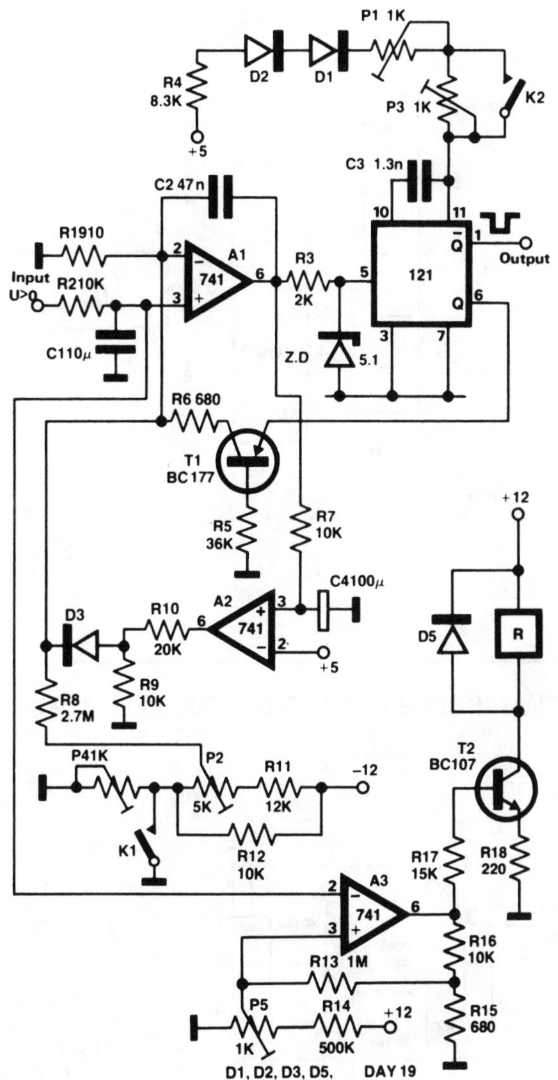

The output voltage of a thermocouple is converted into frequency, measured by a digital frequency meter. The measuring set, connected with Ni-NiCr thermocouple, permits you to measure the temperatures within the range of 5 to 800°C with ±1°C error. The output thermocouple signal is proportional to the temperature difference between the hot junction and the thermostat. Kept at 0°C, the thermostat drives the voltage-to-frequency converter, changing the analog input signal into the output frequency with the conversion ratio adjusted so that the frequency is equal to the measured temperature in Celsius degrees (e.g., for 350°C the frequency value is 350 Hz).

ELECTRONIC ENGINEERING

Fig. 20-1

BASIC DIGITAL THERMOMETER (KELVIN SCALE)

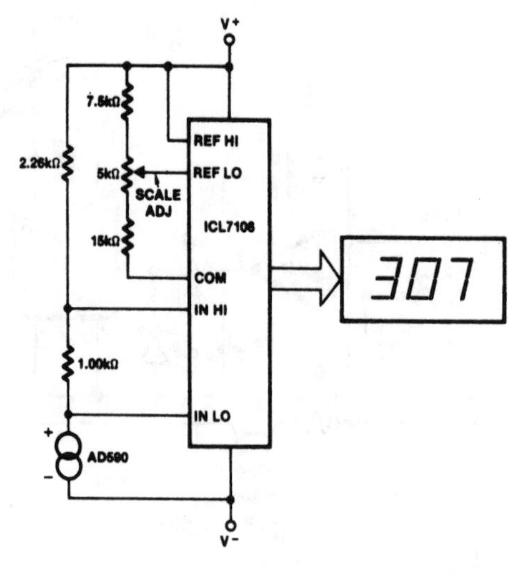

The Kelvin scale version reads from 0 to 1999°K theoretically, and from 223°K to 473°K actually. The 2.26-kΩ resistor brings the input within the ICL7106 V_{CM} range: two general-purpose silicon diodes or an LED can be substituted.

INTERSIL *Fig. 20-2*

BASIC DIGITAL THERMOMETER (KELVIN SCALE WITH ZERO ADJUST)

This circuit allows zero adjustment as well as slope adjustment. The ICL8069 brings the input within the common-mode range, while the 5-kΩ pots trim any offset at 218°K (−55°C), and set scale factor.

INTERSIL *Fig. 20-3*

THERMOCOUPLE AMPLIFIER

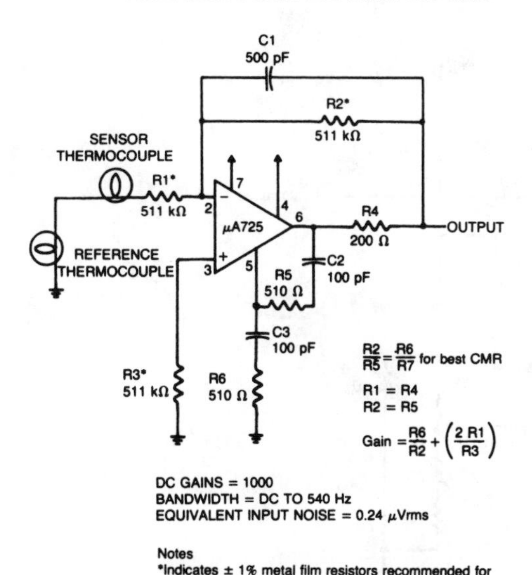

FAIRCHILD CAMERA & INSTRUMENT *Fig. 20-4*

REMOTE TEMPERATURE SENSOR

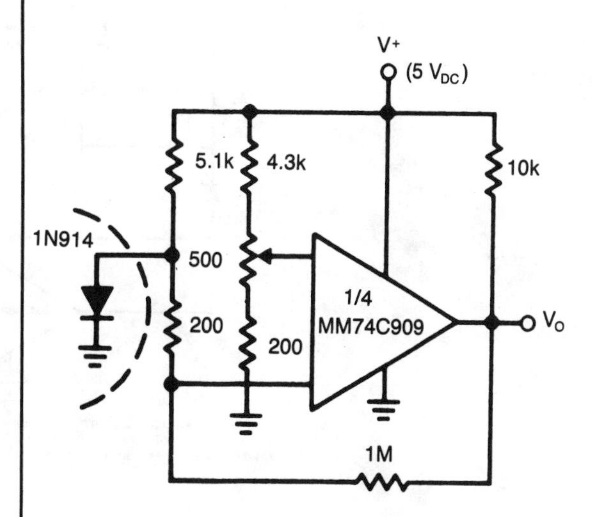

NATIONAL SEMICONDUCTOR *Fig. 20-6*

OPTICAL PYROMETER

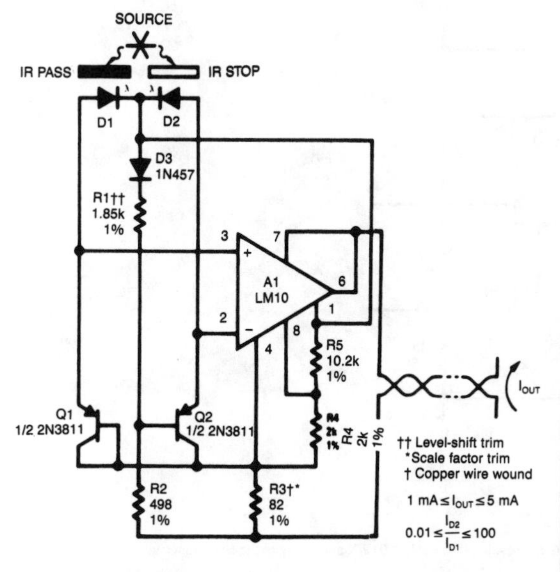

NATIONAL SEMICONDUCTOR *Fig. 20-5*

SIMPLE DIFFERENTIAL TEMPERATURE SENSOR

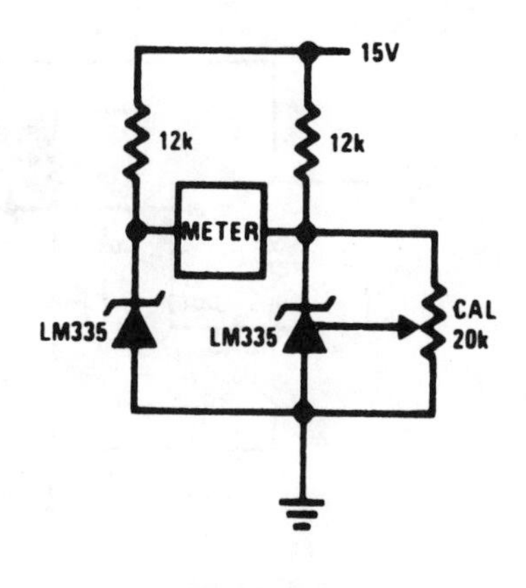

NATIONAL SEMICONDUCTOR *Fig. 20-7*

VARIABLE-OFFSET THERMOMETER

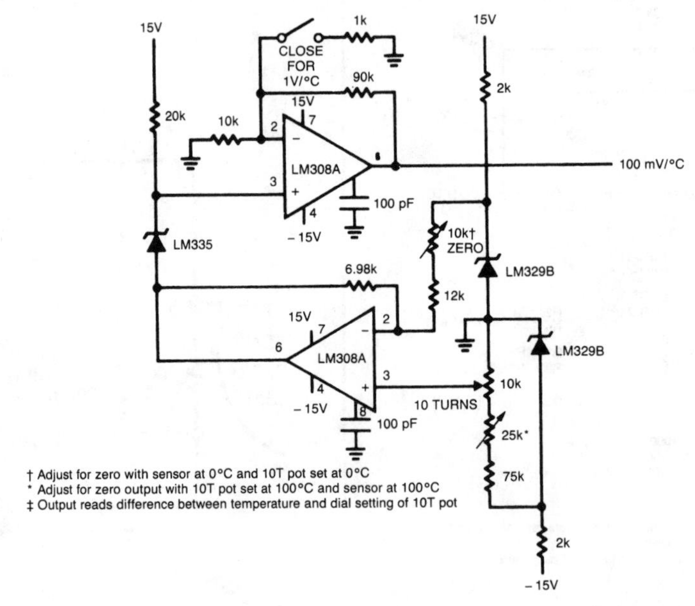

NATIONAL SEMICONDUCTOR

Fig. 20-8

DIFFERENTIAL THERMOMETER

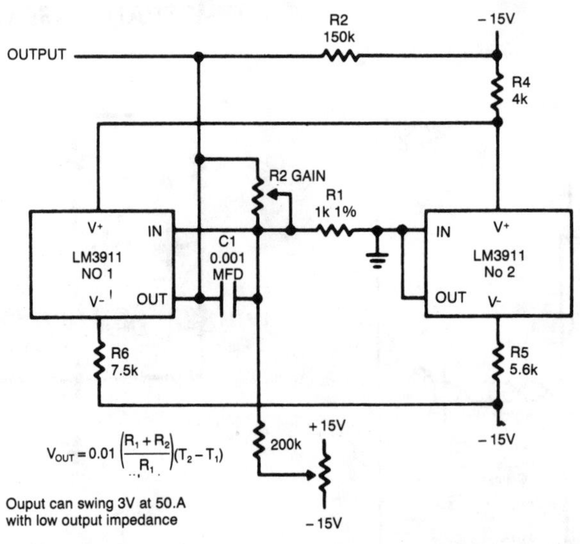

$$V_{OUT} = 0.01 \left(\frac{R_1 + R_2}{R_1}\right)(T_2 - T_1)$$

Ouput can swing 3V at 50.A
with low output impedance

**The 0.01 in the above equation is in units of V/K or V/C, and is a result of the basic 0.01V/K sensitivity of the transducer

Fig. 20-9

ISOLATED TEMPERATURE SENSOR

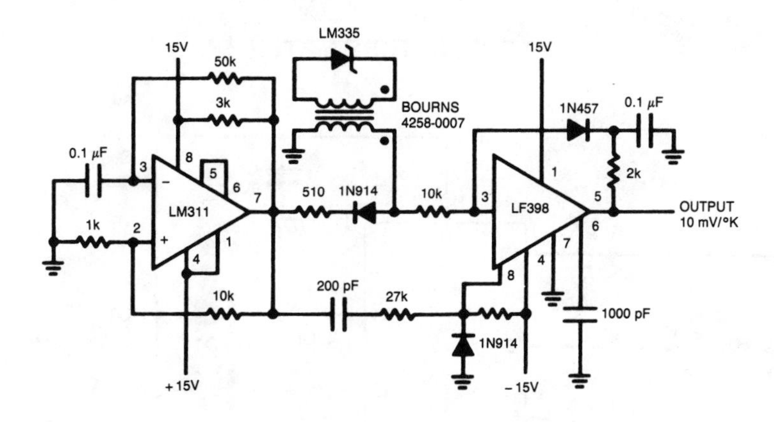

NATIONAL SEMICONDUCTOR *Fig. 20-10*

DIGITAL THERMOMETER

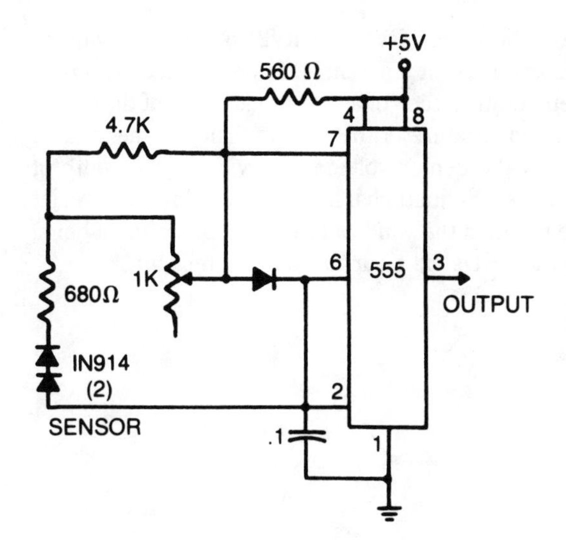

The sensor consists of two series-connected 1N914s, part of the circuit of a 555 multivibrator. Wired as shown, the output pulse rate is proportional to the temperature of the diodes. This output is fed to a simple frequency-counting circuit.

RADIO-ELECTRONICS *Fig. 20-11*

TEMPERATURE METER

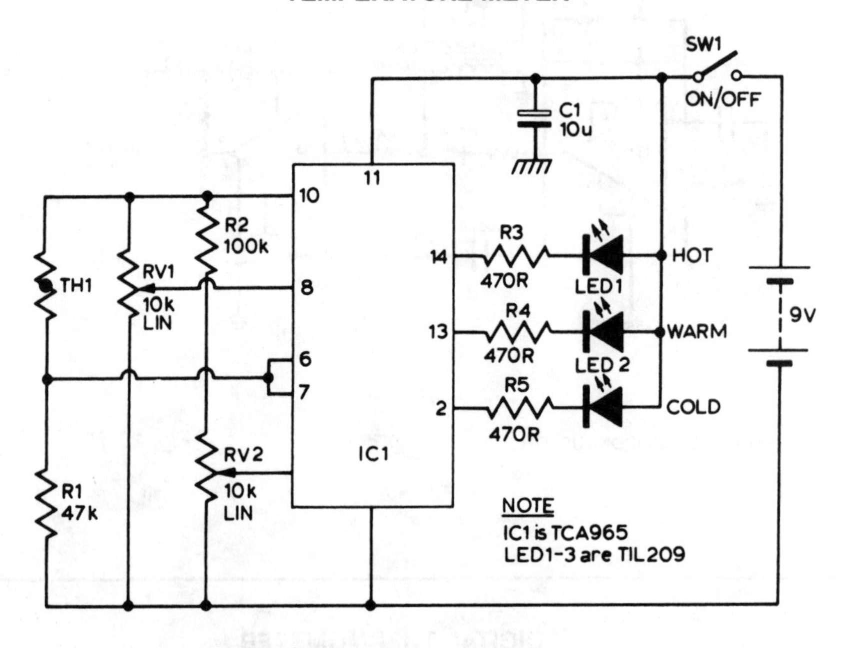

ELECTRONICS TODAY INTERNATIONAL

Fig. 20-12

TCA965 window-discriminator IC allows the potentiometers RV1 and RV2 to set up a window height and window width, respectively. R1 and thermistor TH1 for a potential divider connected across the supply lines. R1 is chosen such that at ambient temperature the voltage at the junction of these two components will be approximately half supply. As the temperature of the sensor changes, the voltage will change. RV1 will set the point that corresponds to the center voltage of a window the width of which is set by RV2. The switching points of the IC feature a Schmitt characteristic with low hysteresis. The outputs of IC1 indicate whether the input voltage is within the window or outside by virtue of being either too high or too low. The outputs of IC1 drive the LEDs via a current-limiting resistor.

0 to 63°C TEMPERATURE SENSOR

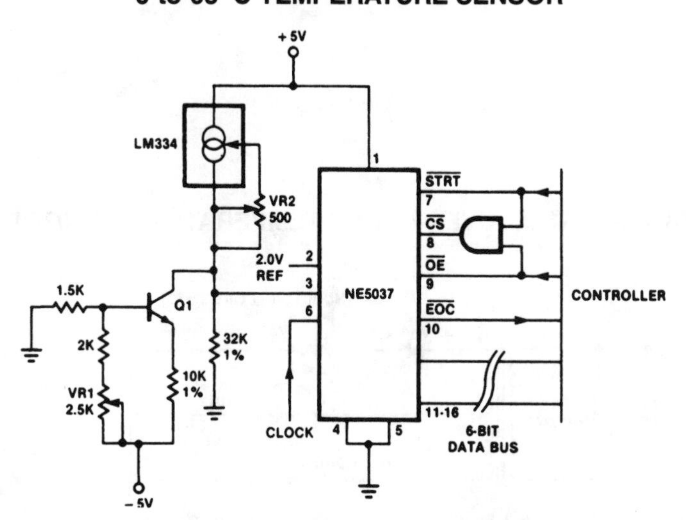

SIGNETICS *Fig. 20-13*

The temperature sensor provides an input to pin 3 of the NE5037 of 32 mV/°C. This 32 mV is the value of one LSB for the NE5037. The LM334 is a three-terminal temperature sensor and provides a current of 1 μA for each degree Kelvin. The 32-kΩ resistor provides the 32 mV for each microamp through it, while the transistor bleeds off 273 μA of the temperature sensor (LM334) current. This bleeding lowers the reading by 273 K, thus converting from Kelvin to Celsius. To read temperature, conversion is started by sending a momentary low signal to pin 7 of the NE5037. When pin 10 of the NE5037 becomes low, conversion is complete and a low is applied to pin 9 of the NE5037 to read data on pins 11 through 16. Notice that this temperature data is in straight binary format. The controller can be a microprocessor in a temperature-control application, or discrete circuitry in a simple temperature-reporting application.

ISOLATED TEMPERATURE SENSOR

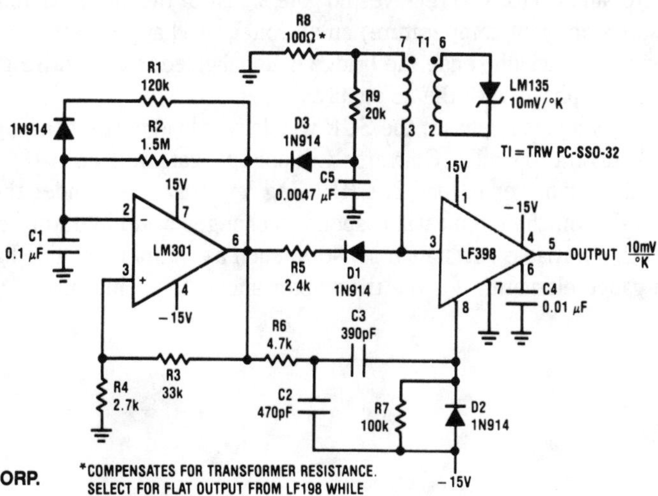

LINEAR TECHNOLOGY CORP. *Fig. 20-14*

DUAL-OUTPUT OVER-UNDER TEMPERATURE MONITOR

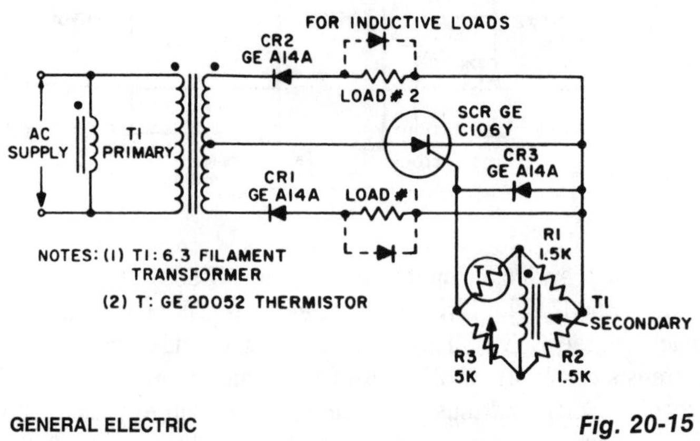

GENERAL ELECTRIC

Fig. 20-15

This circuit is ideal for use as an over-under temperature monitor, where its dual output feature can be used to drive high and low temperature indicator lamps, relays, etc. T1 is a 6.3-V filament transformer whose secondary winding is connected inside of a four-arm bridge. When the bridge is balanced, ac output is zero, and C5 (or C7) receives no gate signal. If the bridge is unbalanced by raising or lowering the thermistor's ambient temperature, an ac voltage will appear across the SCR's gate cathode terminals. Depending on which sense the bridge is unbalanced, the positive gate voltage will be in phase with, or 180° out of phase with the ac supply.

If the positive gate voltage is in phase, the SCR will deliver load current through diode CR1 to load 1, diode CR2 blocking current to load 2. Conversely, if positive gate voltage is 180° out of phase, diode CR2 will conduct and deliver power to load 2, CR1 being reverse biased under these conditions. With the component values shown, the circuit will respond to changes in temperature of approximately 1 to 2°C. Substitution of other variable-resistance sensors, such as cadmium sulfide light-dependent resistors (LDR) or strain-gauge elements, for the thermistor shown is permissible.

THERMOCOUPLE AMPLIFIER WITH COLD-JUNCTION COMPENSATION

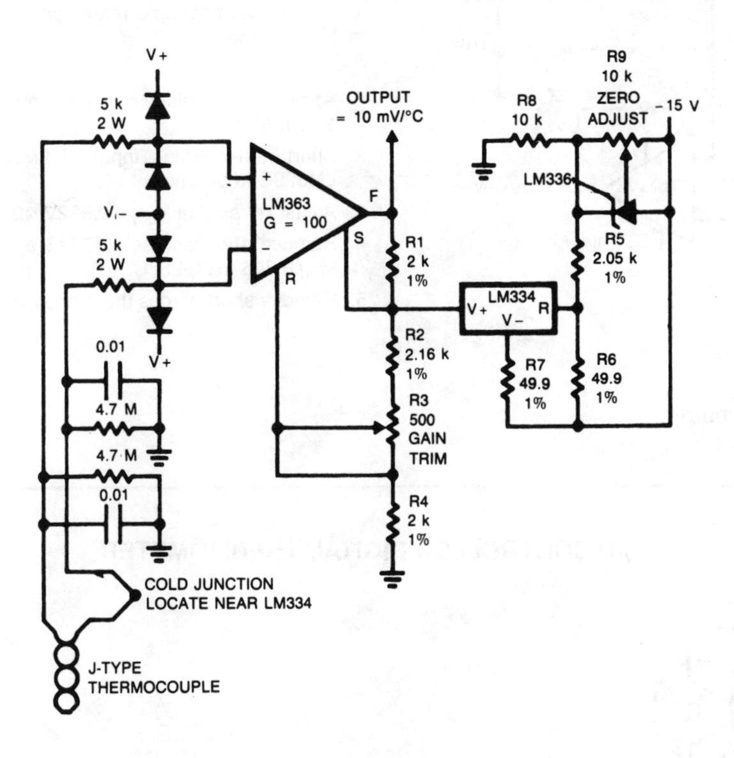

NATIONAL SEMICONDUCTOR **_Fig. 20-16_**

Input-protection circuitry allows thermocouple to short to 120 Vac without damaging the amplifier. Calibration:

1. Apply a 50-mV signal in place of the thermocouple. Trim R3 for $V_{OUT} = 12.25$ V.
2. Reconnect the thermocouple. Trim R9 for correct output.

CENTIGRADE-CALIBRATED THERMOCOUPLE THERMOMETER

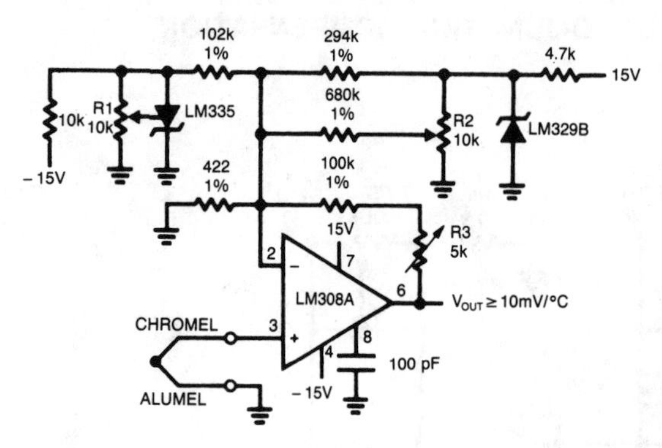

Terminate thermocouple reference junction in close proximity to LM335.

Adjustments:

1. Apply signal in place of thermocouple and adjust R3 for a gain of 245.7.
2. Short non-inverting input of LM308A and output of LM329B to ground.
3. Adjust R1 so that $V_{OUT} = 2.982V@25°C$.
4. Remove short across LM329B and adjust R2 so that $V_{OUT} = 246\ mV@25°C$.
5. Remove short across thermocouple.

NATIONAL SEMICONDUCTOR

Fig. 20-17

μP-CONTROLLED DIGITAL THERMOMETER

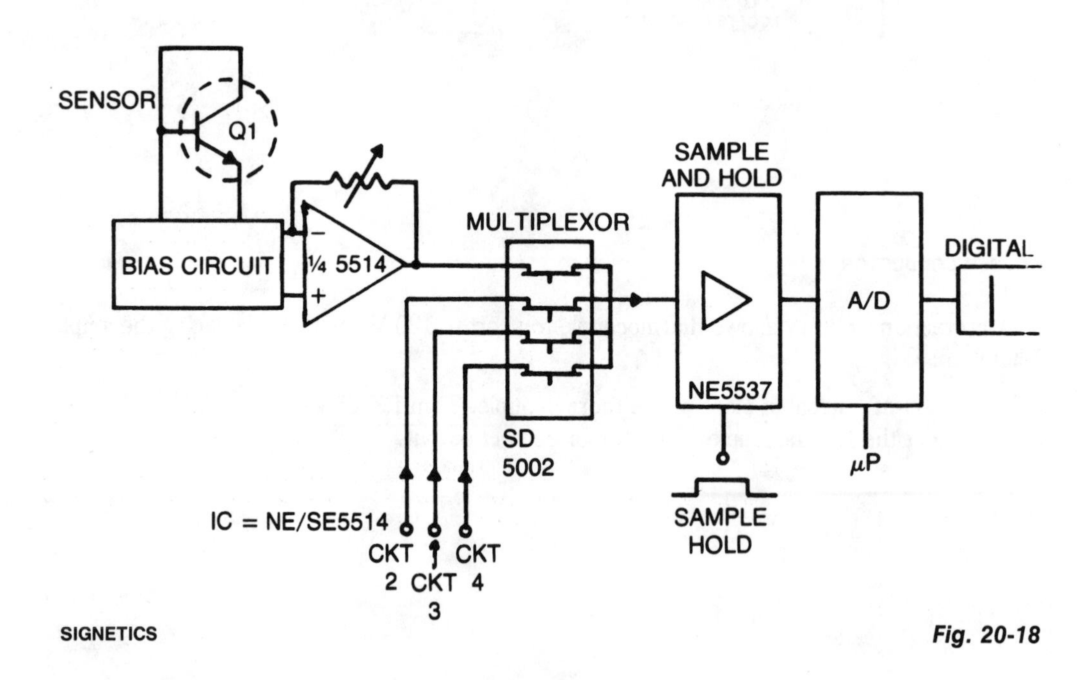

SIGNETICS

Fig. 20-18

IC TEMPERATURE SENSOR

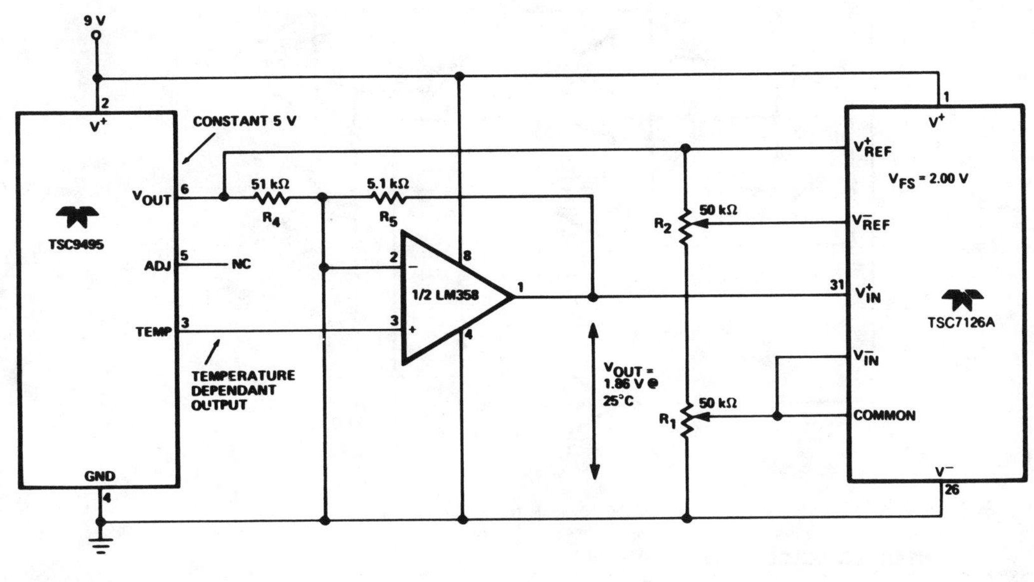

TELEDYNE

Fig. 20-19

PRECISION TEMPERATURE TRANSDUCER WITH REMOTE SENSOR

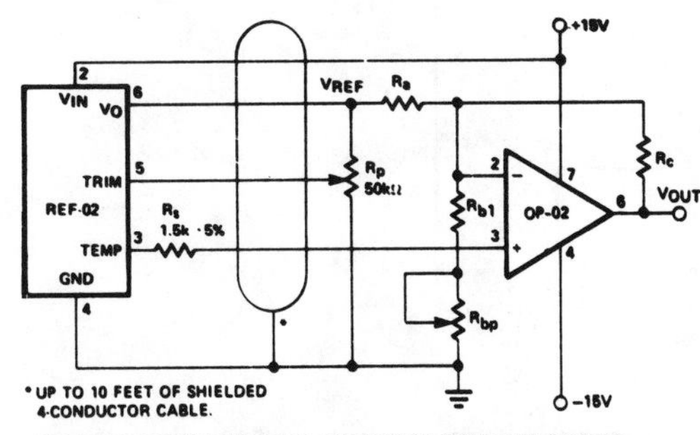

* UP TO 10 FEET OF SHIELDED 4-CONDUCTOR CABLE.

FOR THEORY OF OPERATION AND CALIBRATION PROCEDURE CONSULT APPLICATION NOTE 18, "THERMOMETER APPLICATIONS OF THE REF-02".

RESISTOR VALUES

TCV_{OUT} SLOPE (S)	10mV/°C	100mV/°C	10mV/°F
TEMPERATURE RANGE	−55°C to +125°C	−55°C to +125°C	−67°F to +257°C
OUTPUT VOLTAGE RANGE	−0.55V to +1.25V	−5.5V to +12.5V	−0.67V to +2.57V
ZERO SCALE	0V@0°C	0V@0°C	0V@0°F
R_a (±1% resistor)	9.09kΩ	15kΩ	7.5kΩ
R_{b1} (±% resistor)	1.5kΩ	1.82kΩ	1.21kΩ
R_{bp} (Potentiometer)	200Ω	500Ω	200Ω
R_c (±1% resistor)	5.11kΩ	84.5kΩ	8.25kΩ

* For 125°C operation, the op amp output must be able to swing to +12.5V, increase V_{IN} to +18V from +15V if this is a problem.

PRECISION MONOLITHICS

Fig. 20-20

5-V POWERED, LINEARIZED PLATINUM RTD SIGNAL CONDITIONER

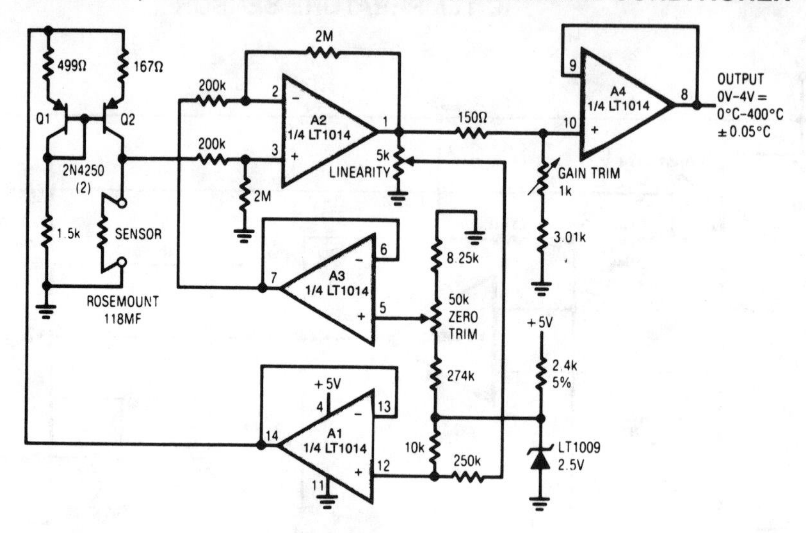

ALL RESISTORS ARE TRW-MAR-6 METAL FILM.
RATIO MATCH 2M–200K ± 0.01%.
TRIM SEQUENCE:
SET SENSOR TO 0° VALUE.
ADJUST ZERO FOR 0V OUT.

LINEAR TECHNOLOGY CORP.

Fig. 20-21

HIGH/LOW TEMPERATURE SENSOR

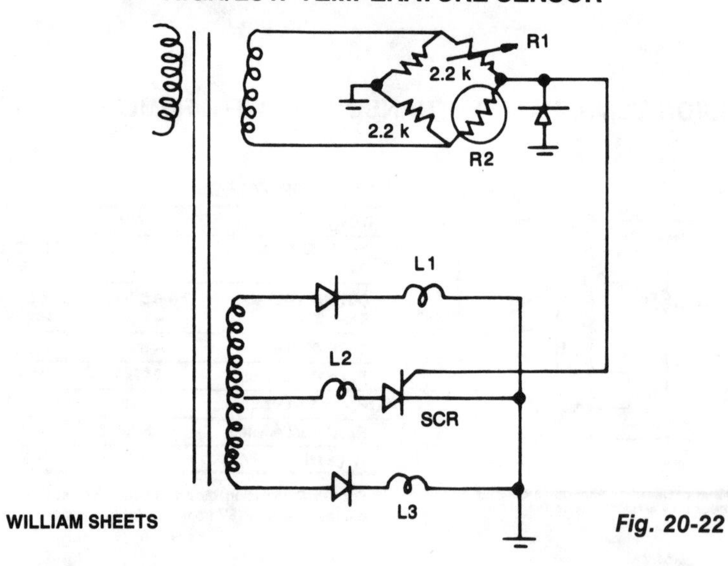

WILLIAM SHEETS

Fig. 20-22

Resistors R1, R2, and the two 2.2-kΩ resistors form a bridge circuit. R2 is a thermistor, and R1 sets the temperature at which L2 lights. Lower or higher temperatures light L1 or L3 to indicate an over- or under-temperature condition.

CURVATURE-CORRECTED PLATINUM RTD THERMOMETER

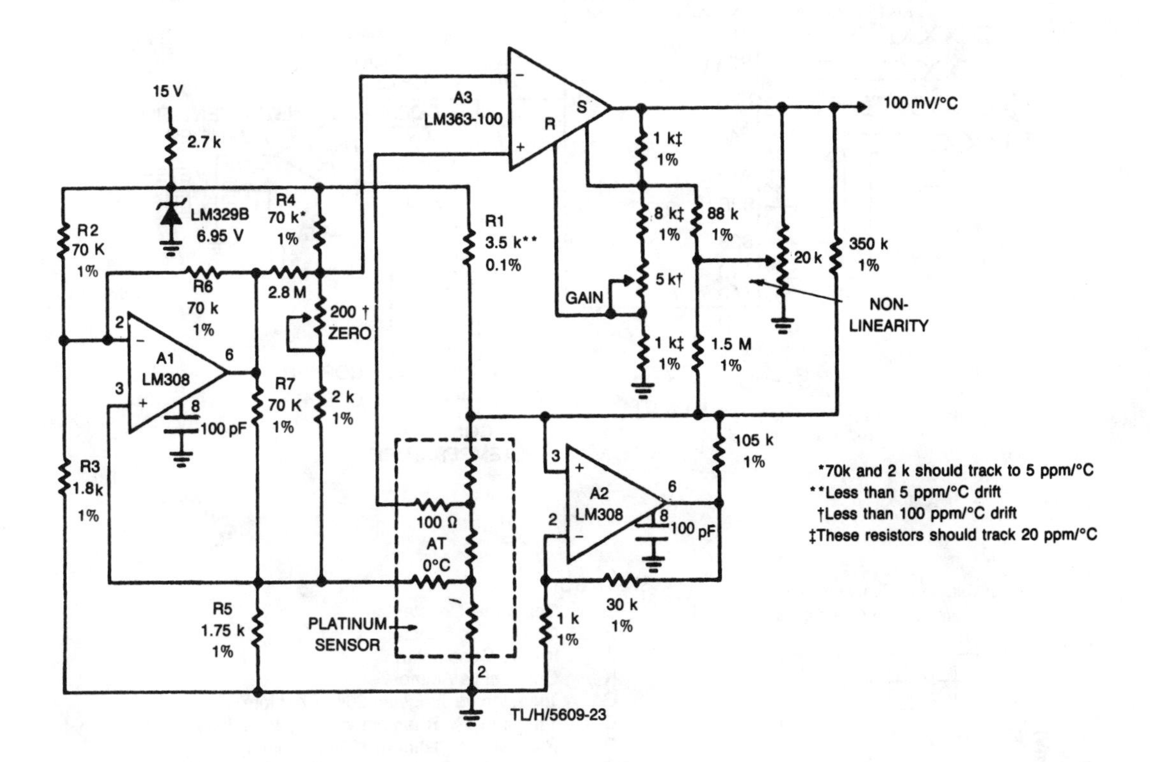

NATIONAL SEMICONDUCTOR ***Fig. 20-23***

This thermometer is capable of 0.01 °C accuracy over −50 to +150 °C. A unique trim arrangement eliminates cumbersome trim interactions so that zero gain, and nonlinearity correction can be trimmed in one even trip. Extra op amps provide full Kelvin sensing on the sensor without adding drift and offset terms found in other designs. A1 is configured as a Howland current pump, biasing the sensor with a fixed current. Resistors R2, R3, R4, and R5 form a bridge driven into balance by A1. In balance, both inputs of A2 are at the same voltage. Because $R_6 = R_7$, A1 draws equal currents from both legs of the bridge. Any loading of the R4/R5 leg by the sensor would unbalance the bridge; therefore, both bridge taps are given to the sensor open-circuit voltage and no current is drawn.

THERMOCOUPLE MULTIPLEX SYSTEM

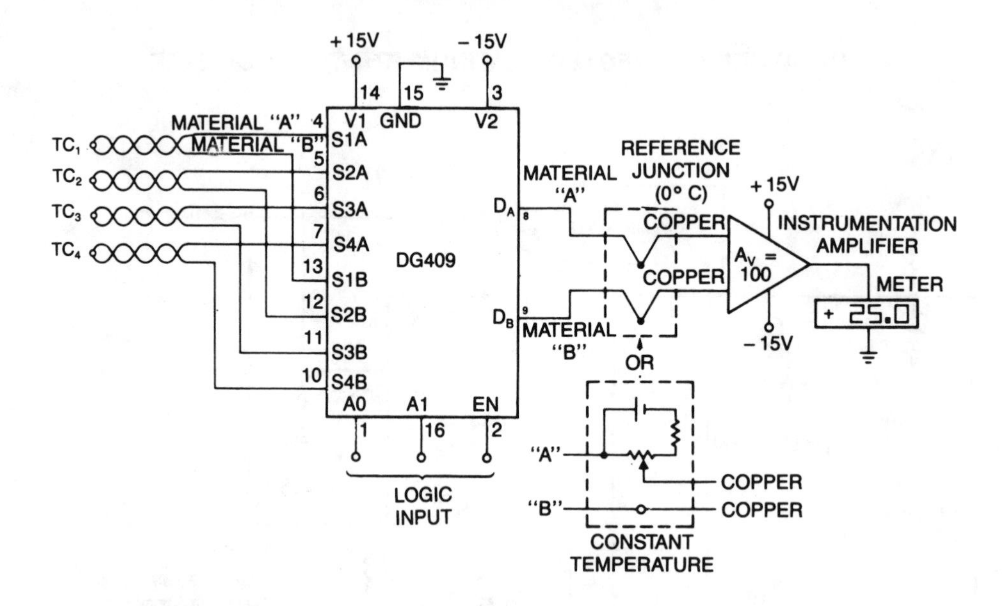

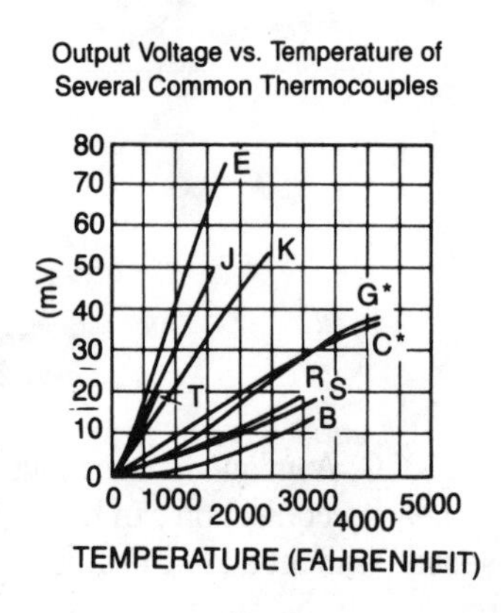

ANSI SYMBOL	
T	Copper vs Constantan
E	Chromel vs Constantan
J	Iron vs Constantan
K	Chromel vs Alumel
G*	Tungsten vs Tungsten 26% Rhenium
C*	Tungsten 5% Rhenium vs Tungsten 26% Rhenium
R	Platinum vs Platinum 13% Rhodium
S	Platinum vs Platinum 10% Rhodium
B	Platinum 6% Rhodium vs Platinum 30% Rhodium

*Not ANSI Symbol

SILICONIX

Fig. 20-24

To decouple the sensors from the meter amplifier, either a reference junction at 0°C or a bucking voltage set at room temperature can be used. The latter method is simpler, but is sensitive to changes in ambient temperature. The table above shows the output voltage vs temperature of several common types of thermocouples.

TEMPERATURE SENSOR AND DVM INTERFACE

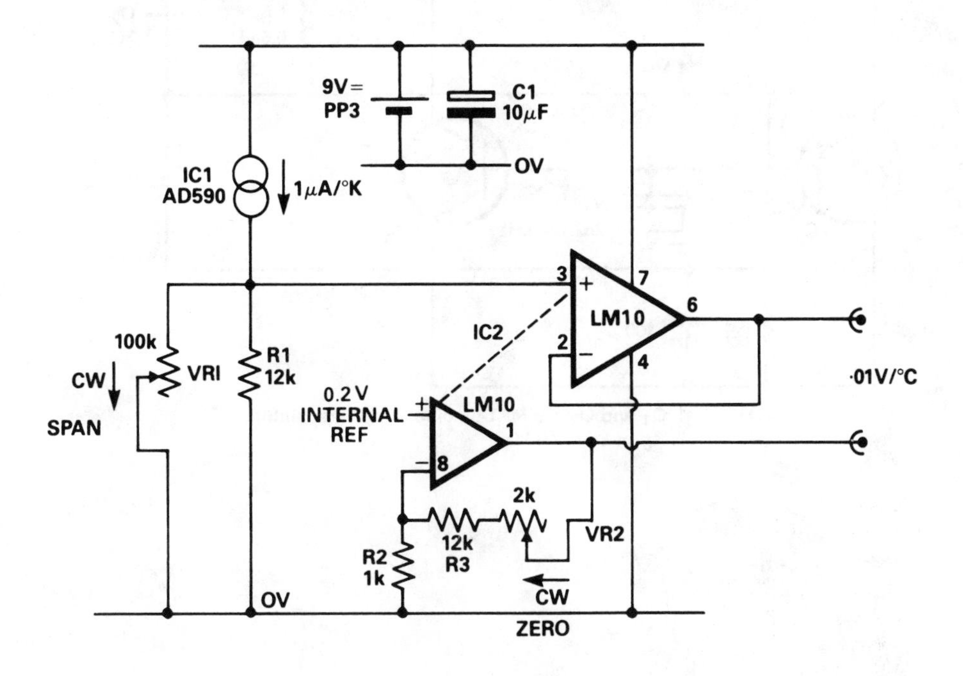

ELECTRONIC ENGINEERING

Fig. 20-25

The DVM gives a direct indication of the temperature of the sensor in degrees Centigrade. The temperature sensor IC1 gives a nominal 1 μA per degree Kelvin, which is converted to 10 mV per degree Kelvin by R1 and VR1. IC2 is a micropower, low-input drift op amp with internal voltage reference and amplifier. The main op amp in IC1 is connected as a voltage follower to buffer the sensor voltage at R1.

The second amplifier in IC1 is used to amplify the 0.2 V internal reference up to 2.73 V in order to offset the 273° below 0°C. The output voltage of the unit is the differential output of the two op amps and is thus equal to 0.01 V per °C.

IMPLANTABLE, INGESTIBLE ELECTRONIC THERMOMETER

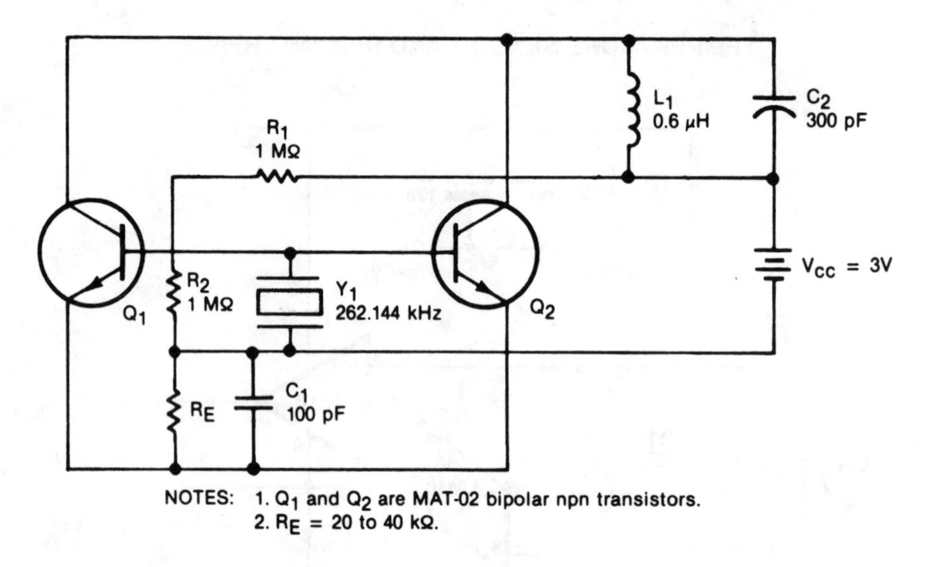

NASA

Fig. 20-26

This oscillator circuit includes a quartz crystal that has a nominal resonant frequency of 262,144 Hz and is cut in the orientation that gives a large linear coefficient of frequency variation with the temperature. In this type of circuit, the oscillation frequency is controlled primarily by the crystal—as long as the gain-bandwidth product is at least four times the frequency. In this case, the chosen component values yield a gain-bandwidth product of 1 MHz. Inductor L1 can be made very small: 100 to 200 turns with a diameter of 0.18 in. (4.8 mm) and a length of 0.5 in. (12.7 mm). Although the figure shows two transistors in parallel, one could be used to reduce power consumption or three could be used to boost the output. The general oscillator circuit can be used to measure temperatures from −10 to +140°C. A unit made for use in the human body from about 30 to 40°C operates at 262,144 ±50 Hz with a frequency stability of 0.1 Hz and a temperature coefficient of 9 Hz/°C.

TEMPERATURE-MEASURING ADD-ON FOR A DMM

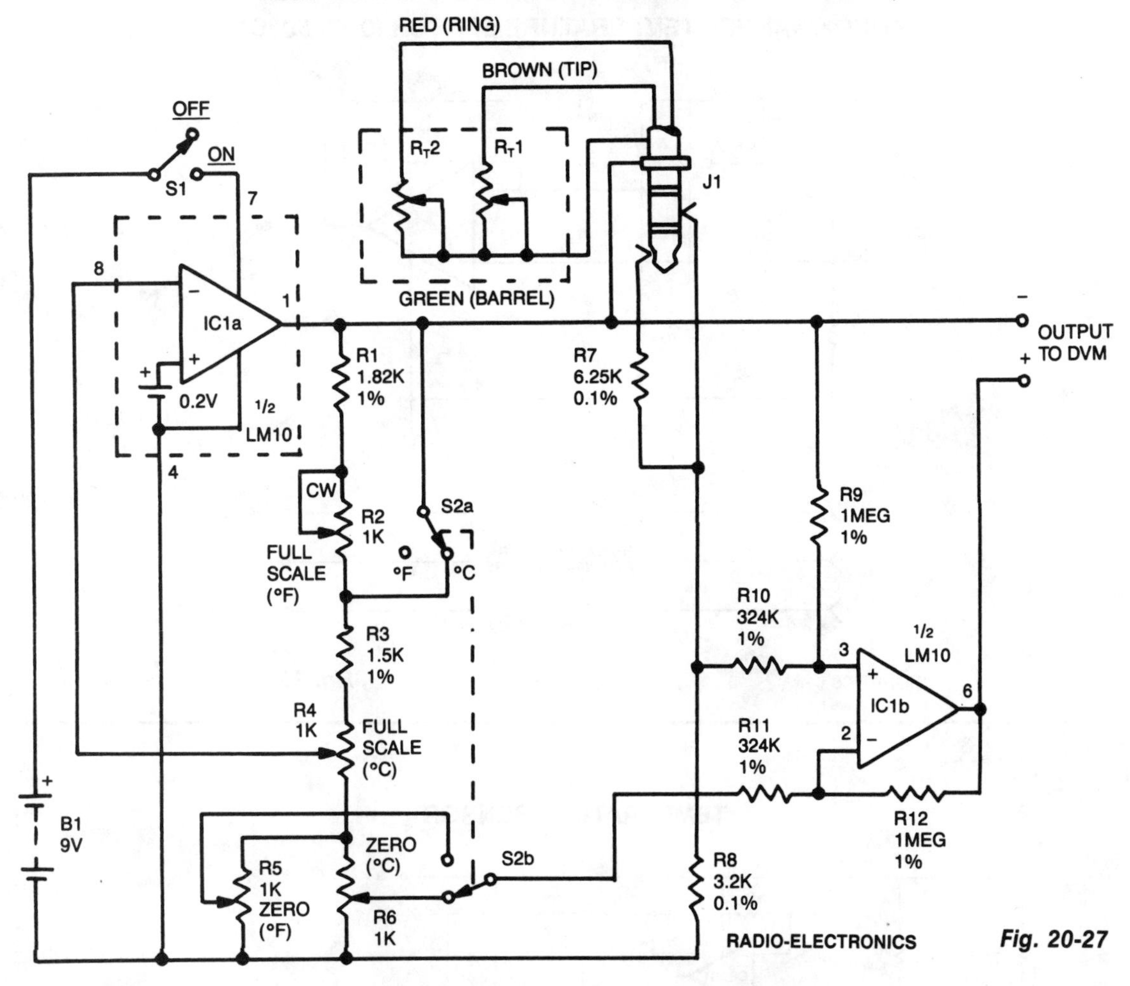

Fig. 20-27

The DVM-to-temperature adapter is built around a single IC, National's LM10. That micropower IC contains a stable 0.2-V reference, a reference amplifier and a general-purpose op amp. The circuit is designed for a linear temperature range of 0 to 100°C (32 to 212°F). The 0.2-V reference and reference amplifier provide a stable, fixed-excitation voltage to the Wheatstone bridge. The voltage is determined by a feedback network consisting of R1 through R6. Switch S2a configures the feedback to increase the voltage from 0.6 V on the Celsius range to 1.08 V on the Fahrenheit range. These differences compensate for the fact that one degree Fahrenheit produces a smaller resistance change than does one degree Celsius.

Resistors R1 through R16 also form the fixed leg of the Wheatstone bridge, nulling the bridge output at zero degrees, because 0°C is different from 0°F, S2b is used to select the appropriate offset.

The LM10's op amp, along with R9 through R12, form a differential amplifier that boosts the bridge output to 10 mV per degree. Since a single supply is used, and since the output must be able to swing both positive and negative, the output is referenced to the bridge supply voltage, rather than to the common supply.

FOUR-CHANNEL TEMPERATURE SENSOR (0 TO 50°C)

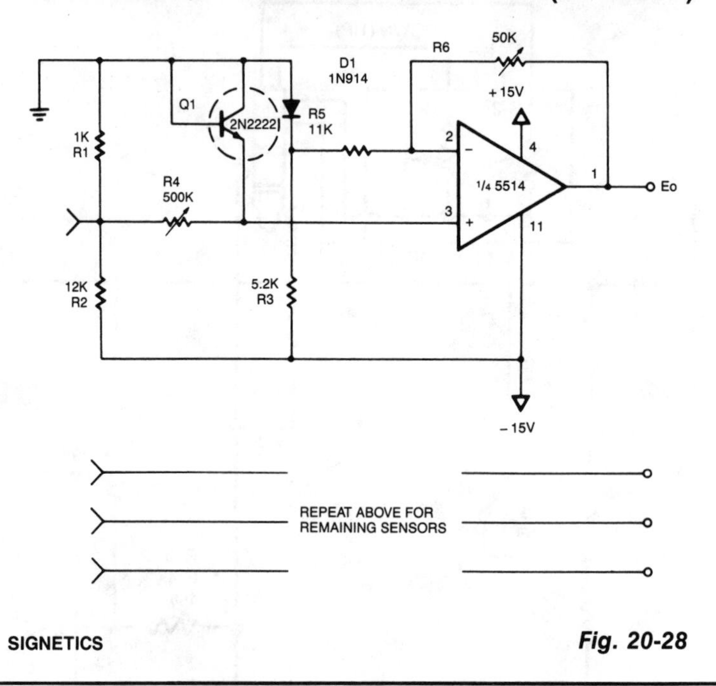

SIGNETICS

Fig. 20-28

TEMPERATURE SENSOR

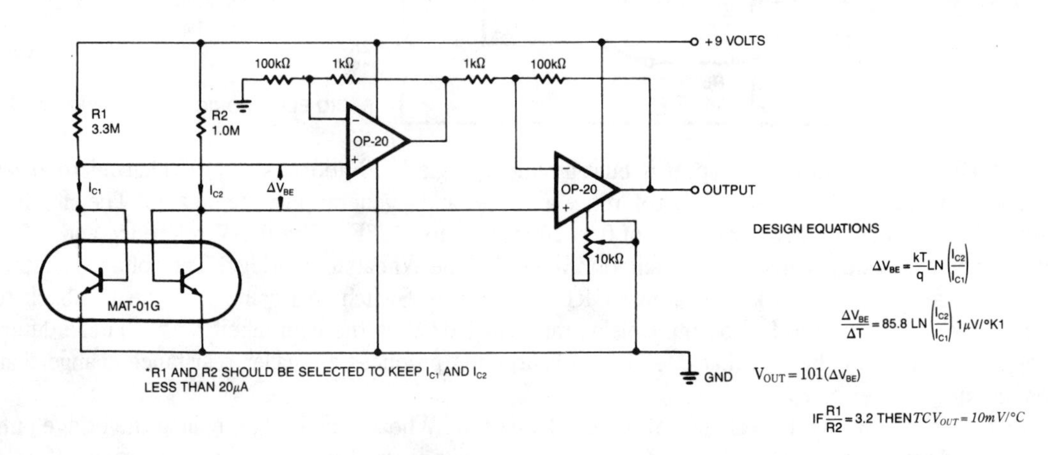

$$\Delta V_{BE} = \frac{kT}{q} LN \left(\frac{I_{C2}}{I_{C1}}\right)$$

$$\frac{\Delta V_{BE}}{\Delta T} = 85.8 \; LN \left(\frac{I_{C2}}{I_{C1}}\right) 1\mu V/°K1$$

$$V_{OUT} = 101(\Delta V_{BE})$$

IF $\frac{R1}{R2} = 3.2$ THEN $TCV_{OUT} = 10mV/°C$

PRECISION MONOLITHICS

Fig. 20-29

BASIC DIGITAL THERMOMETER (CELSIUS AND FAHRENHEIT)

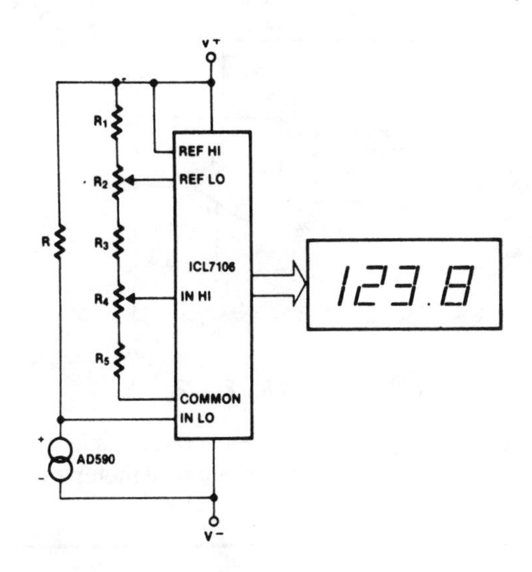

	R	R_1	R_2	R_3	R_4	R_5
°F	9.00	4.02	2.0	12.4	10.0	0
°C	5.00	4.02	2.0	5.11	5.0	11.8

INTERSIL *Fig. 20-30*

Maximum reading on the Celsius range is 199.9°C, limited by the (short-term) maximum allowable sensor temperature. Maximum reading on the Fahrenheit range is 199.9°F (93.3°C), limited by the number of display digits. V_{REF} for both scales is 500 mV.

FAHRENHEIT THERMOMETER

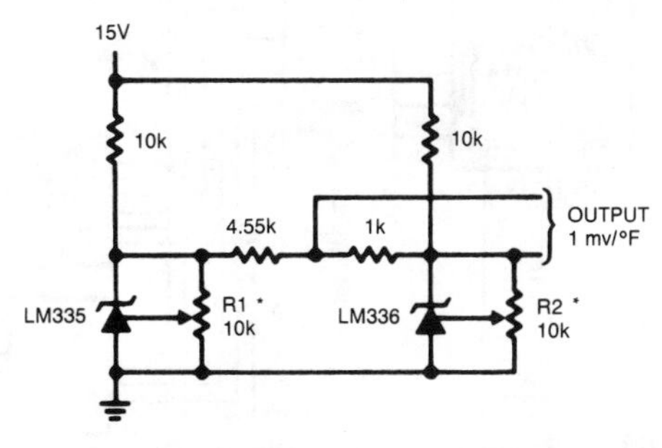

*To calibrate adjust R2 for 2.554V across LM336.
Adjust R1 for correct output.

NATIONAL SEMICONDUCTOR *Fig. 20-31*

DIFFERENTIAL THERMOMETER

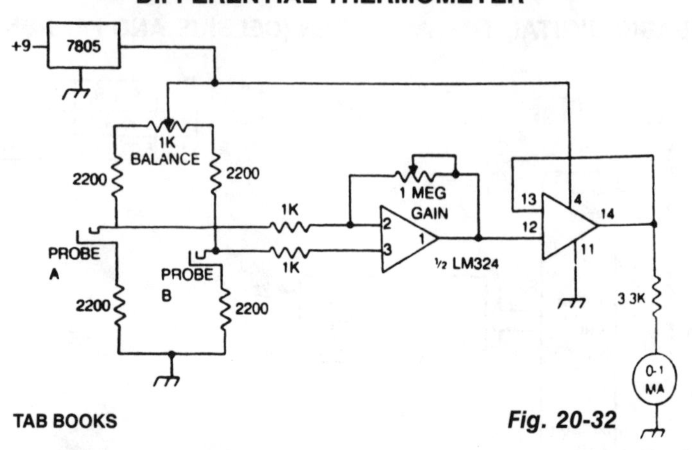

TAB BOOKS

Fig. 20-32

The differential thermometer uses two probes and shows the temperature difference between them, rather than the exact temperature. The thermometer uses a conventional meter as an indicator, and it covers a total range of 20° – 10° low to 10° high.

TEMPERATURE-REPORTING DIGITAL THERMOMETER

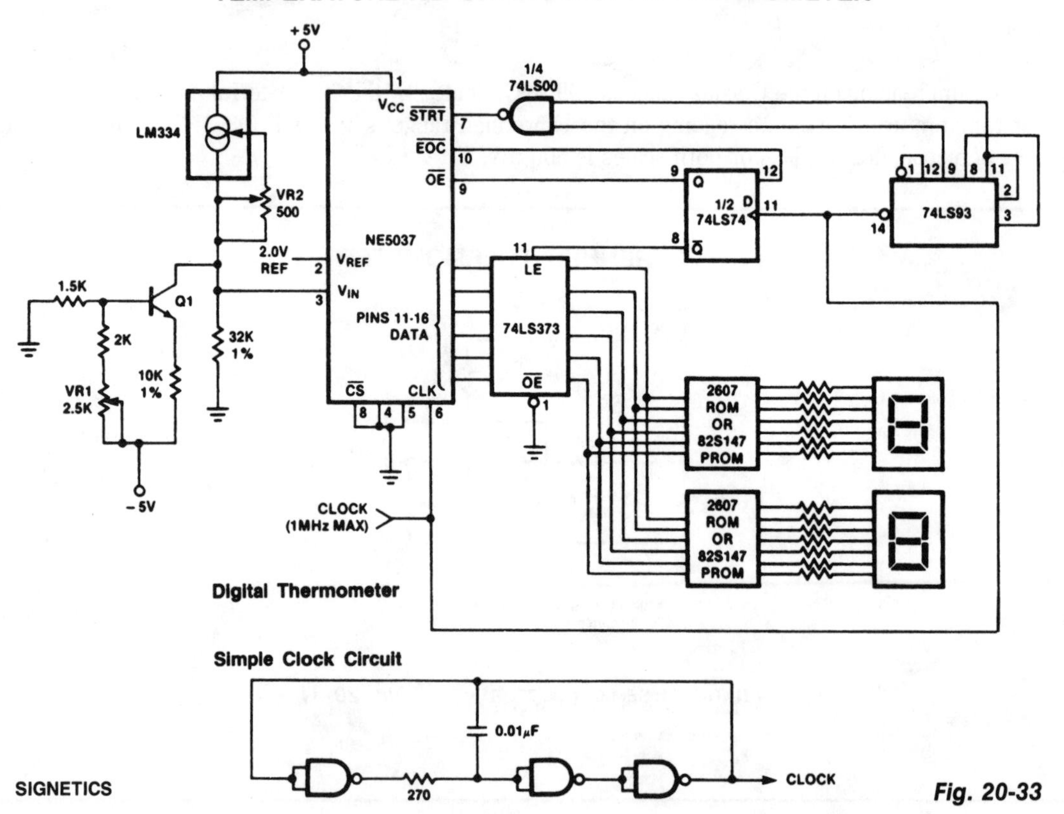

SIGNETICS

Fig. 20-33

The ROMS or PROMs must have the correct code for converting the data from the NE5037—used as address for the ROMs or PROMs—to the appropriate segment driver codes. The displayed amount could easily be converted to degrees Fahrenheit, °F, by the controller of (0 to 63° temperature sensor) or through the (P)ROMs. When doing this, a third (hundreds) digit (P)ROM and display will be needed for displaying temperatures above 99°F. An expensive clock can be made from NAND gates or inverters, as shown.

ELECTRONIC THERMOMETER

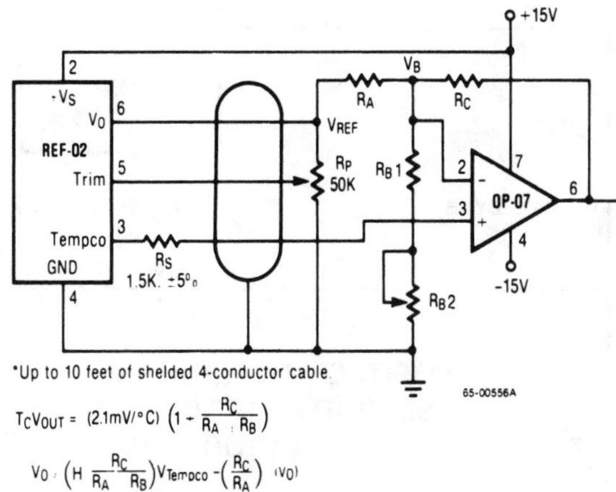

Resistor Values

TCVOUT Slope(s)	10mV/°C	100mV/°C	10mV/°F
Temperature Range	-55°C to +125°C	-55°C to +125°C	-65°F to +257°F
Output Voltage Range	-0.55V to +1.25V	-5.5V to +12.5V	-0.67V to +2.57V
Zero Scale	0V at 0°C	0V at 0°C	0V at 0°F
RA (±1% Resistor)	9.09KΩ	15KΩ	8.25KΩ
RB1 (±1% Resistor)	1.5KΩ	1.82KΩ	1.0KΩ
RB2 (Potentiometer)	200Ω	500Ω	200Ω
RC (±1% Resistor)	5.11KΩ	84.5KΩ	7.5KΩ

RAYTHEON CO. SEMICONDUCTOR DIVISION

Fig. 20-34

This circuit uses the +5-V reference output and the op amp to level shift and amplify the 2.1-mV/°C Tempco output into a voltage signal dependent on the ambient temperature. Different scaling can be obtained by selecting appropriate resistors from the table giving output slopes calibrated in degrees Celsius or degrees Fahrenheit. To calibrate, first measure the voltage on the Tempco pin, V_{TEMPCO}, and the ambient room temperature, T_A in °C. Put those values into the following equation:

$$\frac{V_{TEMPCO} \text{ (in mV)}}{(S)\ (T_A+273)}$$

Where S = Scale factor for your circuit selected from the table in mV. Then, turn the circuit power off, short V_{OUT} at pin 6 of the REF-02 to ground, and while applying exactly 100.00 mV to the op amp output, adjust R_{B2} so that $V_B = (X)$ (100 mV). Now remove the short and the 100-mV source, reapply circuit power and adjust R_P so that the op-amp output voltage equals (T_A) (S). The system is now exactly calibrated.

GROUND-REFERRED CENTIGRADE THERMOMETER I

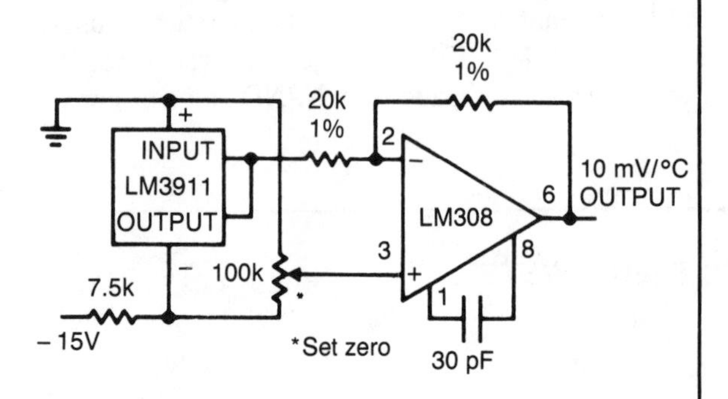

NATIONAL SEMICONDUCTOR *Fig. 20-35*

TEMPERATURE SENSOR

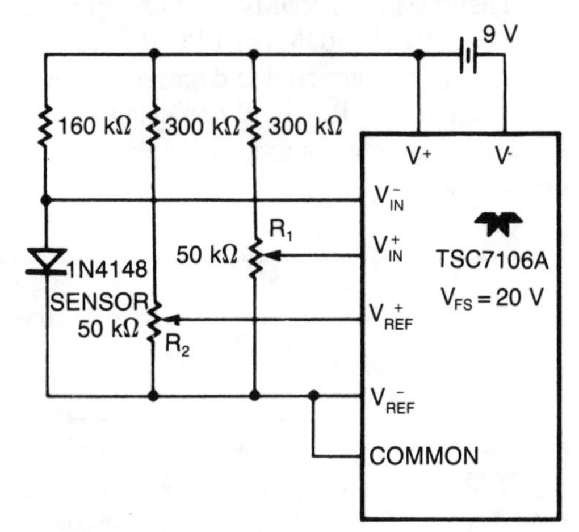

TELEDYNE *Fig. 20-37*

GROUND-REFERRED CENTIGRADE THERMOMETER II

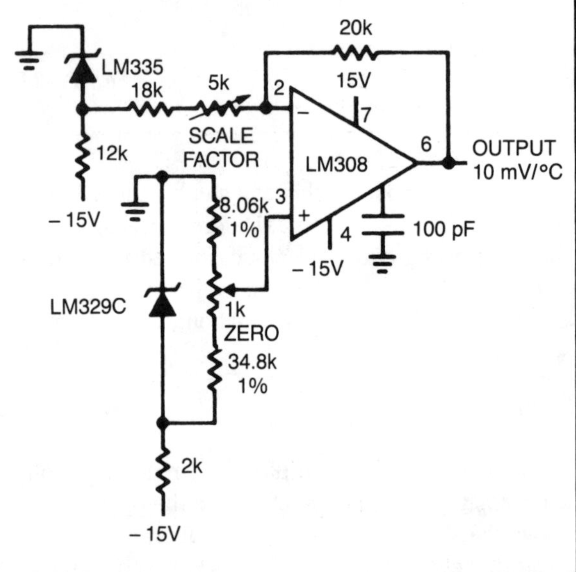

NATIONAL SEMICONDUCTOR *Fig. 20-36*

POSITIVE TEMPERATURE SENSOR COEFFICIENT RESISTOR

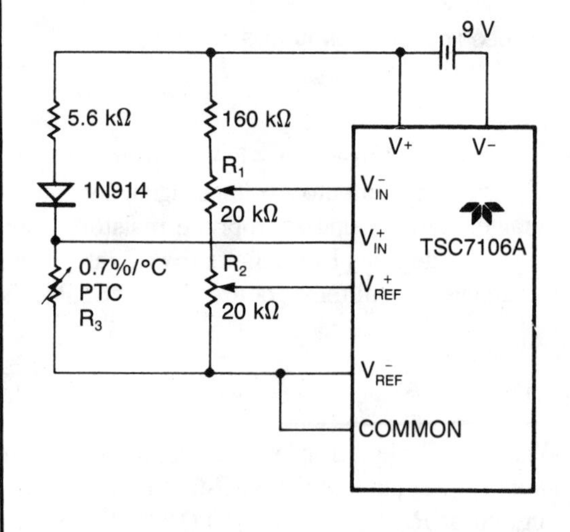

TELEDYNE *Fig. 20-38*

DIFFERENTIAL TEMPERATURE SENSOR

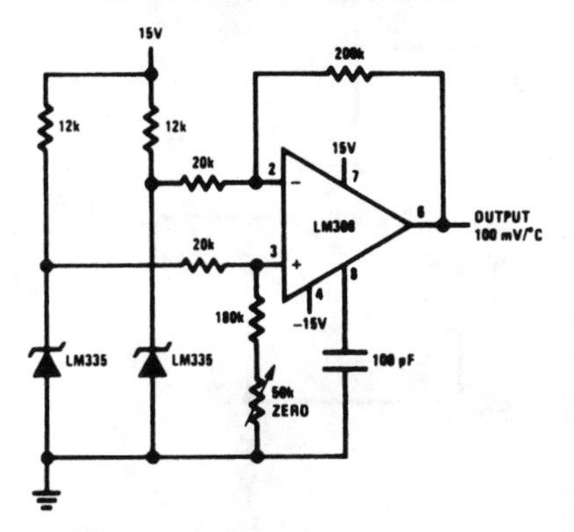

NATIONAL SEMICONDUCTOR *Fig. 20-39*

CENTIGRADE THERMOMETER

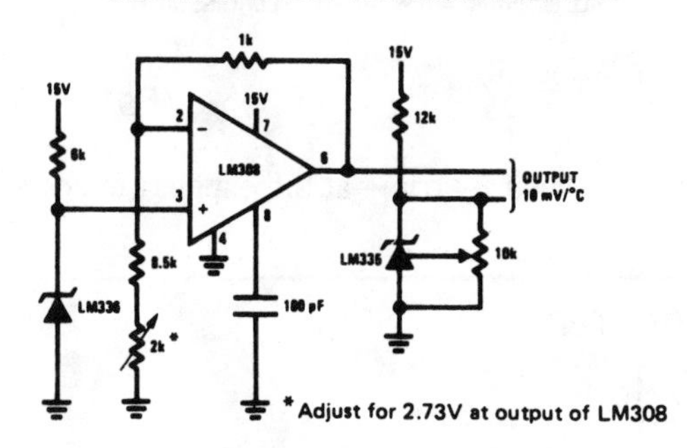

*Adjust for 2.73V at output of LM308

NATIONAL SEMICONDUCTOR *Fig. 20-40*

METER THERMOMETER WITH TRIMMED OUTPUT

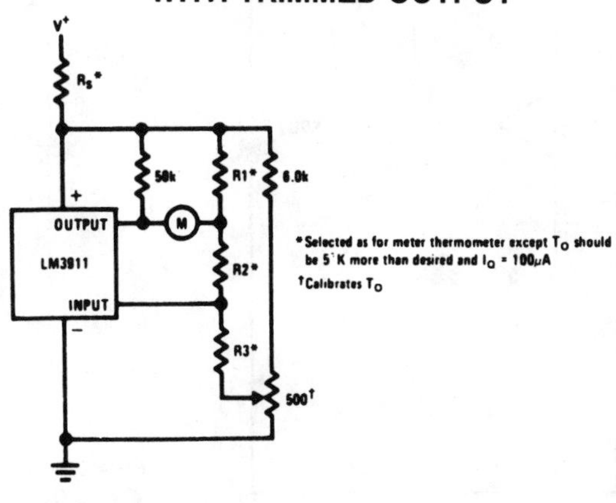

*Selected as for meter thermometer except T_O should be 5°K more than desired and I_Q = 100μA

†Calibrates T_O

NATIONAL SEMICONDUTOR *Fig. 20-41*

KELVIN THERMOMETER WITH GROUND-REFERRED OUTPUT

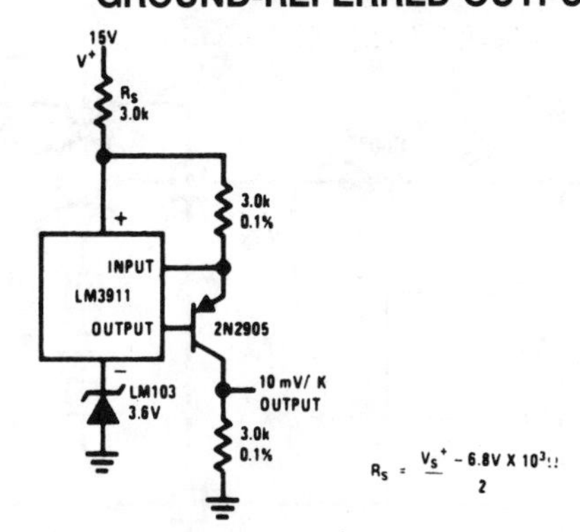

$R_S = \frac{V_S^+ - 6.8V}{2} \times 10^3$

NATIONAL SEMICONDUCTOR *Fig. 20-42*

LOWER POWER THERMOMETER

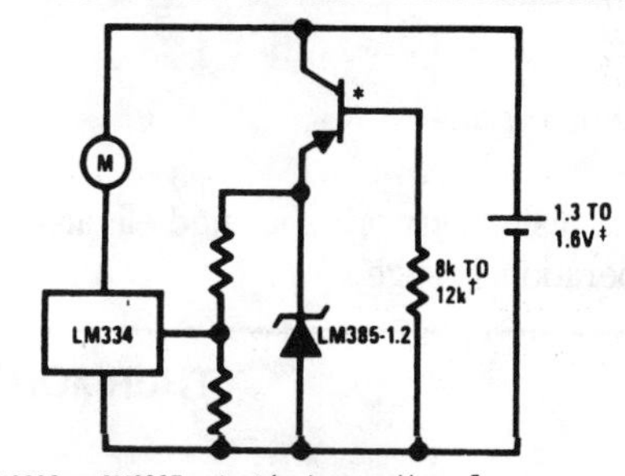

* 2N3638 or 2N2907 select for inverse $H_{FE} \cong 5$

† Select for operation at 1.3V ‡ $I_Q \cong 600$ μA to 900 μA

NATIONAL SEMICONDUCTOR *Fig. 20-43*

0 TO 50°F THERMOMETER

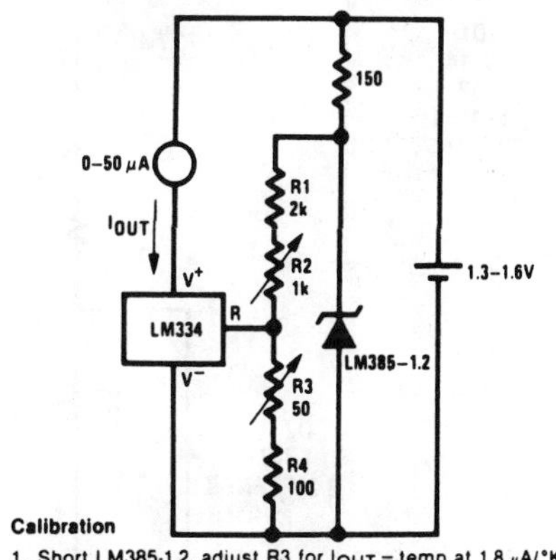

Calibration

1. Short LM385-1.2, adjust R3 for I_{OUT} = temp at 1.8 μA/°K
2. Remove short, adjust R2 for correct reading in °F

NATIONAL SEMICONDUCTOR *Fig. 20-44*

SIMPLE LINEAR THERMOMETER

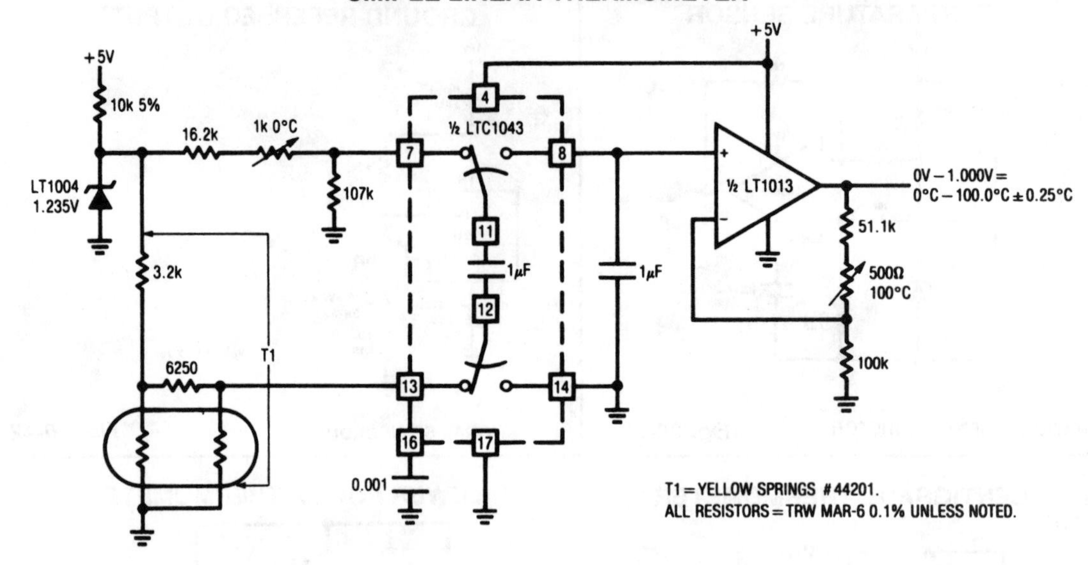

LINEAR TECHNOLOGY CORP. *Fig. 20-45*

The thermistor network specified eliminates the need for a linearity trim—at the expense of accuracy and operational range.

THERMOMETER ADAPTER

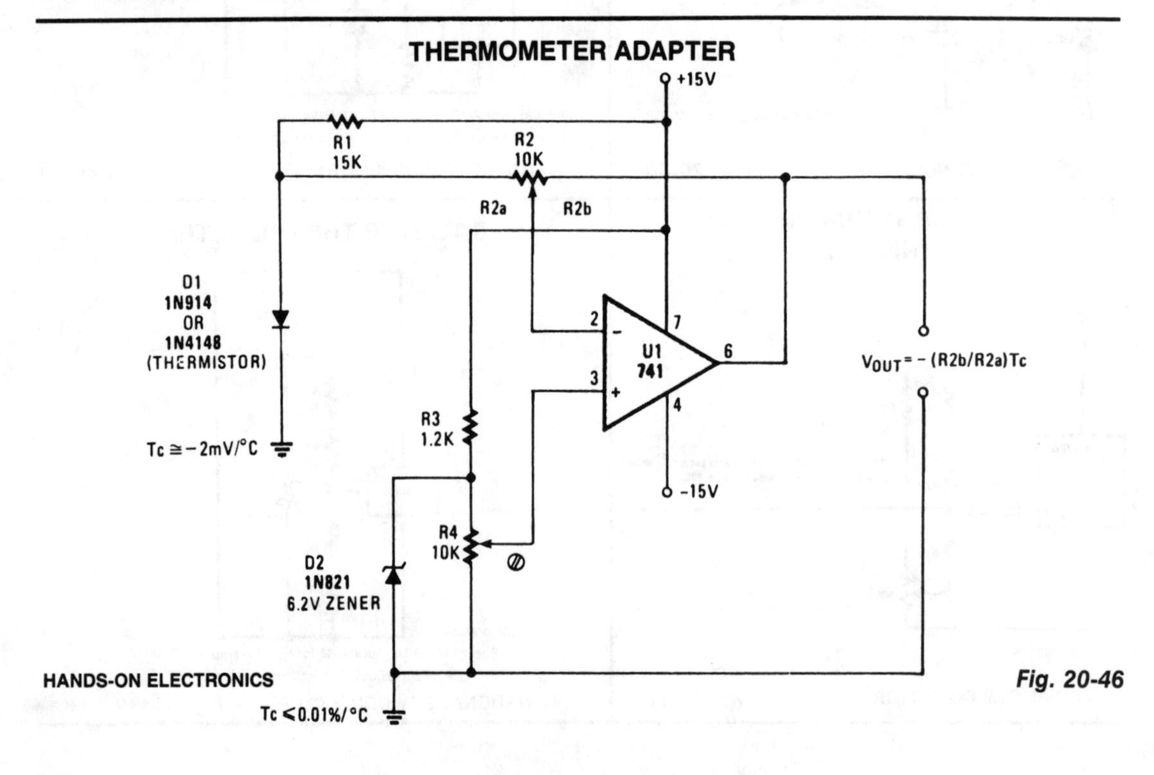

HANDS-ON ELECTRONICS *Fig. 20-46*

THERMOMETER ADAPTER *(Cont.)*

A simple op amp and silicon diode are the heart of the temperature-to-voltage converter that will permit you to use an ordinary voltmeter—either analog or digital—to measure temperature. User adjustments make it possible for a reading of either 10 mV or 100 mV to represent 1°F or C.

Temperature sensor D1 is a 1N4148 silicon diode. It has a temperature coefficient of −2 mV/°C. U1, a 741 op amp, is connected as a differential amplifier. A voltage divider consisting of R3 and Zener diode D2 provides a 6.2-V reference voltage. D2 is shunted by potentiometer R4, so that the offset can be adjusted to align the output voltage with either the Celsius or Fahrenheit scale, as desired.

Gain control R2 is adjusted so that the output of the op amp is in the scale or voltage range of the meter being used. R4, the offset-adjust control, is then adjusted so that the output voltage represents either degrees F or C. The thermometer adapter can be calibrated by adjusting R4 while the probe sensor is at a known temperature.

0 TO 100°C THERMOMETER

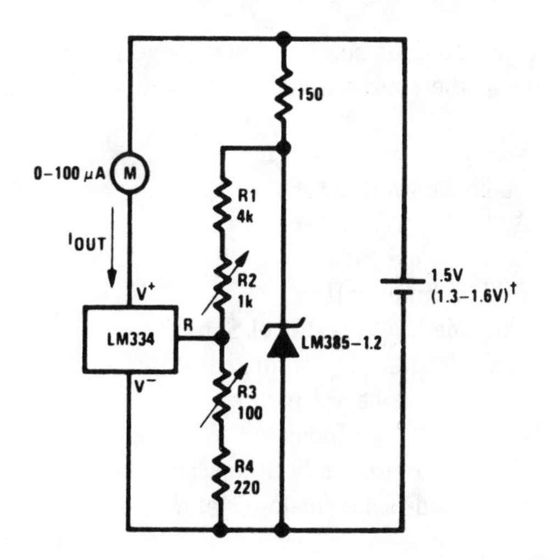

Calibration
1. Short LM385-1.2. adjust R3 for I_{OUT} = temp at 1 μA/°K
2. Remove short. adjust R2 for correct reading in centigrade

† I_Q at 1.3V ≥ 500 μA
I_Q at 1.6V ≥ 2.4 μA

NATIONAL SEMICONDUCTOR

Fig. 20-47

GROUND-REFERRED FAHRENHEIT THERMOMETER

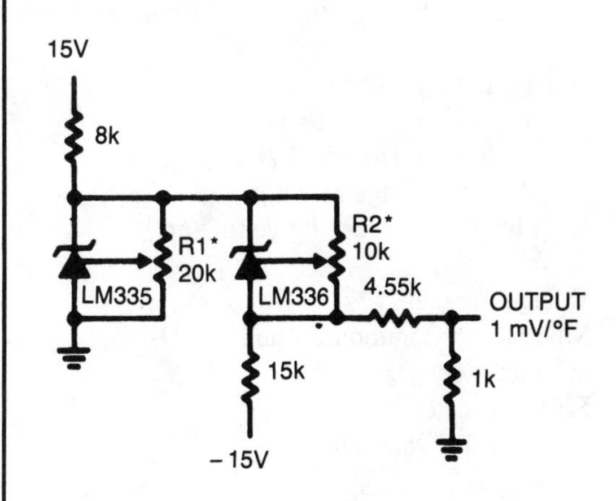

* Adjust R2 for 2.554V across LM336.
Adjust R1 for correct output.

NATIONAL SEMICONDUCTOR

Fig. 20-48

21

Voltage-Indicator/Monitor Circuits

The sources of the following circuits are contained in the Sources section, which begins on page 217. The figure number in the box of each circuit correlates to the source entry in the Sources section.

Voltage-Level Indicator
$4\frac{1}{2}$-Digit DVM
Full-Scale Four-Decade $3\frac{1}{2}$-Digit DVM
Over/Under Voltage Monitor
High-Input Resistance dc Voltmeter
dc Voltmeter
Voltage Freezer
Multiplexed Common-Cathode LED-Display ADC
ac Voltmeter
FET Voltmeter I
Sensitive RF Voltmeter
Voltage Monitor
Audio Millivoltmeter I
Audio Millivoltmeter II
Low-Voltage Indicator
FET Voltmeter II
Simplified Voltage-Level Sensor
Peak Program Detector
Wide-Range ac Voltmeter
Visable Voltage Indicator
Bipolar Reference Source
Expanded-Scale Analog Meter

VOLTAGE-LEVEL INDICATOR

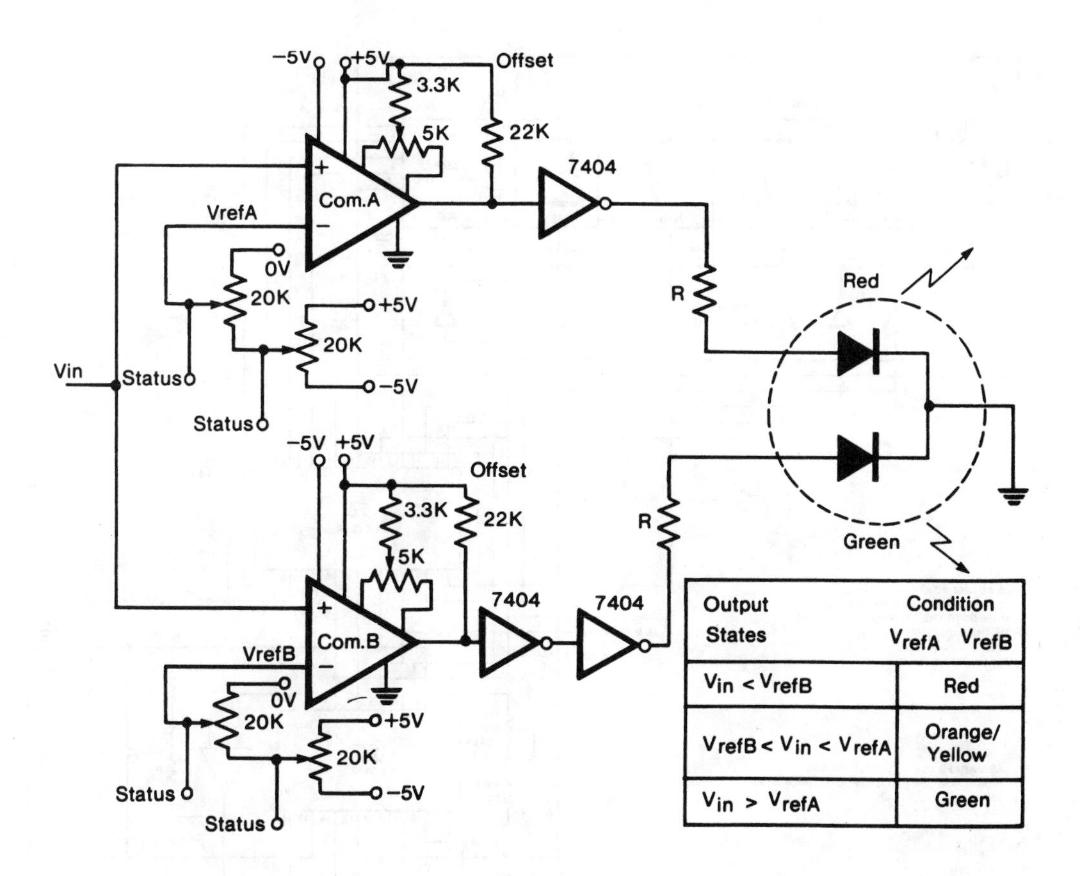

Output States	Condition V_{refA} V_{refB}
$V_{in} < V_{refB}$	Red
$V_{refB} < V_{in} < V_{refA}$	Orange/ Yellow
$V_{in} > V_{refA}$	Green

ELECTRONIC ENGINEERING ***Fig. 21-1***

A tricolor LED, acts as the visual indicator of the voltage level. The voltage to be measured is connected to the two comparators in parallel. The first 20-kΩ trimmer defines a voltage between ±5 V and this becomes the full-scale value of the reference voltage. The second trimmer is a fine adjustment to give any reference voltage between 0 V and the full-scale voltage. Thus, it is possible to select both positive and negative reference voltages. During the initialization procedure, a voltage, equal to the reference voltage of each comparator, is connected to the input terminal, and the offset-balance potentiometer is adjusted to give a reading between the high and low output voltage levels. The inverter following comp A ensures that, whatever the input voltage, at least one diode is lit. The two inverters following comp B leave the voltage largely unchanged, but provide the current necessary to illuminate the diode. The value of the resistance should be chosen so that the current through any single diode does not exceed the specified limit, usually 30 mA. The LED contains a red and a green diode with a common cathode. When both diodes are lit, a third color, orange, is emitted. With V_{refA} greater than V_{refB}, the output states given in the diagram apply.

4½-DIGIT DVM

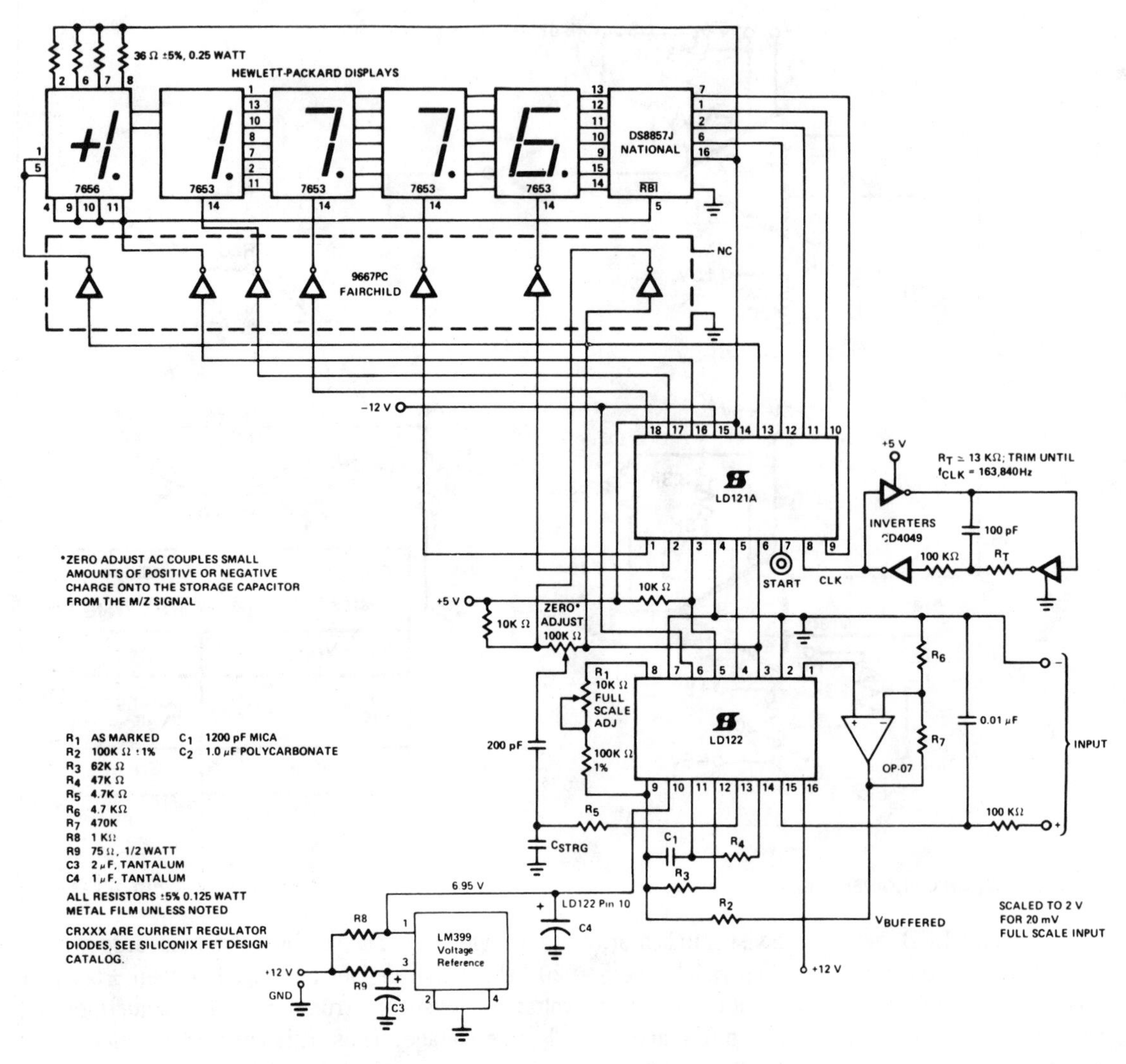

SILICONIX *Fig. 21-2*

- 1-μV resolution
- Overrange blinking
- 0 to 19.999-mV input voltages
- Zero adjust-to-null offset introduced by PC board leakage and the comparator.

FULL-SCALE FOUR-DECADE 3 1/2-DIGIT DVM

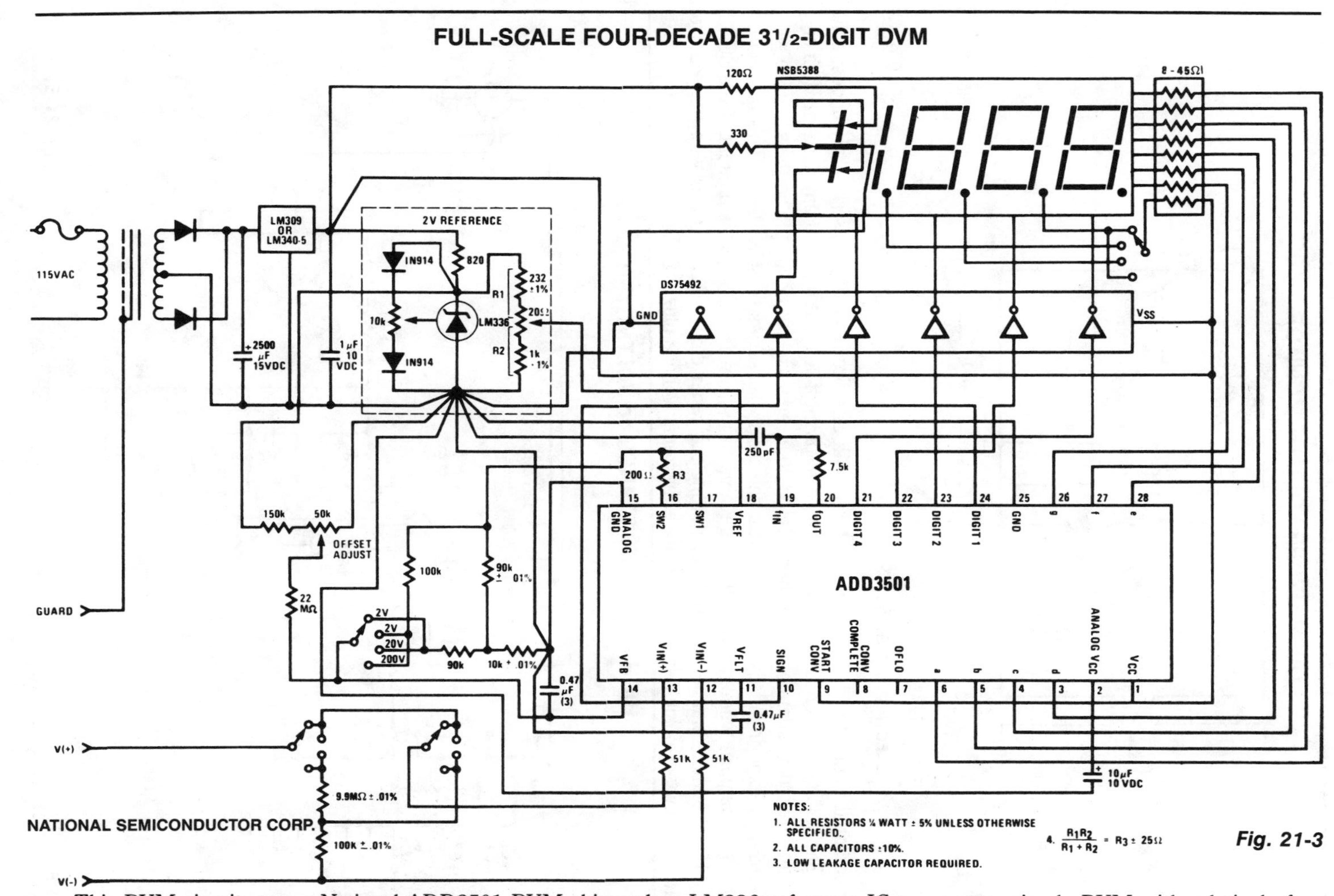

NATIONAL SEMICONDUCTOR CORP.

Fig. 21-3

This DVM circuit uses a National ADD3501 DVM chip and an LM336 reference IC to create a simple DVM with relatively few components. When making a single-range panel meter, the range-switching components can be left out, as required.

OVER/UNDER VOLTAGE MONITOR

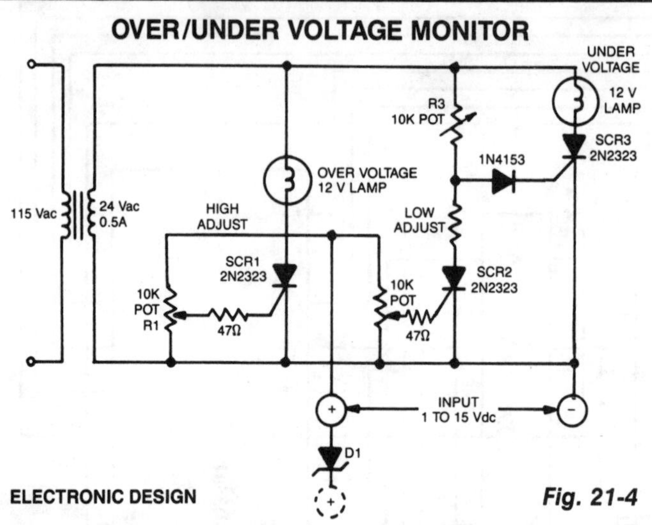

ELECTRONIC DESIGN *Fig. 21-4*

Any potential from 1 to 15 V can be monitored with this circuit. Two lamps alert any undesirable variation. The voltage differential from lamp turn-on to turn-off is about 0.2 V at any setting. High and low set points are independent of each other. The SCRs used in the circuit should be the sensitive gate type. R3 must be experimentally determined for the particular series of SCRs used. This is done by adjusting R3 to the point where the undervoltage lamp turns on when no signal is present at the SCR2 gate. Any 15-V segment can be monitored by putting the zener diode, D1, in series with the positive input lead. The low set-point voltage will then be the zener voltage plus 0.8 V.

HIGH-INPUT RESISTANCE dc VOLTMETER

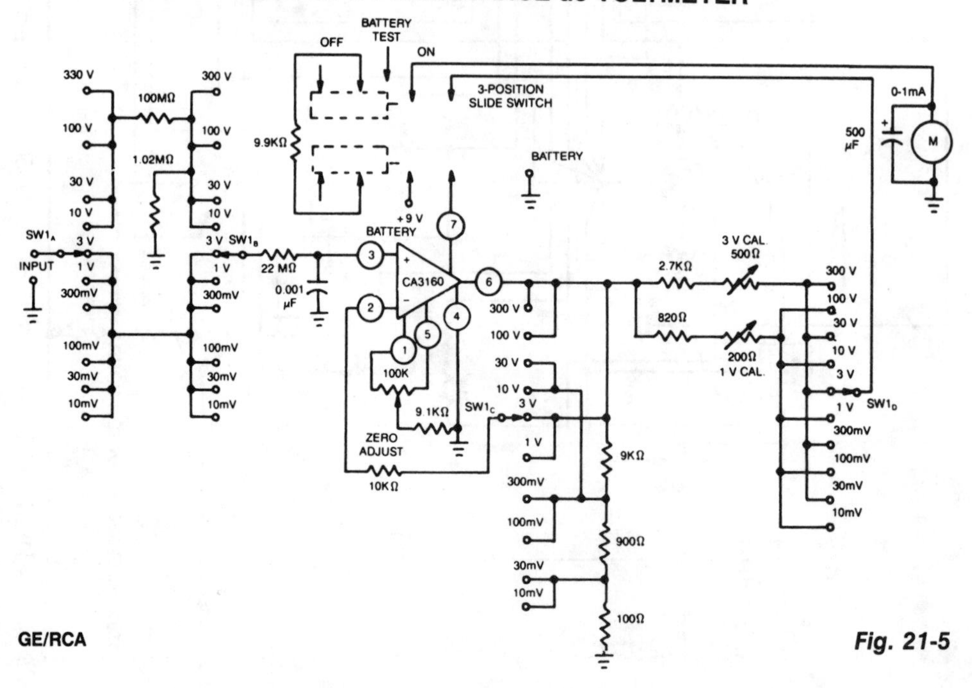

GE/RCA *Fig. 21-5*

HIGH-INPUT RESISTANCE dc VOLTMETER (*Cont.*)

This voltmeter exploits a number of the CA3160 BiMOS op amp's useful characteristics. The available voltage ranges from 10 mV to 300 V. Powered by a single 8.4-V mercury battery, this circuit, with zero input, consumes approximately 500 μA. Thus, at full-scale input, the total supply current will increase by 1000 μA.

dc VOLTMETER

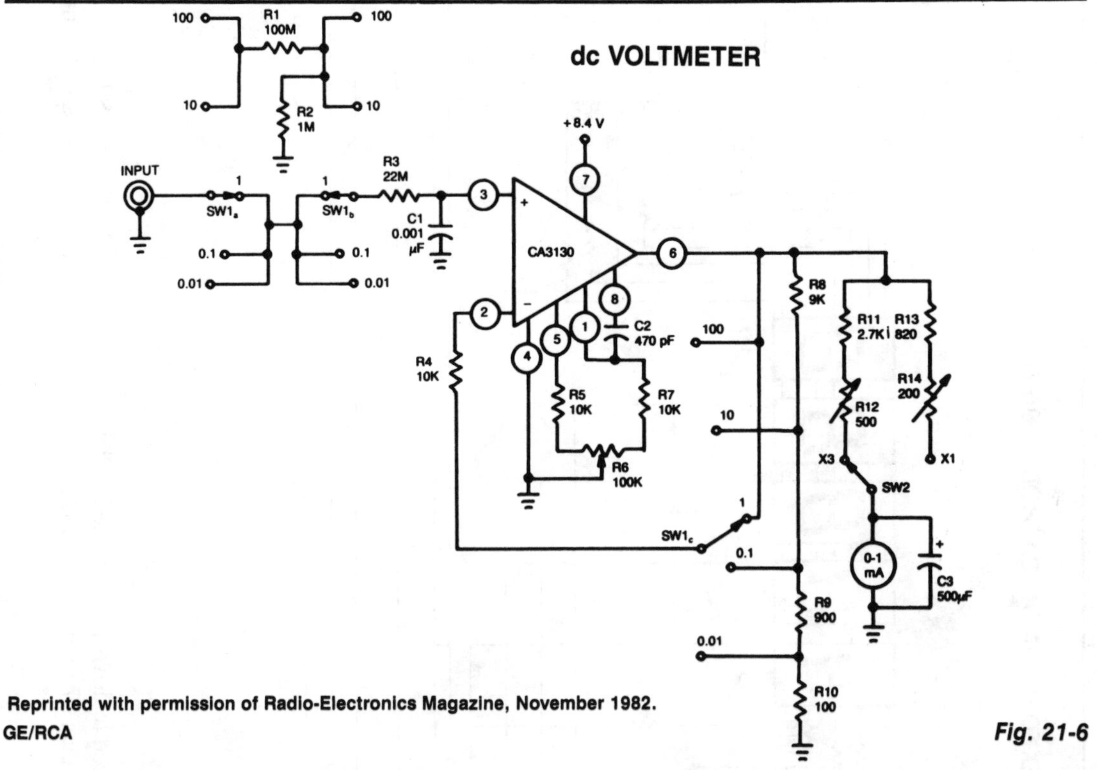

Reprinted with permission of Radio-Electronics Magazine, November 1982.

GE/RCA

Fig. 21-6

This dc voltmeter, with high input resistance, uses a CA3130 BiMOS op amp and measures voltages from 10 mV to 300 V. Resistors R12 and R14 are used individually to calibrate the meter for full-scale deflection. Potentiometer R6 is used to null the op amp and meter on the 10-mV range by shorting the input terminals, then adjusting R6 for the first indication of upscale meter deflection.

VOLTAGE FREEZER

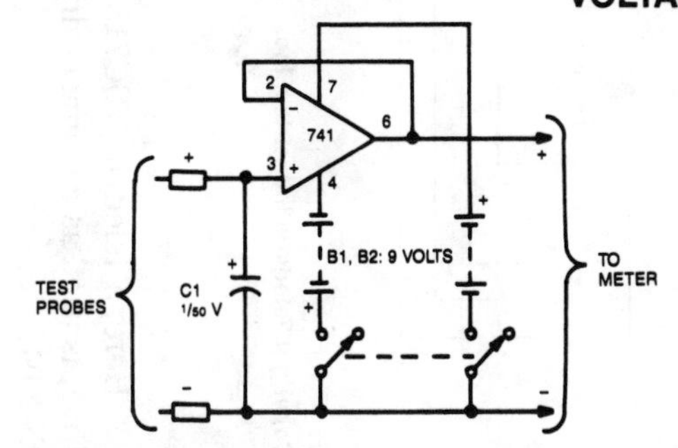

This circuit reads and stores voltages, thus freezing the meter reading even after the probes are removed. The op amp is configured as a unity-gain voltage follower, with C1 situated at the input to store the voltage. For better performance, use an LF13741 or a TL081 op amp in place of the 741. These two are JFET devices and offer a much higher input impedance than the 741.

RADIO-ELECTRONICS

Fig. 21-7

MULTIPLEXED COMMON-CATHODE LED-DISPLAY ADC

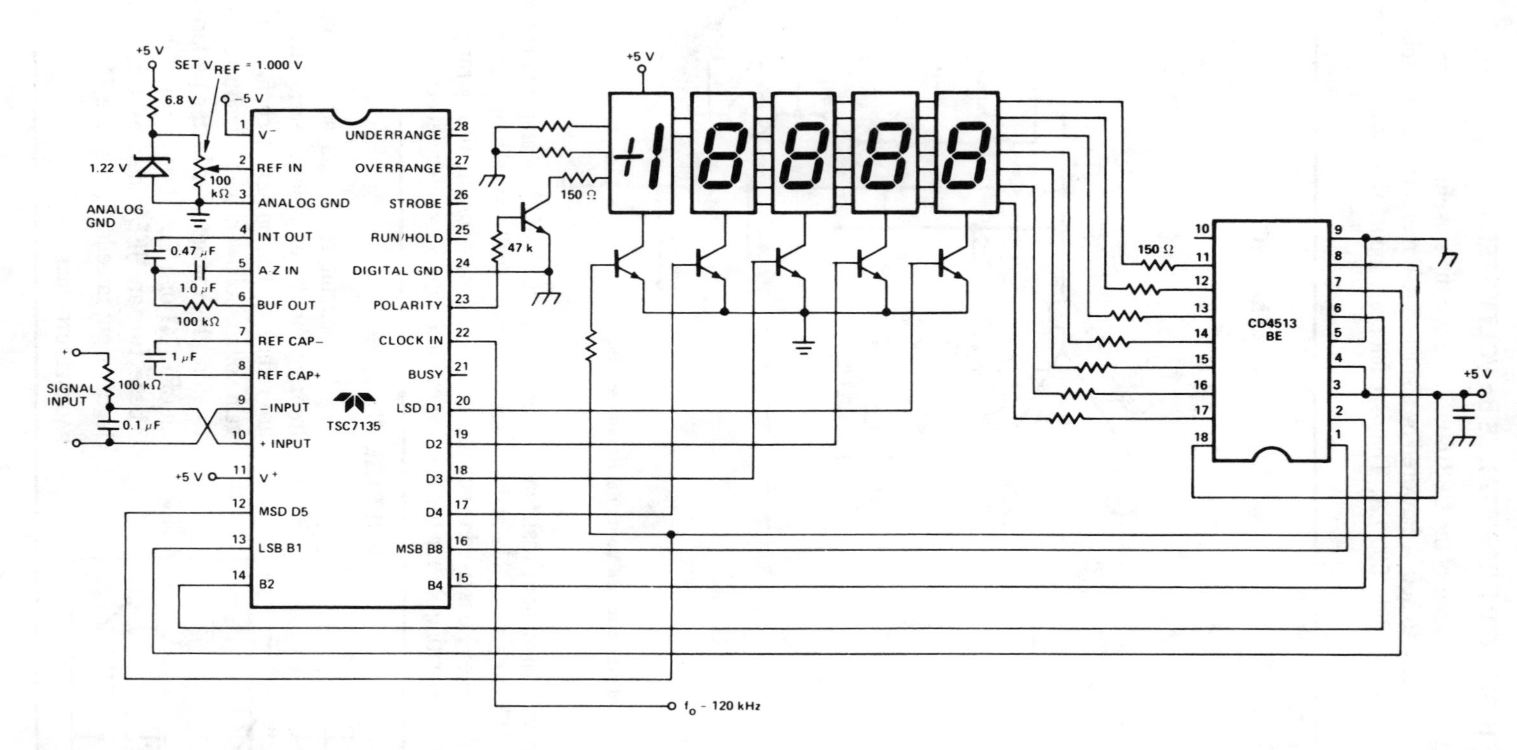

Fig. 21-8

Here, a Teledyne TSC7135 DVM chip is used to drive a multiplexed 5-digit display. A CD4513BE CMOS IC, for common cathode drive, is used as a segment driver selected by pins 17 to 20 of the DVM chip. The transistors can be any suitable npn type, such as 2N3904, etc.

ac VOLTMETER

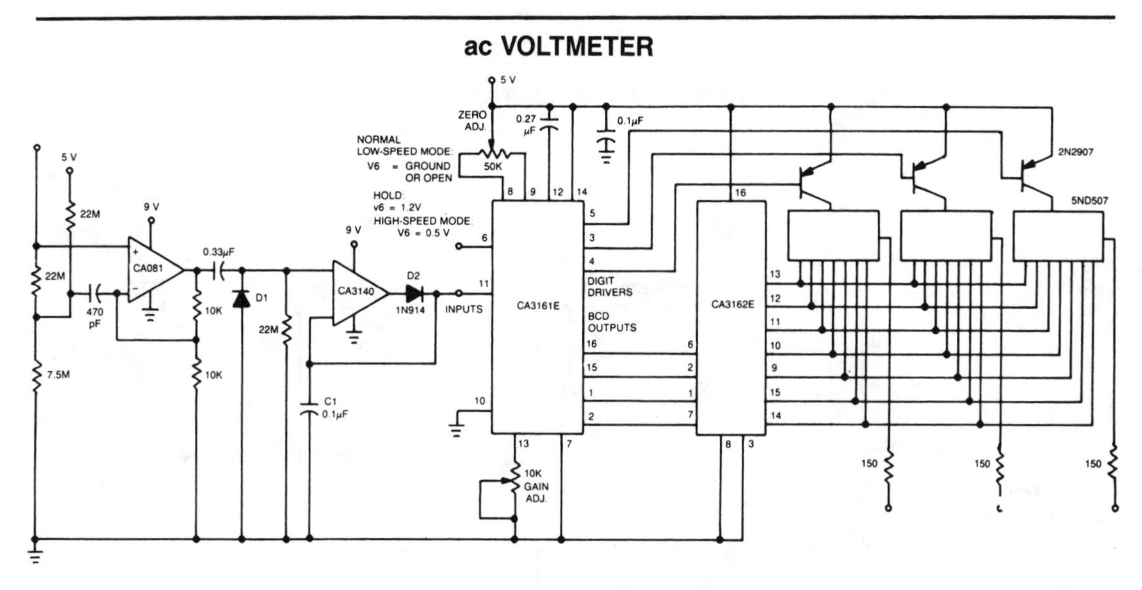

GE/RCA **_Fig. 21-9_**

CA081 and CA3140 BiMOS op amp offer minimal loading on the circuits being measured. The wide bandwidth and high slew rate of the CA081 allow the meter to operate up to 0.5 MHz.

FET VOLTMETER I

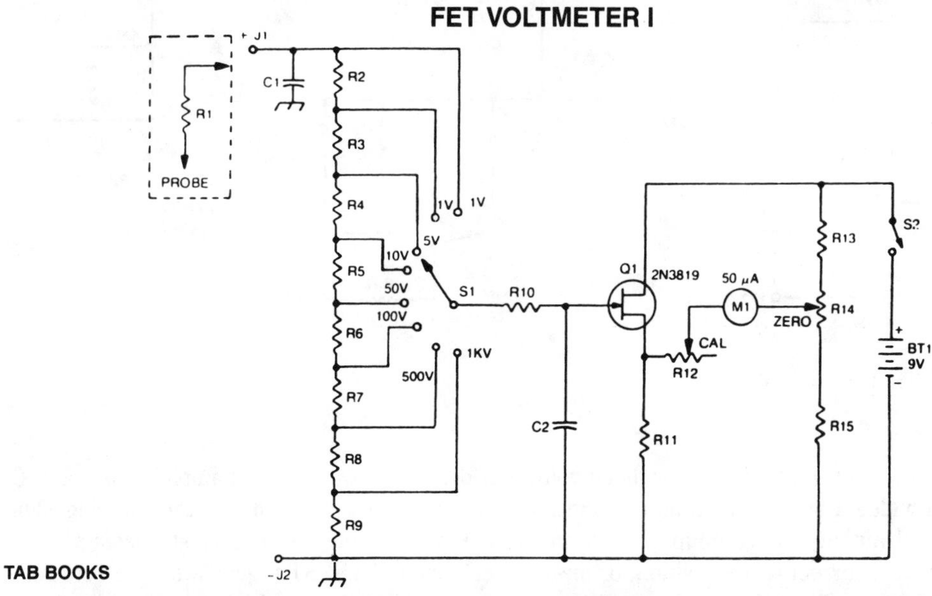

TAB BOOKS **_Fig. 21-10_**

A 2N3819 FET provides a *solid-state VOM*. The 2N3819 acts as a *cathode follower* in a VOM. The bias offset (meter null) obtained with R14 and R12 sets full-scale calibration. R2 through R9 should total about 10 MΩ. R10 is a protective resistor, and C2 provides ac bypassing to limit RF and noise pickup.

SENSITIVE RF VOLTMETER

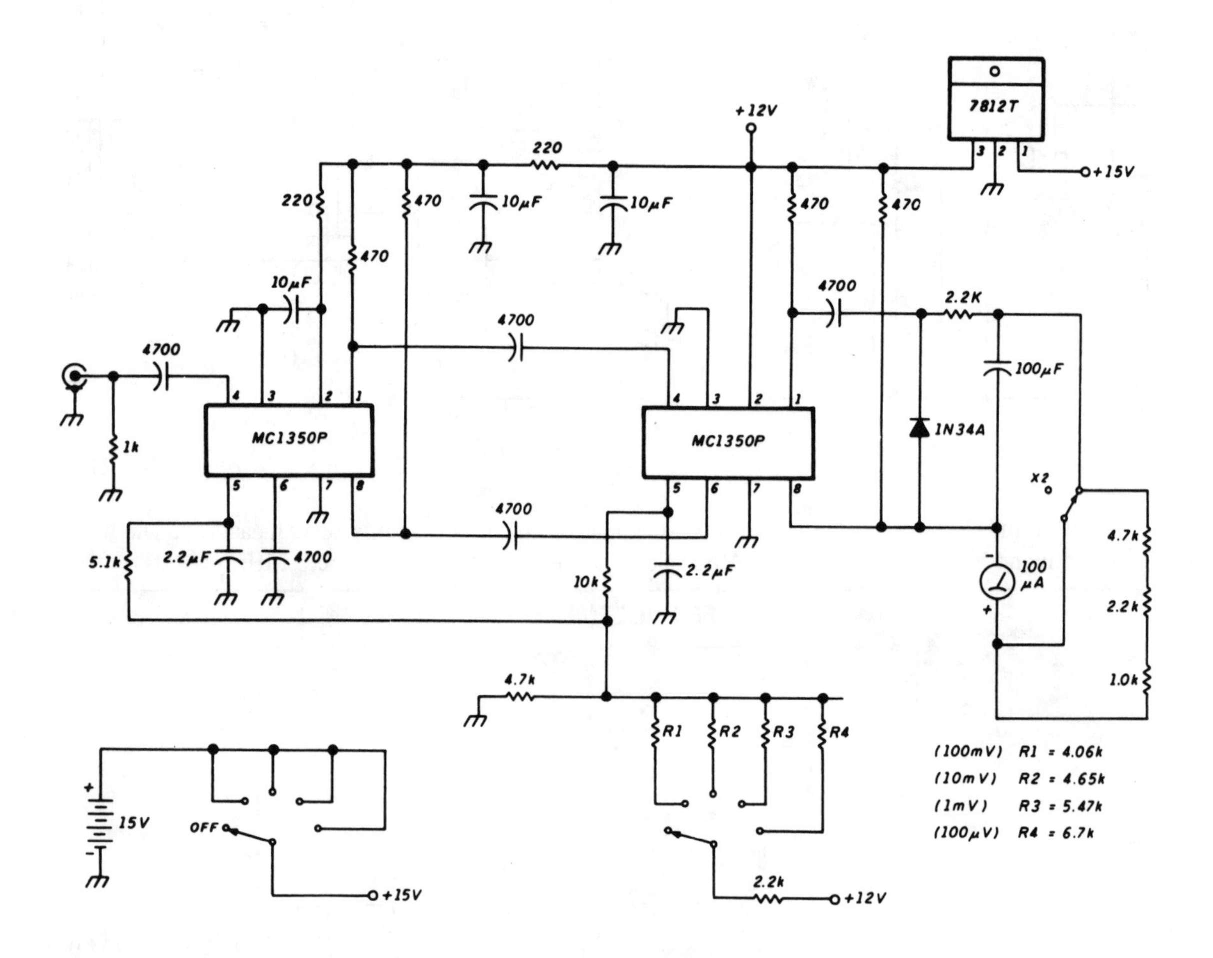

HAM RADIO ***Fig. 21-11***

This schematic shows a peak-reading diode voltmeter driven by two stages of amplification. A 100-μF capacitor provides a fairly large time constant, which results in satisfactory meter damping. The limited differential output voltage coupled with an overdamped meter prevents most *needle pinning* when you select an incorrect range position, or make other errors. An SPST toggle switch selects additional series resistance. This X2 function gives some more overlap of the sensitivity ranges. The resistance values shown are correct for use with a 100-μA meter with 1500-Ω internal resistance.

VOLTAGE MONITOR

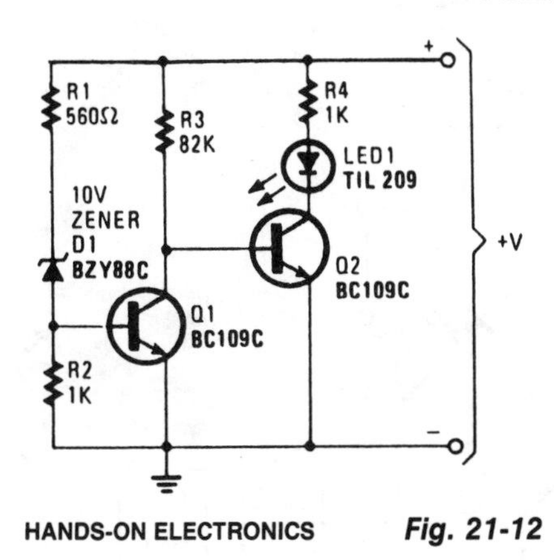

HANDS-ON ELECTRONICS *Fig. 21-12*

If the battery voltage exceeds about 11 V, current flows through R1, D1, and R2. The voltage produced as a result of current flow through R2 is sufficient to bias transistor Q1 into conduction. That places the collector voltage of Q1 virtually at ground. Therefore, Q2, driven from the collector of Q1, is cut off, LED1 and current-limiting resistor R4 are connected in the collector circuit of Q2. With Q2 in the cut-off state, the LED does not light. Should Q1's base voltage drop below approximately 0.6 V, Q1 turns off, biasing Q2 on, and illuminating LED1 to indicate that the battery voltage has fallen below the 11-V threshold level.

AUDIO MILLIVOLTMETER I

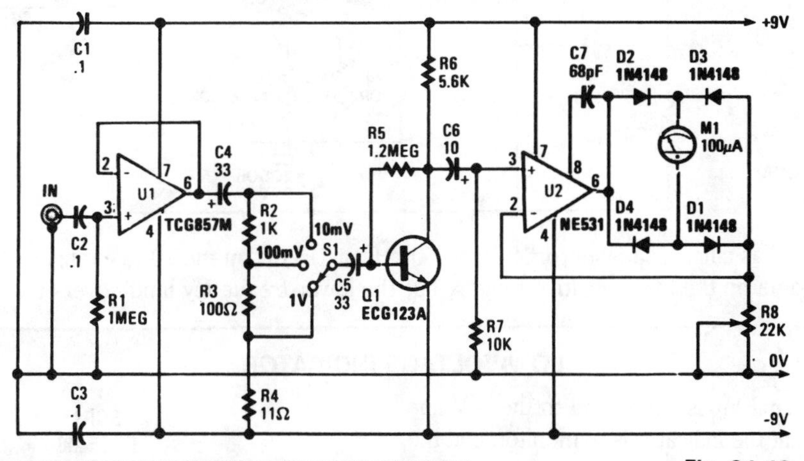

POPULAR ELECTRONICS/HANDS-ON ELECTRONICS *Fig. 21-13*

Capacitor C4 couples the output of U1 to a simple attenuator, which is used to provide a loss of 0 dB, 20 dB, or 40 dB, depending on the setting of range switch S1. The circuit's sensitivity is 10-V rms for full-scale deflection, so the attenuator gives additional ranges of 100 mV and 1 V rms. The attenuator output is connected through capacitor C5 to common-emitter amplifier Q1, which has a high-voltage gain of 40 dB.

To get linear scaling on the meter, we have to use an active-rectifier circuit built around U2. That IC is connected so that its noninverting input is biased to the 0-V bus via R7. Capacitor C6 couples the output of Q1 to the noninverting input of U2; C7 is the compensation capacitor for U2.

The voltage gain of U2 is set by the difference in resistance between the output and the inverting input, and between the inverting input and the ground bus. One resistance is made up of the diode-bridge rectifier D1 through D4, the other by resistor R8. This circuit has a nearly flat frequency response to about 200 kHz.

AUDIO MILLIVOLTMETER II

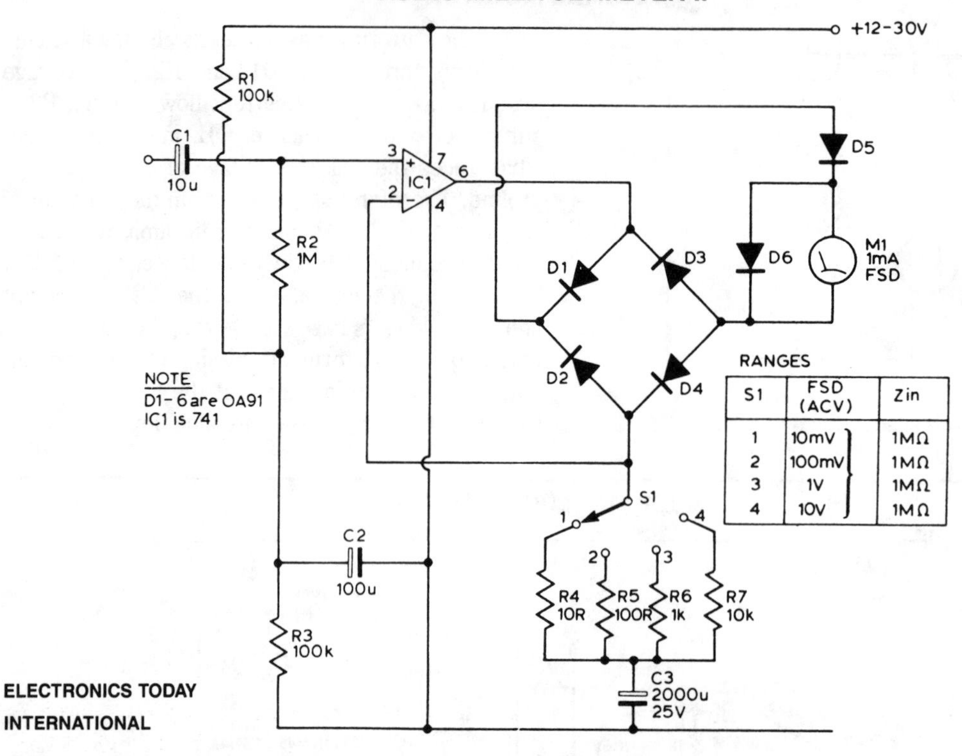

S1	FSD (ACV)	Zin
1	10mV	1MΩ
2	100mV	1MΩ
3	1V	1MΩ
4	10V	1MΩ

ELECTRONICS TODAY INTERNATIONAL

Fig. 21-14

This circuit has a flat response from 8 Hz to 50 kHz at −3 dB on the 10-mV range. The upper limit remains the same on the less sensitive ranges, but the lower frequency limit covers under 1 Hz.

LOW-VOLTAGE INDICATOR

Input terminal V_{IN} is connected to the +V line of the circuit that the indicator is to monitor, and the grounds of both circuits are connected together. The position of potentiometer R1's wiper determines Q1's base voltage. As long as the transistor gets enough bias voltage to remain on, the low voltage at the collector will keep the SCR from firing. As the battery voltage starts to fall, the transistor's base voltage will fall as well. When Q1 turns off (V_{IN} drops), the collector voltage increases. That voltage provides enough gate drive to turn on the SCR, which turns on the LED. The LED could also be a buzzer or almost any other type of warning device.

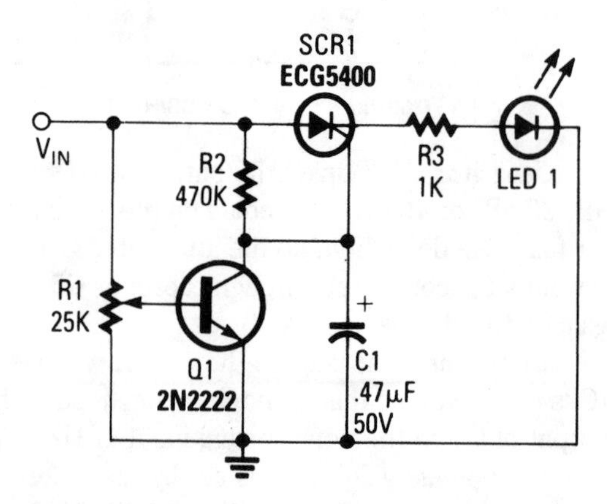

Fig. 21-15

FET VOLTMETER II

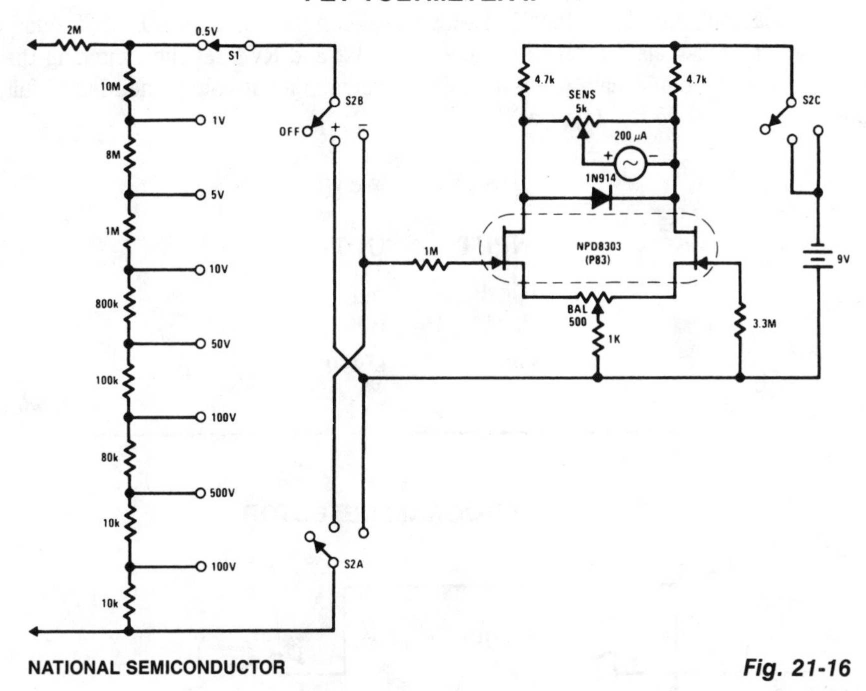

NATIONAL SEMICONDUCTOR *Fig. 21-16*

This FETVM replaces the function of the VTVM and rids the instrument of the usual line cord. In addition, FET drift rates are far superior to vacuum-tube circuits, allowing a 0.5-V full-scale range, which is impractical with most vacuum tubes. The low-leakage, low-noise NPD8303 is ideal for this application.

SIMPLIFIED VOLTAGE-LEVEL SENSOR

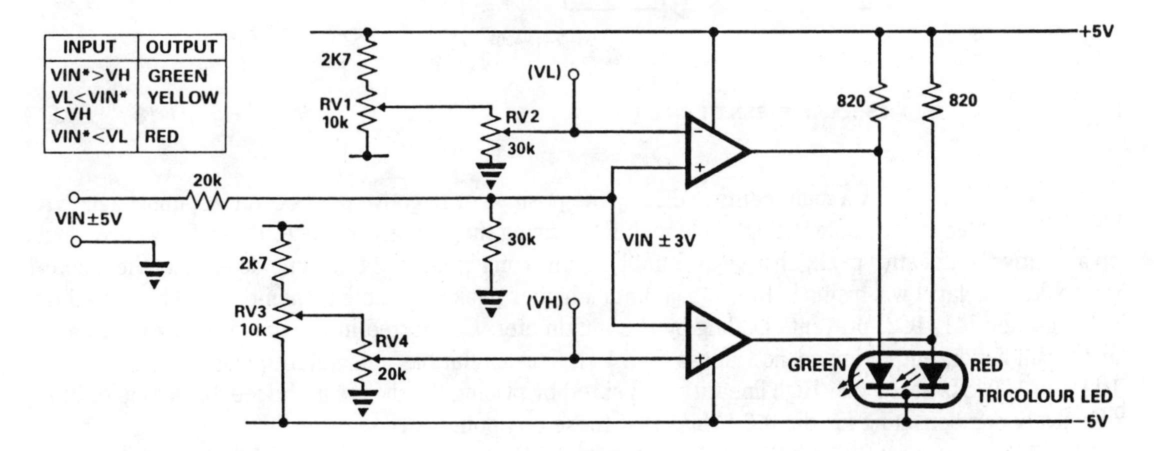

ELECTRONIC ENGINEERING *Fig. 21-17*

SIMPLIFIED VOLTAGE-LEVEL SENSOR (*Cont.*)

This circuit uses only one IC, either 1, LM393 dual comparator or 1/2, LM339 quad comparator. RV1 and RV3 set the full-scale reference voltage, and RV2 and RV4 set the switching thresholds to a value between 0 V and the full-scale reference. The change in input voltage needed to fully switch the output state is less than 0.05 mV (typical).

An alternative is:

INPUT	OUTPUT
V_{IN}, V_H	red
V_H, V_{IN}, V_L	yellow
V_{IN}, V_L	green

PEAK PROGRAM DETECTOR

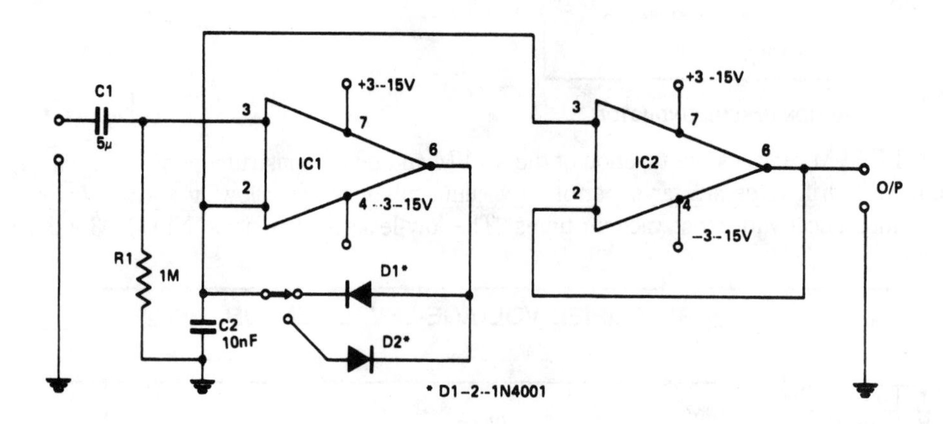

ELECTRONICS TODAY INTERNATIONAL

Fig. 21-18

This circuit will allow a multimeter to display the positive or negative peaks of an incoming signal. A 741, IC1, is used in the noninverting mode with R1 defining the input impedance. D1 or D2 will conduct on a positive or negative peak, charging C2 until the inverting input is at the same dc level as the incoming peak. This level will maintain the voltage until a higher peak is detected, then this will be stored by C2. Another 741, IC2, prevents loading by the multimeter. Connected in the noninverting mode as a unity-gain buffer, output impedance is less than 1 Ω. This circuit has a useful frequency response from 10 Hz to 100 kHz at ±1 dB. High linearity is ensured by placing the diodes in the feedback loop of IC1, effectively compensating for the 0.6 V bias that these components require.

WIDE-RANGE ac VOLTMETER

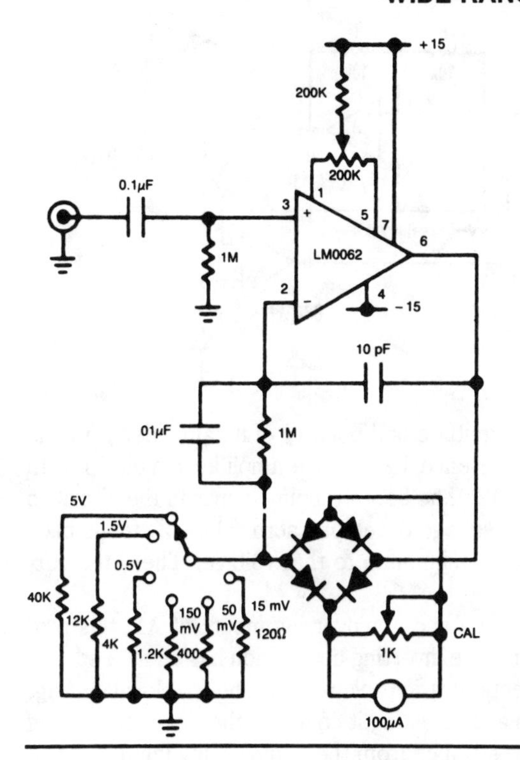

In this circuit, a diode bridge is used as a meter rectifier. The offset voltage is compensated for by the op amp because the bridge is in the feedback network.

NATIONAL SEMICONDUCTOR *Fig. 21-19*

VISABLE VOLTAGE INDICATOR

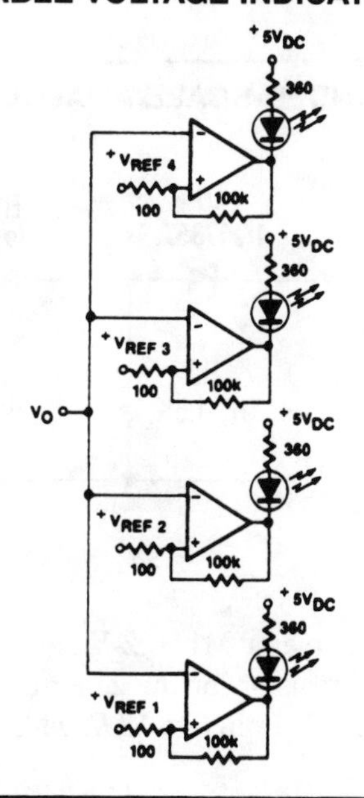

SIGNETICS *Fig. 21-20*

BIPOLAR REFERENCE SOURCE

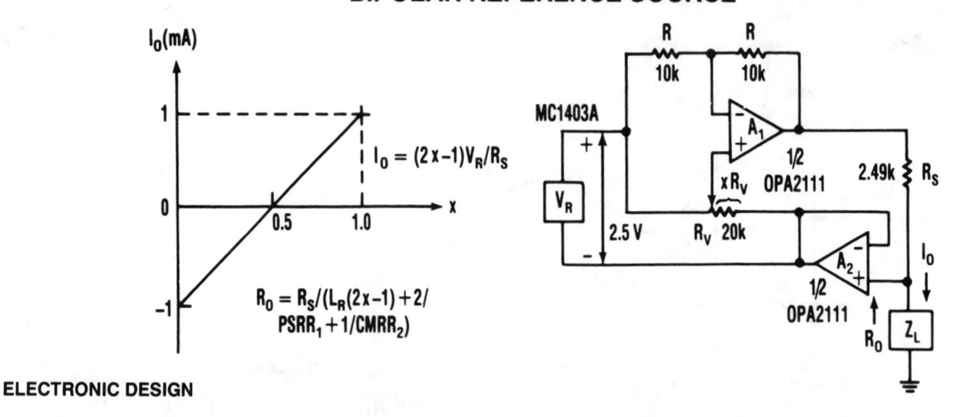

ELECTRONIC DESIGN *Fig. 21-21*

This current source has continuous control of the magnitude and polarity of its amplifier gain and needs only one voltage reference. The circuit includes reference V_R, voltage-amplifier circuit A1 with gain-setting resistor R_S, and bootstrap-follower amplifier A2. The bootstrapping converts the circuit to a current source and allows the load to be grounded. Any voltage developed across load Z_L feeds back to the reference and voltage amplifier, making their functions immune to that voltage. Then, the current-source circuitry floats, instead of the load.

The voltage reference is connected to both the inverting and noninverting inputs of A1; this provides a balanced combination of positive and negative gain. The inverting connection has equal feedback resistors, R, for a gain of -1, and the noninverting connection varies according to the fractional setting, X, of potentiometer R_V. X controls the noninverting gain and adjusting it counters the effect of some of the inverting gain. The value of X is the portion of R_V's resistance from the noninverting input of A1 to the temporarily grounded output of A2. Between potentiometer extremes, the current varies with X ± 1 mA.

EXPANDED-SCALE ANALOG METER

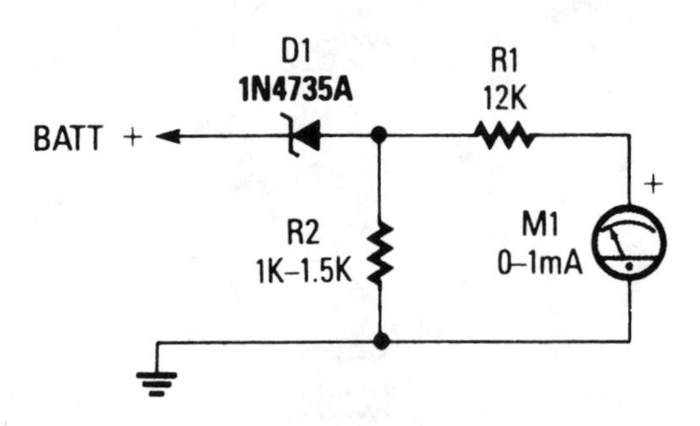

POPULAR ELECTRONICS *Fig. 21-22*

The circuit consists of 0- to 1-mA meter M1, 6.2-V zener diode D1, and 12-kΩ, 1% resistor R1. R2 is included in the circuit as a load resistor for the zener diode. The value of R_2 isn't critical; use a value of 1000 to 1500 Ω. The meter reads from 6 to 18 V, which is perfect for checking a car's charging system.

Sources

Chapter 1

Fig. 1-1. Electronic Engineering, 10/70, p. 17.

Fig. 1-2. Electronic Engineering, Mid5/78, p. 11.

Fig. 1-3. Siliconix Incorporated, Silicon Analog Switch & IC Product Data Book, 1/82, p. 6-19.

Fig. 1-4. Reprinted with the permission of National Semiconductor Corp., Linear Databook, 1982, p. 3-109.

Fig. 1-5. Reprinted with the permission of National Semiconductor Corp., Linear Databook, 1982, p. 3-109.

Fig. 1-6. Electronic Engineering, 2/85, p. 45.

Fig. 1-7. Electronics Today International, 1/75, 6/79, p. 103.

Fig. 1-8. Ham Radio, 9/82, p. 78.

Fig. 1-9. Courtesy of Texas Instruments Inc., Optoelectronics Databook, 1983-84, p. 15-5.

Fig. 1-10. 73 Amateur Radio, 2/79, p. 78.

Fig. 1-11. Electronics Australia, 2/76, p. 91.

Fig. 1-12. 73 Amateur Radio, 2/79, p. 78.

Fig. 1-13. Courtesy of William Sheets.

Fig. 1-14. TAB Books, 101 Sound, Light, and Power IC Projects.

Fig. 1-15. Moli Energy Limited, Publication MEL-126.

Fig. 1-16. Linear Technology Corp., Linear Databook, 1986, p. 2-104.

Fig. 1-17. Motorola, TMOS Power FET Ideas, 1985, p. 7.

Fig. 1-18. Modern Electronics, 9/78, p. 37.

Fig. 1-19. Modern Electronics, 9/78, p. 37.

Chapter 2

Fig. 2-1. 303 Dynamic Electronic Circuits, TAB Books, No. 1060, p. 290.

Fig. 2-2. 73 Amateur Radio, 2/79, p. 79.

Fig. 2-3. Reprinted from Electronics, 5/73, p. 96. Copyright 1973, McGraw-Hill Inc. All rights reserved.

Fig. 2-4. Precision Monolithics Inc., 1981 Full Line Catalog, p. 8-31.

Fig. 2-5. Teledyne Semiconductor, Databook, p. 9.

Fig. 2-6. Siliconix Inc., Siliconix Analog Switch & IC Product Data Book, 1/82, p. 6-4.

Fig. 2-7. Signetics, Analog Data Manual, 1982, p. 8-14.

Fig. 2-8. Courtesy of Motorola Inc., Linear Integrated Circuits, 1979, p. 6-17.

Fig. 2-9. Courtesy of Motorola Inc., Linear Integrated Circuits, 1979, p. 7-8.

Fig. 2-10. Wireless World, 12/74, p. 504.

Fig. 2-11. Courtesy of Motorola Inc., Linear Integrated Circuits, 1979, p. 6-123.

Fig. 2-12. Precision Monolithics Inc., 1981 Full Line Catalog, p. 8-12.

Fig. 2-13. Signetics, Analog Data Manual, 1982, p. 3-38.

Fig. 2-14. Harris Semiconductor, Linear & Data Acquisition Products, p. 2-46.

Fig. 2-15. Harris Semiconductor, Application Note 509.

Fig. 2-16. Precision Monolithics Inc., 1981 Full Line Catalog, p. 8-31.

Fig. 2-17. Courtesy of Motorola Inc., Linear Integrated Circuits, 1979, p. 7-8.

Chapter 3

Fig. 3-1. Courtesy of William Sheets.

Fig. 3-2. GE/RCA, BiMOS Operational Amplifiers Circuit Ideas, 1987, p. 17.

Fig. 3-3. Signetics, 1987 Linear Data Manual Vol. 2: industrial, 2/87, p. 4-78.

Fig. 3-4. Linear Technology, 1986 Linear Databook, p. 2-45.

Fig. 3-5. Texas Instruments, Linear and Interface Circuits Applications, Vol. 1, 1985, p. 3-3, 3-4.

Fig. 3-6. Siliconix, Small-Signal FET Data Book, 1/86, p. 7-29.

Fig. 3-7. Linear Technology Corp., Linear Databook, p. 2-83.

Fig. 3-8. Harris, Analog Product Data Book, 1988, p. 10-183.

Fig. 3-9. Reprinted with permission from Electronic Design. Copyright 1989, Penton Publishing.

Fig. 3-10. Signetics, 1987 Linear Data Manual, Vol. 2: Industrial, 2/87, 5-367.

Fig. 3-11. Linear Technology Corp., Linear Databook, p. 2-101.

Chapter 4

Fig. 4-1. Electronic Engineering, 2/85, p. 34.

Fig. 4-2. Texas Instruments, Linear and Interface Circuit Applications, Vol. 1, 1985, p. 7-21.

Fig. 4-3. Intersil, Applications Handbook, 1988, p. 3-138.

Fig. 4-4. Hands-On Electronics, Fall 19984, p. 68.

Fig. 4-5. Radio-Electronics, 10/77, p. 72.

Chapter 5

Fig. 5-1. Gernsback Publications Inc., 42 New Ideas, 1984, p. 18.

Fig. 5-2. Hands-On Electronics/Popular Electronics, 1/89, p. 59.

Fig. 5-3. Reprinted from EDN, 3/74, (C) 1989 Cahners Publishing Co., a division of Reed Publishing USA.

Fig. 5-4. Intersil, Component Data Catalog, 1987. p. 14-49.

Fig. 5-5. R-E Experimenters Handbook, 1987, p. 151.

Fig. 5-6. Reprinted from EDN, 11/10/88, (C) 1989 Cahners Publishing Co., a division of Reed Publishing USA.

Chapter 6

Fig. 6-1. Reprinted with permission of National Semiconductor Corp., Linear Databook, 1982, p. 3-123.

Fig. 6-2. NASA Tech Briefs, 7-8/86, p. 37.

Fig. 6-3. GE/RCA, BiMOS Operational Amplifiers Circuit Ideas, 1987, p. 14.

Fig. 6-4. Reprinted with permission from General Electric Semiconductor Department, GE Semiconductor Handbook, Third Edition, p. 305.

Fig. 6-5. Reprinted with permission of National Semiconductor Corp., Transistor Databook, 1982, p. 11-35.

Fig. 6-6. GE/RCA, BiMOS Operational Amplifiers Circuit Ideas, 1987, p. 17.

Fig. 6-7. Linear Technology Corp., Linear Databook, 1986, p. 2-85.

Fig. 6-8. Linear Technology Corp., Linear Databook, 1986, p. 2-57.

Fig. 6-9. Linear Technology Corp., Linear Applications Handbook, 1987, p. AN3-13.

Fig. 6-10. Intersil, Component Data Catalog, 1987, p. 7-4.

Fig. 6-11. Reprinted with permission of National Semiconductor Corp., Application Note AN-71, p. 5.

Chapter 7

Fig. 7-1. Ham Radio, 8/81, p. 27.

Fig. 7-2. Ham Radio, 8/81, p. 28.

Fig. 7-3. CQ, 1/87, p. 36.

Fig. 7-4. Ham Radio, 8/81, p. 27.

Fig. 7-5. Ham Radio, 8/81, p. 26.

Fig. 7-6. Ham Radio, 8/81, p. 26.

Fig. 7-7. Ham Radio, 6/77, p. 42.

Fig. 7-8. Ham Radio, 8/81, p. 27.

Chapter 8

Fig. 8-1. GE/RCA, BiMOS Operational Amplifiers Circuit Ideas, 1987, p. 14.

Fig. 8-2. Reprinted with permission of National Semiconductor Corp., Linear Databook, 1982, p. 171.

Fig. 8-3. Reprinted with permission of National Semiconductor Corp., Linear Databook, 1982, p. 9-171.

Fig. 8-4. Electronics Today International, 3/78, p. 50.

Fig. 8-5. Reprinted from Electronics, 12/74, p. 105. Copyright 1974, McGraw-Hill Inc. All rights reserved.

Fig. 8-6. Intersil, Intersil Data Book, 5/83, p. 6-34.

Fig. 8-7. Courtesy of Motorola Inc., Linear Interface Integrated Circuits, 1979, p. 5-102.

Fig. 8-8. Intersil, Intersil Data Book, 5/83, p. 6-52.

Fig. 8-9. Electronic Engineering, 9/84, p. 30.

Fig. 8-10. Reprinted with permission of National Semiconductor Corp., Linear Databook, 1982, p. 9-188.

Fig. 8-11. Reprinted with permission of National Semiconductor Corp., Linear Databook, 1982, p. 9-172.

Fig. 8-12. Electronics Today International, 10/82, p. 80.

Fig. 8-13. GE/RCA, BiMOS Operational Amplifiers Circuit Ideas, 1987, p. 12.

Chapter 9

Fig. 9-1. Ham Radio, 9/86, p. 67.

Fig. 9-2. TAB Books, The Giant Book of Electronics Projects, No. 1367, p. 114.

Fig. 9-3. TAB Books, The Giant Book of Electronics Projects, No. 1367, p. 114.

Fig. 9-4. 73 Amateur Radio.

Fig. 9-5. Hands-On Electronics, 3/87, p. 27.

Fig. 9-6. Ham Radio, 1/85, p. 51.

Fig. 9-7. Hands-On Electronics, 8/87, p. 65.

Fig. 9-8. Courtesy of William Sheets.

Fig. 9-9. Courtesy of William Sheets.

Fig. 9-10. Hands-On Electronics, Fact Card No. 57.

Fig. 9-11. 73 Amateur Radio, 10/83, p. 53.

Fig. 9-12. Practical Wireless, 5/85, p. 37.

Fig. 9-13. Courtesy of William Sheets.

Fig. 9-14. Hands-On Electronics, 3/87, p. 27.

Fig. 9-15. Modern Electronics, 2/78, p. 47.

Fig. 9-16. TAB Books, The Giant Book of Electronics Projects, No. 1367, p. 480.

Chapter 10

Fig. 10-1. 73 Amateur Radio, 6/83, p. 106.

Fig. 10-2. TAB Books, 104 Weekend Electronic Projects, No. 1436, p. 166.

Fig. 10-3. Intersil, Intersil Data Book, 5/83, p. 6-49.

Fig. 10-4. TAB Books, The Giant Book of Electronics Projects, No. 1367, p. 109.

Fig. 10-5. EXAR, Telecommunications Databook, 1986, p. 11-38.

Chapter 11

Fig. 11-1. 73 Amateur Radio, 7/77, p. 35.

Fig. 11-2. Electronics Today International, 6/76, p. 40.

Fig. 11-3. Elector Electronics, 7-8/87 Supplement, p. 36.

Fig. 11-4. Reprinted from EDN, 5/74, (C) 1989 Cahners Publishing Co., a division of Reed Publishing USA.

Fig. 11-5. Courtesy of Texas Instruments Inc., Optoelectronics Databook, 1983-84, p. 15-11.

Fig. 11-6. Reprinted with permission of National Semiconductor Corp., Linear Databook, 1982, p. 9-172.

Fig. 11-7. Courtesy of Texas Instruments Inc., Optoelectronics Databook, 1983-84, p. 15-11.

Chapter 12

Fig. 12-1. Machine Design, 9/80, p. 126.

Fig. 12-2. Machine Design, 9/80, p. 127.

Fig. 12-3. Reprinted with permission of National Semiconductor Corp., Linear Databook, 1982, p. 9-191.

Fig. 12-4. Reprinted with permission of National Semiconductor Corp., Data Conversion/Acquisition Databook, 1980, p. 3-91.

Fig. 12-5. Reprinted with permission of National Semiconductor Corp., Hybrid Products Databook, 1982, p. 1-89.

Fig. 12-6. Reprinted with permission of National Semiconductor Corp., Data Conversion/Acquisition Databook, 1980, p. 13-50.

Chapter 13

Fig. 13-1. Courtesy of Fairchild Camera & Instrument Corp., Linear Databook, 1982, p. 5-25.

Fig. 13-2. Precision Monolithics Inc., 1981 Full Line Catalog, p. 10-8.

Fig. 13-3. Electronics Today International, 7/75, p. 40.

Fig. 13-4. Radio-Electronics, 1/80, p. 68.

Fig. 13-5. Signetics, Analog Data Manual, p. 9-40.

Fig. 13-6. Electronic Engineering, 6/87, p. 28.

Fig. 13-7. Popular Electronics, 8/69, p. 71.
Fig. 13-8. Electronics Today International, 1/76, p. 52.
Fig. 13-9. Electronics Today International, 1/76, p. 47.
Fig. 13-10. Reprinted with permission of National Semiconductor Corp., Linear Databook, 1982, p. 9-140.
Fig. 13-11. Signetics, Analog Data Manual, 1983, p. 9-38.
Fig. 13-12. Reprinted with permission of National Semiconductor Corp., Linear Databook, 1982, p. 9-187.
Fig. 13-13. Reprinted with permission of National Semiconductor Corp., National Semiconductor CMOS Databook, 1981, p. 8-124.
Fig. 13-14. Intersil, Data Book, 1978.
Fig. 13-15. Reprinted with permission of National Semiconductor Corp., Linear Databook, 1982, p. 3-86.
Fig. 13-16. TAB Books, Third Book of Electronic Projects, No. 1446, p. 40.
Fig. 13-17. Electronics Today International, 8/73, p. 82.
Fig. 13-18. Electronic Design, 10/73, p. 114.
Fig. 13-19. Electronic Engineering, 7/85, p. 44.
Fig. 13-20. Hands-On Electronics, 5-6/86, p. 63.
Fig. 13-21. TAB Books, 303 Dynamic Electronic Circuits, No. 1060, p. 153.
Fig. 13-22. Electronics Today International, 10/78, p. 97.
Fig. 13-23. Reprinted with permission of National Semiconductor Corp., Transistor Databook, 1982, p. 11-49.
Fig. 13-24. Electronics Today International, 7/85, p. 44.
Fig. 13-25. Popular Electronics, 1/82, p. 76.
Fig. 13-26. Hands-On Electronics/Popular Electronics, 12/88, p. 61.
Fig. 13-27. Siliconix, Integrated Circuits Data Book, 3/85, p. 10-137.
Fig. 13-28. 73 Amateur Radio, 8/88, p. 24.
Fig. 13-29. Linear Technology Corp., Linear Databook, 1986, p. 2-96.
Fig. 13-30. NASA, NASA Tech Briefs, 1/89, p. 19.
Fig. 13-31. Reprinted from EDN, 5/83, (C) 1989 Cahners Publishing Co., a division of Reed Publishing USA.
Fig. 13-32. Intersil, Component Data Catalog, 1987, p. 14-70.
Fig. 13-33. Intersil, Component Data Catalog, 1987, p. 14-121.
Fig. 13-34. GE, Optoelectronics, Third Edition, Ch. 6, p. 11.
Fig. 13-35. Hands-On Electronics, 7-8/86, p. 51.
Fig. 13-36. Elektor Electronics, 7-8/87 Supplement, p. 23.
Fig. 13-37. GE/RCA, BiMOS Operational Amplifiers Circuit Ideas, 1987, p. 20.
Fig. 13-38. Electronic Engineering, 9/78, p. 20.
Fig. 13-39. Reprinted from EDN, 5/2/85, (C) 1989 Cahners Publishing Co., a division of Reed Publishing USA.
Fig. 13-40. Popular Electronics, 10/89, p. 104.
Fig. 13-41. Hands-On Electronics, Winter 1985, p. 31.
Fig. 13-42. Ham Radio, 4/86, p. 24.
Fig. 13-43. Gernsback Publications Inc., 42 New Ideas, 1984, p. 24.
Fig. 13-44. Ham Radio, 12/88, p. 19.
Fig. 13-45. Texas Instruments, Linear and Interface Circuits Applications, 1987, p. 12-5.
Fig. 13-46. Radio-Electronics, 9/87, p. 32.
Fig. 13-47. GE/RCA, BiMOS Operational Amplifiers Circuit Ideas, 1987, p. 12.
Fig. 13-48. Electronic Engineering, 11/86, p. 34.
Fig. 13-49. Popular Electronics, 10/89, p. 84.
Fig. 13-50. Teledyne Semiconductor, Data Acquisition IC Handbook, p. 15-11.
Fig. 13-51. GE, Optoelectronics, Third Edition, Ch. 6, p. 112.
Fig. 13-52. Reprinted from EDN, 3/23/85, (C) 1989 Cahners Publishing Co., a division of Reed Publishing USA.
Fig. 13-53. 73 Amateur Radio, 6/89, p. 44.
Fig. 13-54. Reprinted from EDN, 3/79, (C) 1989 Cahners Publishing Co., a division of Reed Publishing USA.

Chapter 14

Fig. 14-1. Texas Instruments, Linear and Interface Circuits Applications, 1985, Vol. 1, p. 7-13.
Fig. 14-2. Reprinted from EDN, 3/82, (C) 1989 Cahners Publishing Co., a division of Reed Publishing USA.
Fig. 14-3. Siliconix, Integrated Circuits Data Book, 3/85, p. 10-62.
Fig. 14-4. Elektor Electronics. 7-8/87 Supplement, p. 50.
Fig. 14-5. Electronic Design, 5/79, p. 102.
Fig. 14-6. Reprinted from Electronics, 7/72, p. 77. Copyright 1972, McGraw-Hill Inc. All rights reserved.
Fig. 14-7. Texas Instruments, Linear and Interface Circuits Applications, 1985, Vol. 1, p. 4-2.

Fig. 14-8. Signetics, 1987 Linear Data Manual, Vol. 2: Industrial, 2/87, p. 7-62.
Fig. 14-9. CQ, 11/83, p. 72.
Fig. 14-10. Electronics Today International, 7/77, p. 77.
Fig. 14-11. Radio-Electronics, 10/84, p. 32.
Fig. 14-12. 73 Amateur Radio, 8/88, p. 47.
Fig. 14-13. Radio-Electronics, 7/70, p. 36.
Fig. 14-14. Courtesy of William Sheets.
Fig. 14-15. Electronics Today International, 1978.
Fig. 14-16. CQ, 11/83, p. 72.
Fig. 14-17. Ham Radio, 12/88, p. 26.
Fig. 14-18. 73 Amateur Radio, 12/76, p. 170.

Chapter 15

Fig. 15-1. Reprinted with the permission of National Semiconductor Corp., Linear Databook, 1982, p. 9-205.
Fig. 15-2. Reprinted with the permission of National Semiconductor Corp., Linear Databook, 1982, p. 9-191.
Fig. 15-3. Courtesy of Texas Instruments Inc., Linear Control Circuits Data Book, Second Edition, p. 374.
Fig. 15-4. Reprinted with the permission of National Semiconductor Corp., Application Note 222.
Fig. 15-5. Courtesy of Motorola Inc., Motorola Semiconductor Library, Vol. 6, Series B, p. 8-58.

Chapter 16

Fig. 16-1. Hands-On Electronics, 12/87, p. 73.
Fig. 16-2. Ham Radio, 7-89, p. 20.
Fig. 16-3. Reprinted from EDN, 10/16/86, (c) 1989 Cahners Publishing Co., a division of Reed Publishing USA.
Fig. 16-4. Electronics Today International, 5/77, p. 37.
Fig. 16-5. Linear Technology Corp., Linear Applications Handbook, 1987, p. AN9-9.
Fig. 16-6. Microwaves and RF, 3/86, p. 143.
Fig. 16-7. Reprinted from EDN, 6/72, (c) 1989 Cahners Publishing Co., a division of Reed Publishing USA.
Fig. 16-8. Reprinted from EDN, 7/7/88, (c) 1989 Cahners Publishing Co., a division of Reed Publishing USA.
Fig. 16-9. Reprinted from EDN, 3/73, (c) 1989 Cahners Publishing Co., a division of Reed Publishing USA.
Fig. 16-10. Harris, Analog Product Data Book, 1988, p. 10-182.
Fig. 16-11. Canadian Projects Number 1, p. 86.
Fig. 16-12. Electronics Today International, 11/75, p. 74.
Fig. 16-13. Ham Radio, 2/73, p. 56.
Fig. 16-14. Linear Technology Corp., Linear Applications Handbook, 1987, p. AN8-4.
Fig. 16-15. Siliconix, Mospower Applications Handbook, p. 6-62.
Fig. 16-16. Electronics Today International, 3/81, p. 19.
Fig. 16-17. 101 Electronic Projects, 1975, p. 47.
Fig. 16-18. 73 Amateur Radio, 10/83, p. 66.
Fig. 16-19. Electronics Today International, 6/79, p. 103.
Fig. 16-20. Electronics Today International, 1/76, p. 44.
Fig. 16-21. Reprinted from Electronics, 7/76, p. 121. Copyright 1976, McGraw-Hill Inc. All rights reserved.
Fig. 16-22. Electronics Today International, 1/76, p. 52.
Fig. 16-23. Electronics Today International, 1/76, p. 51.
Fig. 16-24. Ham Radio, 8/80, p. 18.
Fig. 16-25. 73 Amateur Radio, 4/88, p. 20.
Fig. 16-26. Reprinted from EDN, 12/8/88, (c) 1989 Cahners Publishing Co., a division of Reed Publishing USA.
Fig. 16-27. Popular Electronics, Fact Card No. 100.
Fig. 16-28. Hand-On Electronics, 9/87, p. 96.

Chapter 17

Fig. 17-1. Teledyne, Semiconductor Databook, p. 11.
Fig. 17-2. Reprinted from Electronics 4/76, p. 104. Copyright McGraw-Hill Inc. All rights reserved.
Fig. 17-3. Electronics Today International, 11/78, p. 68.
Fig. 17-4. CQ, 6/78, p. 33.
Fig. 17-5. 73 Amateur Radio.
Fig. 17-6. CQ, 6/78, p. 32.
Fig. 17-7. Signetics, 1987 Linear Data Manual Vol. 2: Industrial, 2/87, p. 7-66 and 7-67.
Fig. 17-8. NASA, Tech Briefs, 1/88, p. 18.
Fig. 17-9. Hands-On Electronics/Popular Electronics, 12/88, p. 24.
Fig. 17-10. National Semiconductor Corp., Voltage Regulator Handbook, p. 10-59.
Fig. 17-11. Reprinted by permission from the Aug. 1981 issue of Insulation/Circuits magazine. Copyright 1981, Lake Publishing Corporation, Libertyville, Illinois, 60048-9989, USA.
Fig. 17-12. (c) Siliconix Inc., Application Note AN154.
Fig. 17-13. Electronic Engineering, 7/85, p. 44.

Fig. 17-14. Electronic Engineering, 11/86, p. 34.
Fig. 17-15. Electronic Engineering, 4/86, p. 34.

Chapter 18

Fig. 18-1. Popular Electronics/Hands-On Electronics, 3/89, p. 24.
Fig. 18-2. Popular Electronics/Hands-On Electronics, 12/88, p. 26.

Chapter 19

Fig. 19-1. Reprinted from EDN, 3/3/88, (c) 1989 Cahners Publishing Co., a division of Reed Publishing USA.
Fig. 19-2. Reprinted from EDN, 5/26, (c) 1989 Cahners Publishing Co., a division of Reed Publishing USA.
Fig. 19-3. Electronic Engineering, Applied Ideas, 11/88, p. 28.

Chapter 20

Fig. 20-1. Electric Engineering, 7/84, p. 31.
Fig. 20-2. Intersil Data Book, 5/83, p. 5-71.
Fig. 20-3. Intersil Data Book, 5/83, p. 5-71.
Fig. 20-4. Courtesy of Fairchild Camera & Instrument Corp., Linear Databook, 1982, p. 4-42.
Fig. 20-5. Reprinted with the permission of National Semiconductor Corp., Linear Databook, 1982, p. 3-108.
Fig. 20-6. Reprinted with the permission of National Semiconductor Corp., CMOS Databook, 1981, p. 6-7.
Fig. 20-7. Reprinted with the permission of National Semiconductor Corp., Linear Databook, 1982, p. 9-31.
Fig. 20-8. Reprinted with the permission of National Semiconductor Corp., Data Conversion/Acquisition Databook, 1980, p. 12-10.
Fig. 20-9. Reprinted with the permission of National Semiconductor Corp., Linear Databook, 1982, p. 9-162.
Fig. 20-10. Reprinted with the permission of National Semiconductor Corp., Data Conversion/Acquisition Databook, 1980, p. 12-7.
Fig. 20-11. Radio-Electronics, 3/80, p. 60.
Fig. 20-12. Electronics Today International, 12/78, p. 32.
Fig. 20-13. Signetics, 1987 Linear Data Manual Vol. 2: Industrial, 11/14/86, p. 5-58.
Fig. 20-14. Linear Technology Corp., Linear Databook, 1986, p. 8-43.
Fig. 20-15. GE, SCR Manual, Sixth Edition, 1979, p. 222.
Fig. 20-16. National Semiconductor Corp., 1984 Linear Supplement Databook, p. S1-42.
Fig. 20-17. Reprinted with the permission of National Semiconductor Corp., Data Conversion/Acquisition Databook, 1980, p. 12-9.
Fig. 20-18. Signetics Analog Data Manual, 1982, p. 3-78.
Fig. 20-19. Teledyne, Semiconductor Databook, p. 12.
Fig. 20-20. Precision Monolithics Inc., 1981 Full Line Catalog, p. 10-16.
Fig. 20-21. Linear Technology Corp., Linear Databook, 1986, p. 2-101.
Fig. 20-22. Courtesy of William Sheets.
Fig. 20-23. National Semiconductor Corp., 1984 Linear Supplement Databook, p. S1-41.
Fig. 20-24. Siliconix, Integrated Circuits Data Book, 1988, p. 13-204.
Fig. 20-25. Electronic Engineering, 9/85, p. 30.
Fig. 20-26. NASA, NASA Tech Briefs, 10/87, p. 34.
Fig. 20-27. R-E Experimenters Handbook, 1987, p. 11.
Fig. 20-28. Signetics Analog Data Manual, 1983, p. 10-65.
Fig. 20-29. Precision Monolithics Inc., 1981 Full Line Catalog, p. 6-147.
Fig. 20-30. Intersil, Intersil Data Book, 5/83, p. 5-70.
Fig. 20-31. Reprinted with the permission of National Semiconductor Corp. Linear Databook, 1982, p. 9-29.
Fig. 20-32. TAB Books, 44 Electronics Projects for the Darkroom.
Fig. 20-33. Signetics, 1987 Linear Data Manual Vol. 2: Industrial, 11/14/86, p. 5-58.
Fig. 20-34. Raytheon, Linear and Integrated Circuits, 1989, p. 8-16.
Fig. 20-35. Reprinted with the permission of National Semiconductor Corp., Linear Databook, 1982, p. 9-160.
Fig. 20-36. Reprinted with the permission of National Semiconductor Corp., Linear Databook, 1982, p. 9-31.
Fig. 20-37. Teledyne, Semiconductor Databook, p. 11.

Fig. 20-38. Teledyne, Semiconductor Databook, p. 11.

Fig. 20-39. Reprinted with the permission of National Semiconductor Corp., Linear Databook, 1982, p. 9-31.

Fig. 20-40. Reprinted with the permission of National Semiconductor Corp., Linear Databook, 1982, p. 9-29.

Fig. 20-41. Reprinted with the permission of National Semiconductor Corp.,Linear Databook, 1982, p. 9-160.

Fig. 20-42. Reprinted with the permission of National Semiconductor Corp.,Linear Databook, 1982, p. 9-162.

Fig. 20-43. Reprinted with the permission of National Semiconductor Corp.,Linear Databook, 1982, p. 10-107.

Fig. 20-44. Reprinted with the permission of National Semiconductor Corp.,Linear Databook, 1982, p. 2-46.

Fig. 20-45. Linear Technology Corp., Linear Applications Handbook, 1987, p. AN3-14.

Fig. 20-46. Hands-On Electronics, 9-10/86, p. 32.

Fig. 20-47. Reprinted with the permission of National Semiconductor Corp.,Linear Databook, 1982, p. 2-46.

Fig. 20-48. Reprinted with the permission of National Semiconductor Corp.,Linear Databook, 1982, p. 9-29.

Chapter 21

Fig. 21-1. Electronic Engineering, 10/84, p. 41.

Fig. 21-2. Siliconix, Integrated Circuits Data Book, 3/85, p. 3-21.

Fig. 21-3. National Semiconductor Corp., CMOS Databook, 1981, p. 3-41.

Fig. 21-4. Reprinted with permission from Electronic Design. Copyright 1970, Penton Publishing.

Fig. 21-5. GE/RCA, BiMOS Operational Amplifiers Circuit Ideas, 1987, p. 16.

Fig. 21-6. GE/RCA, BiMOS Operational Amplifiers Circuit Ideas, 1987, p. 16.

Fig. 21-7. Radio-Electronics, 11/82, p. 92.

Fig. 21-8. Teledyne, Teledyne Semiconductor Databook, p. 9.

Fig. 21-9. GE/RCA, BiMOS Operational Amplifiers Circuit Ideas, 1987, p. 15.

Fig. 21-10. TAB Books, Third Book of Electronic Projects, p. 37.

Fig. 21-11. Ham Radio, 7/89, p. 62.

Fig. 21-12. Hands-On Electronics, 4/87, p. 93.

Fig. 21-13. Popular Electronics/Hands-On Electronics, 4/89, p. 25.

Fig. 21-14. Electronics Today International, 10/78, p. 95.

Fig. 21-15. Radio-Electronics, 1/86, p. 104.

Fig. 21-16. National Semiconductor Corp., Transistor Databook, 1982, p. 7-26.

Fig. 21-17. Electronics Engineering, 9/85, p. 25.

Fig. 21-18. Electronics Today International, 6/76, p. 42.

Fig. 21-19. Reprinted with the permission of National Semiconductor Corp.,Linear Databook, 1982, p. 362.

Fig. 21-20. Signetics, 1987 Linear Data Manual Vol. 2: Industrial, 11/14/86, p. 5-269.

Fig. 21-21. Reprinted with permission from Electronic Design. Copyright 1989, Penton Publishing.

Fig. 21-22. Popular Electronics, 6/89, p. 22.

Index

D

E

F

G

H

I

R

S

T

V

W

Z

Other Bestsellers of Related Interest